the

SOVIET UNION

a systematic geography

Leslie Symons
Professor of Geography, University College of Swansea

J. C. Dewdney
Professor of Geography, University of Durham

D. J. M. Hooson
Professor of Geography, University of California

R. E. H. Mellor
Emeritus Professor of Geography, University of Aberdeen

W. W. Newey
Senior Lecturer in Geography, University of Edinburgh

the

SOVIET UNION

a systematic geography

Hodder & Stoughton

LONDON SYDNEY AUCKLAND TORONTO

ROUTLEDGE

NEW YORK

The photograph reproduced on the title page is of a Caucasus mountain settlement.

British Library Cataloguing in Publication Data
The Soviet Union: a systematic geography.—2nd ed.
  1. Soviet Union. Geographical features
  I. Symons, Leslie
  914.7

ISBN 0 340 50300 9
ISBN 415-9035-80 (US edition)

First Published 1983
Second edition 1990

Typeset by Macmillan India Ltd., Bangalore 25
Printed in Great Britain for the educational publishing division of Hodder and Stoughton, Mill Road, Dunton Green, Sevenoaks, Kent by Thomson Litho, East Kilbride.
Published in the United States for Routledge, a division of Routledge, Chapman and Hall, Inc. 29 West 35th Street, New York, NY 10001.

Contents

Preface to the Second Edition

There has been a great reawakening of interest in the Soviet Union since the first edition of this book was published. At that time (1983) the Soviet Union was perceived in the western world and in many countries of the 'third world' as an intermediate or second world dominated by the dogma of a communist ideology which allowed little scope for change or for more understanding relationships with the capitalist or 'free enterprise' countries. This situation inhibited trade and encouraged mutual suspicion between 'the west' and the 'iron curtain' countries of the Soviet Union and the Eastern European bloc.

In 1985 the administration of Mikhail Gorbachev brought to the USSR a sharp wind of change. During the period 1985–8 the groundwork was laid for restructuring the economy ('*perestroika*') and for a new 'openness' or '*glasnost*' in the reporting of events and in general interchange of views with the people and, more especially, the authorities and officials, of other countries. Trade has been encouraged and new laws passed to facilitate, *inter alia*, the development of joint industrial enterprises with foreign countries.

At the time of writing, therefore, the USSR is in ferment economically and politically and discussion is being actively encouraged into the reforms. The revision of this book takes into account these developments where there is clear evidence of effects on the geography of the country. It is, however, necessary to say that at this time, the reforms are still very much at the legal and experimental stage and there is no information yet available to indicate what changes are likely in the geographical and spatial aspects of the country. A parallel might be drawn with the effects that resulted from the entry of the UK into the European Economic Community. It was not possible to foresee in 1972 in detail what this change might bring in the way of economic change or alterations to the landscape. Even now there are many difficulties in assessing what these changes have so far been. It would therefore be wrong to forecast what 'perestroika' might do to the geographical aspects of Soviet industry, population distribution, etc. This revision therefore takes note of the legal changes and incentives newly introduced into farm and factory in the Soviet Union but does not attempt to anticipate their spatial significance.

It has, however, been possible to make major changes in the chapters dealing with the biosphere and mineral development where numerous changes have been occurring throughout the past decade and these chapters have been completely rewritten. Elsewhere all material has been updated and statistics revised to the last available figures published, normally the 1987 figures, published in 1988.

The Soviet Union, from its very size and diversity in landscape, peoples, economics and cultures, deserves much more attention than it has hitherto received in geographical studies. This book aims to fill some of the gaps that exist in the material at present easily available to schools and colleges. Though there are a number of excellent textbooks available on the USSR, certain topics have not been covered adequately in them. The present volume provides not only an **up-to-date** account of the basic aspects of the physical and economic geography of the USSR but also deals in detail with topics such as vegetation and water resources not usually adequately covered in textbooks.

The study of the Soviet Union, through the very diversity that it encompasses, presents some difficulties. This diversity, which defies brevity of description in the case of the physical features, poses even more problems in the variety of the human geography. This is best illustrated by the matter of different nationalities, of which over 90 are recognised in the current classification used by the Soviet authorities, with many additional lesser ethnic groups. Each of these nationality groups has its own language but access to material on the Soviet Union is simplified by the status of Russian, the language of the dominant ethnic group, as a *lingua franca*. Soviet atlases and maps show place names in Russian and it is the Russian forms that are used here, transliterated by the standard system used by British and American geographers, reproduced on page viii. A few of these names may appear unfamiliar to the reader, because of the adherence to the transliteration system, e.g. Baykal rather than Baikal, Tadzhikistan instead of Tajikistan. The system is extended to transliteration of other Russian words, *tayga* rather than *taiga*. As this word should be pronounced approximately *taygá*, not 'taeega', as commonly in English usage, the transliteration gives a better idea of correct Russian pronunciation. For simplification, soft and hard signs have been omitted throughout. Where

English forms of names are in general use, e.g. Moscow (*Moskva*), Georgia (*Gruzinskaya Respublika*), these are used, and where alternative names, such as Turkmeniya and Turkmenistan are equally acceptable, the authors' usage has not been standardised by the editor. With these exceptions the spellings will be found to coincide with the Times Atlas as this uses the same transliteration system.

Capital letters are used for regions defined administratively e.g. North-west, North Caucasus economic regions, while north-west Russia, north Caucasus, etc. indicate geographical areas not so defined.

All the contributors to this volume have specialised in the geography of the Soviet Union. John Dewdney, Professor of Geography at the University of Durham (Chapters 2, 6 and 12), has studied, especially, Soviet demography and regional development; David Hooson, Professor of Geography at the University of California at Berkeley (Chapter 3), is the author of regional and methodological studies in particular; R. E. H. Mellor, Emeritus Professor of Geography at the University of Aberdeen (Chapters 10 and 11), has specialised in the economic and urban geography of the region; and Walter W. Newey, Senior Lecturer in Geography at the University of Edinburgh (Chapters 4, 5 and 8), specialises in biogeography. Leslie Symons (editor and contributor of Chapters 1, 7 and 9) is Professor of Geography in the Centre of Russian and East European Studies and the Department of Geography at the University College of Swansea.

Acknowledgements

The editor and authors gratefully acknowledge the help of colleagues in the Department of Geography, University College of Swansea. Mr. Guy B. Lewis and Mrs. Nicola Jones drew the maps from material supplied by the authors and based mainly on the following sources: *Atlas SSSR, Atlas razvitiya khozyaystva i kul'tury SSSR, Atlas sel'skogo khozyaystva SSSR, Fiziko-geograficheskiy atlas mira* and Soviet statistical publications, with supplementary information derived from other sources referred to in the text with *Soviet Geography, Review and Translation* particularly valuable for a wide range of articles and comments from geographers of both the Soviet Union and the West. The photographs on page 33, 66, 79, 100, 115, 126, 132, 136, 148, 154, 160, 184, 232, 239 (upper) and page 260 are reproduced by courtesy of the Novosti Press Agency (A.P.N.) and on page 219 by Mr. John Massey Stewart. Other photographs are by Dr. T. E. Armstrong (page 49), Dr. E. M. Bridges (page 264) and Mrs. Wendy Playfoot (pages 48 and 59) and the remaining photographs, including the cover picture, are by the editor. Special thanks are due to Mrs. E. Baker and Mrs. G. Symons for much of the work in preparing the index.

Every effort has been made to contact the holders of copyright material but if any have been inadvertently overlooked the publisher will be pleased to make the necessary alterations at the first opportunity.

Transliteration system

Russian	English rendering	Russian	English rendering
Аа	a	Рр	r
Бб	b	Сс	s
Вв	v	Тт	t
Гг	g	Уу	u (pronounced o͞o)
Дд	d	Фф	f
Ее	ye	Хх	kh
Ёё	yo (short o)	Цц	ts
Жж	zh	Чч	ch
Зз	z	Шш	sh
Ии	i	Щщ	shch
Йй	y	Ъъ	" (hard sign, not pronounced)
Кк	k	Ыы	y
Лл	l	Ьь	' (soft sign)
Мм	m	Ээ	e (eh)
Нн	n	Юю	yu
Оо	o	Яя	ya
Пп	p		

Maps and Diagrams

Photographs

Introduction

The Union of Soviet Socialist Republics is, in terms of area, the largest state in the world, covering about one-sixth of the land surface of the earth. For mankind, however, much of this land area is inhospitable and difficult to utilise, and the population (about 282 million in 1987) is largely concentrated in the more favourable parts. It was in one of these more favoured areas, today usually referred to as European Russia, that the Russian state originated and from which explorers, traders, soldiers and settlers moved out into all the other areas and eventually brought them under Russian control.

It is for this reason that the name of Russia is commonly used to apply to the whole of the USSR, but such a use can lead to confusion and is not geographically satisfactory, so 'Russia' will not be used in this widest sense here. At the same time, it is not easy to lay down hard and fast rules about the use of names like 'Russia' and we shall sometimes use it, particularly in its adjectival form, in a rather wider use than if used only for European Russia. This is, indeed, necessary in the case of the pre-communist Russian Empire, while today the largest of the republics that constitute the USSR, or Soviet Union, is called the Russian Soviet Federal Socialist Republic, although it stretches far beyond the traditional limits of European Russia to embrace Siberia and the Soviet Far East (see Fig. A).

Altogether there are 15 Soviet Socialist Republics, including the RSFSR, making up the USSR. The name of each indicates the nationality group for which it is a 'national home' and which makes up the majority of the population, except

that in Kazakhstan the population is only about 30% Kazakh, compared with over 50% Slav. An SSR is intended to give a measure of self-expression to each of the major national groups within the USSR, while smaller national and linguistic groups are recognised by other divisions, notably the Autonomous Soviet Socialist Republic (ASSR) usually an enclave within the great RSFSR, while other divisions are purely administrative and not connected with nationality or language. Further details are given in the table on page 3, Fig. B (page 4) and in Chapter 12.

During the years when revolutionary groups were preparing the way for the overthrow of the Tsar the term 'soviet' (council) became adopted for a revolutionary group or cell, and seizure of power by the Bolshevik Party was achieved largely through the efforts of the members of the soviets of workers, soldiers and sailors. The word thus became an honoured term in the language of the revolution and its adoption in national and regional administrative divisions symbolised the transfer of power. Similarly, the term 'socialist' in the titles signifies the organisation of the state for the common good, rather than for the benefit of royal, aristocratic and other powerful groups.

Soviets at republic and local levels, and the various committees and administrative bodies of all these organisations provide opportunities for members to carry out useful work. Local soviets are elected for two and a half years (formerly two years), their work being largely concerned with educational, health and cultural affairs.

Most of the period since N. S. Khrushchev was ousted from the leadership (abruptly, but peace-

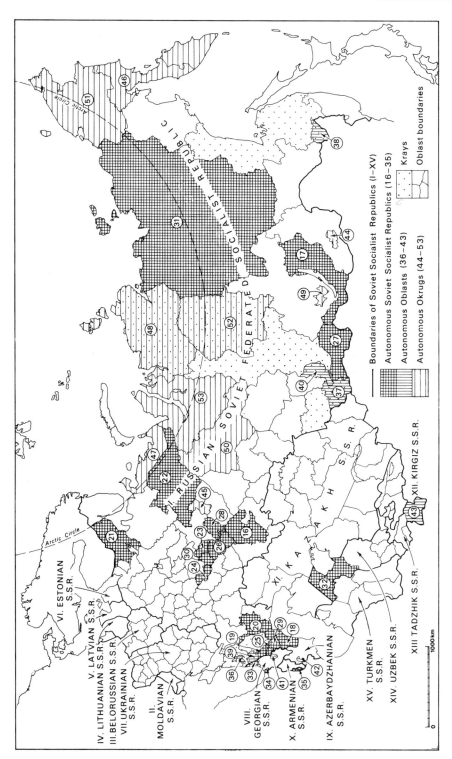

Fig. A Political and administrative divisions of the Soviet state (Numbers key pp. 246–7)

ADMINISTRATIVE DIVISIONS OF THE SOVIET UNION

The Soviet system of local administration involves the following major units:

English name	Russian name	Description
Union of Soviet Socialist Republics (USSR)	Soyuz Sovetskikh Sotsialisticheskikh Respublik, i.e. SSSR (Cyrillic letters CCCP)	Originated as a union of several nominally separate republics, as indicated by its name.
Russian Soviet Federated Socialist Republic (RSFSR)	Rossiyskaya Sovetskaya Federativnaya Sotsialisticheskaya Respublika	Comprises European Russia and other areas where Russians are dominant, i.e. Siberia and the Far East.
Soviet Socialist Republic (SSR)	Sovetskaya Sotsialisticheskaya Respublika	The term is applied to the 14 republics (e.g. Kazakhstanskaya SSR) which, together with the RSFSR, constitute the Soviet Union.
Autonomous Soviet Socialist Republic (ASSR)	Avtonomnaya Sovetskaya Sotsialisticheskaya Respublika	Are contained within the RSFSR or an SSR and represent the homelands of important minority groups, e.g. Tatarskaya ASSR.
Autonomous Oblast, (AOb) and Autonomous Okrug (AOk)	Avtonomnaya Oblast, Avtonomnyy Okrug	Administrative divisions with a limited degree of local autonomy. The bulk of the population belongs to one of the smaller minority groups, e.g. Khakasskaya AOb, Koryakskiy AOk. Prior to 1977, Autonomous Okrugs were known as National Okrugs.
Oblast	Oblast	The basic administrative division of the RSFSR and most SSRs; usually named after its 'capital' and consisting of a town and the surrounding area, e.g. Leningradskaya Oblast.
Kray	Kray	A larger administrative division, found only in the RSFSR, e.g. Primorskiy (Maritime) Kray in the Far East.
For smaller divisions see page 247		
Economic Region	Ekonomicheskiy Rayon	In some cases subdivisions of republics comprising several oblasts; in others they consist of a single republic (e.g. Kazakhstanskiy Ekonomicheskiy Rayon); or may unite several republics, e.g. the Central Asian Economic Region (Sredneaziatskiy Ekonomicheskiy Rayon).

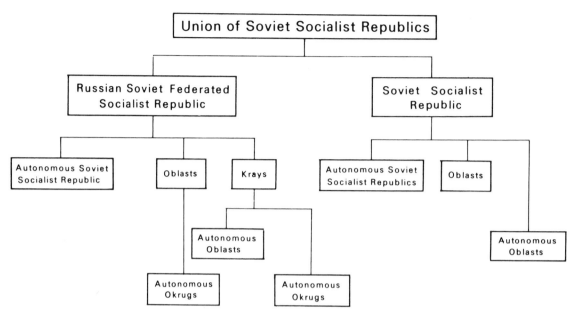

Fig. B Political-administrative structure of the USSR. The term 'Autonomous okrug' was adopted in 1977 for the areas previously known as 'National okrugs'

fully, so that he was able to enjoy a 'normal' retirement) the Soviet Union experienced a long period of conservative rule under L.N. Brezhnev, a period which is now regarded as one of stagnation. After Brezhnev's death, there were two 'caretaker' leaders—Chernenko and Andropov. Then came the appointment in 1985 of Mikhail Gorbachev whose reforming zeal had already been noted by western 'Kremlinologists' but whose immediate declaration of *'perestroika'* (restructuring or reform) of the Soviet economy ushered in a new era in the life of Soviet citizens and relationships with other countries.

The primary aim of perestroika, a word which quickly entered into the languages of all major western countries, was to inculcate into the Soviet economy as much as possible of the dynamism of free enterprise without sacrificing the fundamentally socialist ideals and organisation of the country. On the 1st January, 1988 a wide range of industries, including agriculture, became responsible for managing their own finances, subject to overriding commitments to the state. Some aspects of this economic revolution will be dealt with in appropriate chapters of this book as indicated in the Preface to this edition.

Major political changes were also initiated in 1988 in the direction of much greater democracy at all levels. These should go some way to curbing the power of the Central Committee of the Communist Party, its secretariat, and, in the regions, the local Party officers. Criticism of the Party will be possible at all levels and alternative political parties will be permitted.

A new constitution seeks to provide for freedom of information, conscience and dissent, and to make secure the rule of law within the framework of a one-party state. The Communist Party is at the time of writing the only political party but elections may now be contested by individuals of an infinite number of persuasions. In the 1989 elections for the Congress of Peoples Deputies many of the official Party candidates were rejected by the popular vote. Furthermore only 100 of the 2 250 seats are reserved for the Party while 650 are for nominees of the other national organisations such as trade union groups, women's organisations and other societies. 1 500 members are elected by constituencies and districts throughout the Union. A further provision in the parliamentary election ensured that no candidate polling less than 50% of the vote should be elected without a second poll of the top two candidates or reselection if the candidate had been unopposed.

The Congress elects the 542 members of the Supreme Soviet by secret ballot. The Supreme Soviet has become a true legislature sitting for two four-month terms each year. The President is also limited to two five-year terms, Mr Gorbachev becoming the first to hold this office following the elections in the spring of 1989, while retaining the post of General Secretary of the Party.

Thus open criticism of economic and political, including nationalistic matters, is now accepted and there is scope for substantial divergence of opinion and opposition to the conservative elements in the Communist Party

BIBLIOGRAPHY

Aganbegyan, A. (1988), *Challenge: economics of perestroika*, Hutchinson, London.

Aganbegyan, A. (ed. in chief) (1988), *Perestroika annual*, Futura, London.

Bater, J. H. (1989), The Soviet Scene, *A Geographical Perspective*, Edward Arnold.

Brown, A. *et al.* (eds) (1982), *Cambridge encyclopedia of Russia and the Soviet Union*, Cambridge University Press, Cambridge.

Dellenbrant, J. A. and Hill, R. J. (1989), *Gorbachev and perestroika*, International Library of Studies in Communism, vol. 1, Edward Elgar, Aldershot.

Frolic, B. M. (1972), Decision making in Soviet cities, *American Political Science Review*, **66**, 38–52.

Glebov, O. and Crowfoot, J. (ed. and trans.) (1989), *The Soviet Empire: its nations speak out*, Harwood Academic Publishers, New York.

Hill, R. J. (1985) *Soviet Union, politics, economics and society from Lenin to Gorbachev*, Pinter, London.

Lane, D. (1985) *Soviet economy and society*, Blackwell.

Lydolph, P. E. (1990), *Geography of the USSR*, Misty Valley, Wisconsin.

McCauley, M. (1987), *The Soviet Union under Gorbachev*, Macmillan.

Murarka, D. (1988), *Gorbachev, the limits of power*, Hutchinson

Novak, M (1988), Taking glasnost seriously: towards an open Soviet Union, American Enterprise Institute for Public Policy Research.

Pallot, J. and Shaw, D. (1981), *Planning in the Soviet Union*, Croom Helm, London.

Sacks, M. P. and Pankhurst, J. G. (1988), *Understanding Soviet Society*, Allen and Unwin, London.

Sallnow, J. (1989), *Reform in the Soviet Union: glasnost and the future*, Pinter, London.

Schöpflin, G. (ed.) (1970), *The Soviet Union and Eastern Europe, a handbook*, Anthony Blond, London.

1 The Evolution of the Russian State

No other state has managed to control such a vast, contiguous yet diversified territory as the Soviet Union. The British Empire in its heyday, before granting independence to many of its overseas constituents, was comparable in total area but comprised many scattered overseas colonies held together by seapower. The Russian Empire grew by overland penetration into the vastness of Siberia and Central Asia, which fell to the Tsars almost by default of other powers not contesting ownership. The Soviet Union has consolidated its hold on this vast area. The Soviet constitution provided for constituent republics to secede but such a move was not expected until *perestroika* (reconstruction) and *glasnost* (openness) opened the gates for nationalist ambitions in the late 1980s. Few outsiders had appreciated the important distinction between 'Soviet' and 'Russian'. The protestations came mainly from non-Slav peoples in the Baltic and Transcaucasian areas, only a tiny proportion of the total number of nationalities whose experiences combine to make up the full history of the Soviet Union.

These groups are described in a later chapter, and for a brief summary here of the historical geography of the Soviet Union and of its predecessor, the Russian Empire, it must suffice to summarise the development and geographical diffusion of the dominant ethnic group, the Slavs, who eventually brought the other groups under their control. The Slav civilisation of central and eastern Europe was itself the result of a long period of evolution. From early Paleolithic times there was a gradual northward movement from the Middle East and the Black Sea coasts into the steppe and forest regions. As the ice retreated groups of hunters penetrated into the northern forest zones and an important Mesolithic settlement has been identified by archaeologists as far north as Kunda in the Estonian SSR.

While hunting, fishing and gathering of food remained the basis of life for most communities, agriculture in the form of forest-fallow cultivation with the rearing of livestock began to appear in the Neolithic period (fifth to second millenium BC). This was particularly so in the southern areas where climatic and soil conditions were most favourable and nearness to the hearths of agriculture in the Middle East and central Asia facilitated the transfer of ideas, seeds, plants and livestock. A pastoral–agricultural economy was well developed in the succeeding Bronze Age over a large part of the present-day territories of the USSR. Metal working was particularly well developed in the Caucasus and Transcaucasian areas, and this was also the region in which iron working developed first, but in the Iron Age such crafts spread northward and also eastward into Siberia and encouraged trade as well as agriculture.

From about the eighth century BC new raiders and colonists were appearing along the Black Sea coast and establishing settlements there and trade developed, particularly under the stimulus provided by the Greeks. Then the area fell under Roman domination, but there were also many incursions by a variety of raiders and the towns fell into decay, and migration northward reinforced the communities that had developed in the steppes and forests. It is to this period that the

Soviet archaeologists and historians look for the signs of the first Slavonic groups in central and eastern Europe, notably in the basins of the Vistula and Dnepr and in Volynia, and for the split of the Slavs by about 500 BC into eastern, southern and western groups, with the former becoming much the largest group and providing the basis of the future Russian state.

During the first to ninth centuries AD the east Slav tribes developed agricultural and trading communities along the Dnepr, Desna, Dnestr, Volga and other river valleys. They were subject to raids by marauding tribes including the Goths, Huns, Avars, Khazars and Bulgars from the east and south, while from the north the forest lands were penetrated along the river routes by the Vikings, or Varangians, as the Slavs called them. The Scandinavians became the most consistent raiders and colonists. Trading and intermarrying with the Slavs, they appear to have played an important role in the development of the still separate communities to which the name Rus became attached. By 862 Rurik had established Novgorod as the capital of a small but distinct princedom in the northern forest zone which had become known as Rus. On the southern fringes of the forest-steppe belt, Kiev had evolved as a leading city state and in 882 this fell to Oleg, Rurik's successor. Kievan Rus became the most advanced of the embryos of the Russian state but there were many other communities evolving into small states, many of them coming under the rule of the Kievan princes for a time. In spite of raids by the Pechenegs and others Kievan Rus flourished and the adoption of Christianity late in the tenth century facilitated the forging of links with Constantinople and royal houses throughout Europe.

In Kievan Rus forest-fallow forms of cultivation persisted but there was some transition to permanent fields, and cultivation was aided by a variety of implements—ard irons, plough shares, coulters, sickles and scythes. Archaeological investigations reveal well-made grain pits, with millet as the favoured crop, no doubt because of its drought-resistant qualities appropriate to forest-steppe conditions. Established field systems with wheat and rye (more suitable in northerly latitudes) gradually became common in the central forest areas and even in Novgorod by about the eleventh century. By this time also there was increased reliance on domestic livestock, as op-

posed to hunting, with pigs probably most numerous, but cattle, sheep and goats also common.

Other raiders came to Kiev, which was particularly exposed to the nomads of the steppes and after the city was sacked by the Polovtsy or Cumans it failed to regain its previous eminence and the areas more protected by the forests from marauders gained in strength, notably Vladimir–Suzdal. Exports from the forest lands included furs, honey and wax, while from the farms came flax, hemp, hides, skins, suets, tallow and grains. Local craftsmen developed manufactures based on these raw materials. Under the growing pressure of population, improved field systems developed, and two-field agriculture, in which one field was left fallow after a crop, merged into threefield systems, in which two out of three fields were productive each year. The light *sokha* plough was generally used in the northern districts, where glaciation had deposited numerous boulders, while the *ralo* was used on heavy soils. This more cumbersome plough needed a team of draught animals and may have been one of the factors encouraging the growth of the commune and the gradual introduction of slavery and of serfdom by the more powerful members of the community.

By the eighth to ninth centuries towns were becoming an important factor in economic, social and administrative systems. The need to establish strongholds to control areas over which princes claimed suzerainty, to keep out raiders and to store tribute led to the creation of many towns. The security they offered, as well as trade links, would then lead to the growth of craft industries, trading rows and markets. Churches, cathedrals and monasteries also stimulated urban developments.

THE TATAR PERIOD

The thirteenth century saw a reversal of the progress that was being made throughout the Russian lands, as the Tatar nomads began sustained raiding of the steppes and repeated incursions into the forest zone. These people, also known as Mongol, possessed superior military skill and exploited to the full their horsemanship, which gave them total supremacy in open land. They inflicted a series of defeats on Slav forces from 1223 onward and gradually conquered Transcaucasia and southern Russia, culminating

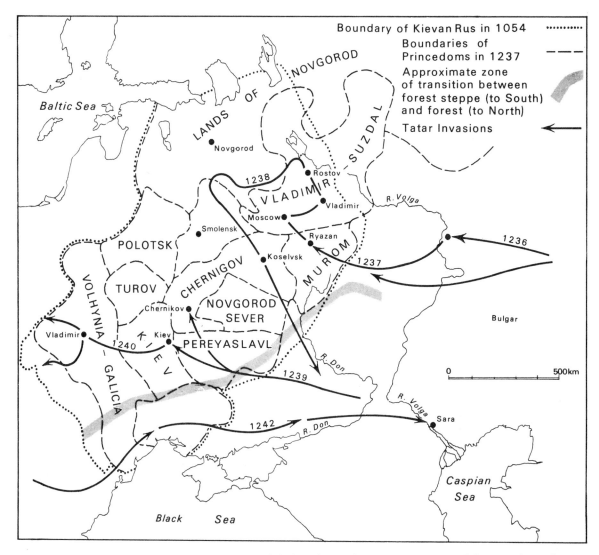

Fig. 1.1 Tatar invasions of the thirteenth century and the boundaries of Kievan Rus in 1054 and the princedoms of 1237. The forests provided some protection from the raiders who easily penetrated the forest steppe from the more southerly treeless steppe

in the fall of Kiev in 1240. The principalities which had emerged in the forest zone as independent of Kiev and Vladimir–Suzdal fell under Tatar control, although in due course their princes were permitted to remain leaders if they collected tribute for the Khan who had established the capital of the Golden Horde at Saray on the Volga.

Novgorod remained independent and had trade links with the Baltic and Finnish lands, trading especially in the furs collected in the forests from as far as, and occasionally beyond, the Ural mountains. However, it had its own campaign to fight against the incursions of the Swedes and Germans who were organised in orders of Knights. In 1242 the victory of Alexander Nevsky in the 'battle on the ice' assured the survival of Novgorod, but it was not entirely free of raids and partial control by the Tatars. During the second half of the thirteenth century, the

Tatar yoke fell heavily on the Russian people as trade, agriculture and crafts were affected by the disruption and exaction of tribute. Nevertheless, firm administration by the Tatars gradually re-established the trade routes and eventually disruption within the Tatar state itself facilitated the rise of new forces within Russia.

It was Moscow, first mentioned in the Annals of 1143, that emerged as the nucleus of the new Russian state. It had no doubt benefited from its sheltered position in the forest lands and from a strategic position in relation to communications. Its merchants had access to river routes in all directions, for here the upper reaches of the Volga make a great loop to the north of the city, and southward the Oka cuts across to join it, and so also gives access to the upper Don and other rivers by easy portages. As in the wooded steppe near Kiev, the soils were relatively easy to cultivate with the available implements, so facilitating the development of agriculture.

Moscow, however, was not free from the attacks of the Tatars and not until 1328, when it acquired the task of collecting the tribute from other principalities for the Tatars, did it begin to assume the role of leadership. In 1326 the seat of the Metropolitan of the Orthodox Church was moved from Kiev to Moscow and when Constantinople fell to the Turks in 1453 its mantle was held to have fallen on Moscow—the 'Third Rome'. Meanwhile, the Tatar leaders had been converted to Islam, which helped to prevent any union between the rivals and stimulated more church support for the Russian princes. In 1380 Dmitry Donskoy defeated the Saray Tatars at Kulikovo on the Don. Two years later Moscow was burnt by Tokhtamysh, but before the end of the century Tokhtamysh's Golden Horde was in turn defeated by a rival eastern prince, Timur the Lame, or Tamerlane. The Golden Horde remained strong enough to besiege Moscow again in 1408, but soon after began to disintegrate and a partial split took place with the formation of the khanates of the Crimea, Kazan, Astrakhan and Siberia. In 1480 a final, unsuccessful, campaign by the Tatars was followed by the denunciation of

St Basil's Cathedral and the Kremlin, Moscow

the 'Yoke' by Ivan III, 'the Great'. Novgorod, Tver, Vyatka and other local princedoms were gradually absorbed into Muscovy.

Moscow was now firmly set on the road of expansion and during the long reign of Ivan IV 'the Terrible', who was crowned 'Tsar' (derived from Caesar) and grand prince of all Russia, the khanates of Kazan and Astrakhan were subjugated and the whole of the Volga area was brought under Russian rule. During this time, Russia was visited by the Englishman, Richard Chancellor, and the Muscovy Company was formed for trade between England and Russia, leading later to the founding of the port of Arkhangelsk (Archangel) on the White Sea.

Thus, by the middle of the sixteenth century, the forest state of Moscow had emerged to undisputed leadership of the east Slavs, and one of the major themes of Russian history and geography was already beginning to develop—the search for outlets to the sea to enable the benefits of trade to be realised. Sweden, Lithuania and Poland blocked the way to the Baltic Sea while, in the steppes to the south, the Tatars were still strong enough to bar Russian access to the Black Sea. As late as 1571 the Crimean Tatars raided Moscow and fortifications were built along the line of the wooded steppe, where forest shelter was available. In this frontier zone there grew up communities of a semi-military nature, formed by people who sought independence from serfdom and oppression in Poland, Lithuania and Muscovy and took the name of Cossacks, meaning 'free warriors'. They are known mainly for the part they later played in the expansion of the Russian empire, being employed by the Tsars as frontiersmen and soldiers to occupy and control conquered areas.

The conquest of Siberia

The Ural mountains and the Ural river, which flows from the mountains to the Caspian Sea, have long been regarded as a border between Europe and Asia. Today, this border has no meaning, but in the first half of the sixteenth century it was a frontier zone beyond the limits of Russian civilisation, developed from Slav, Greek, Roman, Scandinavian and Germanic cultures, with the way barred by the Tatar khanate of Kazan. When Kazan was captured in 1552 the Russians could advance into this vast, almost untapped hunting-ground from which could be obtained great quantities of furs, the greatest source of wealth in Russia.

The Stroganov family, possessors of great estates in Russia, acquired new lands in and beyond the Urals and sent a Cossack army to conquer the Siberian Tatars. This they did, though they lost their leader, Yermak, in battle. Tyumen fort was built in 1585 and soon there was a string of strongholds and trading posts along the valleys of the Tobol, Irtysh and Ob, a vast river system flowing from the mountains and steppes through the forest to the tundra and the Arctic Sea. Now began in earnest the battle against nature in the frozen north and the *tayga*—the great coniferous forest that stretches in a belt some 8000 km across northern Russia and Siberia. Incredible hardships were tolerated by the traders, soldiers and administrators who would trek by river valley and forest

Tamara's Castle, in the glaciated Terek Valley, one of the main routes through the Caucasus mountains, now used by the Georgian Military Highway. Queen Tamara ruled 1184–1213 when Georgia was a powerful and cultured state

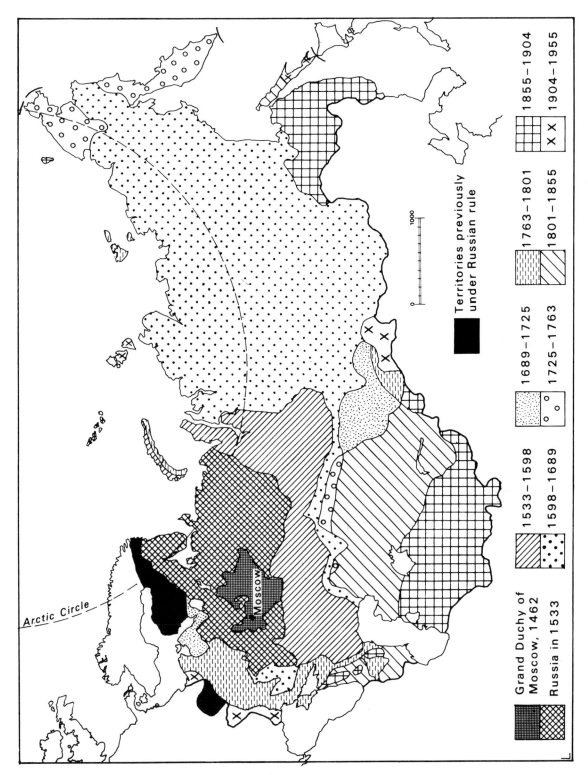

Fig. 1.2 The expansion of the Russian Empire. Some simplification has been necessary in the interests of clarity, especially in respect of gains and losses of territory in the European areas

trail from Moscow or, in the brief summer, would be shipped round from the White Sea to the mouth of the Ob, many of them being ship-wrecked and falling among hostile tribes on the way. From Moscow to a base at Tobolsk or Tyumen could be (depending on route) a journey greater than that from Moscow to London, and from the Siberian base the Russians would then work hundreds of kilometres out into the forest, bargaining with the tribes for furs and wintering in simple huts through temperatures 40°C or more below freezing point.

With the Cossacks often pioneering the routes, they sailed up the rivers, dragged their heavy boats over long portages and took the next favourably oriented valley to pursue the route eastwards. In winter, the frozen rivers and lakes provided travel conditions often better than the summer, when other perils and hindrances from rapids to mosquitoes took their toll. They pene-trated beyond the Yenisey into eastern Siberia and, another thousand miles on, founded Yakutsk in 1632 on the Lena, and, penetrating the last mountain barriers, reached the Pacific Ocean in 1648. On the sea of Okhotsk they set up more trading posts and went on to explore the vast area that still lay between them and the northern seas.

The eighteenth century saw colonisation of Kamchatka, a peninsula which looks insignificant on the map of the USSR, yet is as big as Great Britain. After earlier setbacks, the Russian fleet under Bering proved the existence of the straits named after him and paved the way for later Russian colonisation of Alaska. The Tsars had acquired a vast empire wrapped round a third of the globe. They were not very much interested in it, except for the wealth it could provide—furs, salt, gold and other minerals.

After the strong reign of Ivan the Terrible, Russia suffered severe problems of succession. Ivan had killed his own son in a fit of rage and for thirty years Moscow was the scene of plot and counter-plot to secure the throne. The Poles supported pretenders to the Russian throne and actually took command in Moscow, only to be ejected by a local uprising. Eventually a youth of 16, to whom there were fewer objections than to any other claimant, was elected Tsar. He was Michael Romanov, and thus began the line that was to hold the Russian throne until the re-volution, 304 years later.

A medieval tower in Baku, Azerbaydzhan, used as a refuge when raiders came across the Caspian Sea. Parts of the Soviet city can be seen behind the old Moslem town

RUSSIA UNDER THE ROMANOVS

Despite the great advances to the east, the Romanovs were most concerned with their posi-tion in the west, where the boundaries of Russia met those of other powerful states. It was in western Europe that the great advances in science and technology were being made and it was to the west that Russia turned for help in its own modernisation. Peter I came to the throne in 1689 already aware of western progress, and visited Prussia, Hanover, Holland and England to acquire practical knowledge of new techniques and re-cruit foreign engineers, artisans, surgeons and other specialists. When he returned he had a new city built on the marshy land near the Baltic, on which the Russians had gained a precarious hold

The Bronze Horseman, a striking statue in Leningrad commemorating the foundation of the city, formerly called St Petersburg, by Peter the Great

against Swedish power, and called it his 'window on the west', St. Petersburg. He transferred the capital to the new city in 1712 and from its foundation it was an important port. In a few decades it was a large industrial city, yet distinguished for the charm of its planned layout and the beauty of its buildings. Moscow was to be eclipsed as a capital for over 200 years, but with its central position in European Russia and the momentum of its past it continued to grow, though at a rate reduced by the loss of its functions as a capital.

Peter initiated many reforms including systematic organisation of the provincial governments, subordination of the nobility and clergy, currency reform and educational developments ranging from introduction of a system of elementary education to the foundation of the Academy of Sciences. The effects were in some cases slow to permeate the economy but industrial development was aided by the experts from western countries. Factories were set up, encouraged by the state, not only in Moscow and St. Petersburg but also in the Ural region, where iron ore was available. Shipyards and artillery works were set up or modernised to provide the basis of a reformed army and navy.

Wars were fought against Turkey and Sweden. The Russians had great successes in the north, enabling Russian rule to be extended over Karelia, Estonia and other areas, so consolidating the Russian position on and near the Baltic Sea. Still, however, Russia lacked access to a warmer, icefree sea, and Peter's failures to defeat the Turks left this a problem to be tackled in the latter half of the century by Catherine the Great. For over 30 years under Catherine (1761–96), Russia made appreciable territorial advances, acquiring the Black Sea steppes, absorbing the lands of the Cossacks and, eventually, most of the Ukraine and the Crimea. Although the Bosphoros–Dardanelles passage from the Black Sea to the Mediterranean was to cause trouble with Turkey and the western European powers right through to the twentieth century, Russia now had access to warm-water ports. Odessa, it is true, does freeze up for about a month but St. Petersburg was closed for about five months and Arkhangelsk even longer. Also, the communications with the Mediterranean countries were economically as well as strategically valuable, though not until the Suez canal was opened did the full benefit of the Black Sea possession become evident. Meanwhile, however, foreign trade was facilitated and within the country internal customs duties were abolished and roads improved. A notable achievement was the completion in 1773 of the Yakutsk Track from Irkutsk to Yakutsk, linking with the Yakutsk–Okhotsk track, so strengthening Russian power in the Far East.

On the fluctuating western frontiers also, Russia made progress. In the dismemberment of Poland by the three partitions, 1772, 1793 and 1795, Prussia, Austria and Russia each took over large tracts of territory. Russia also acquired Lithuania and Latvia, with the useful port of Riga, and in 1809, control of Finland, so consolidating her position on the Baltic, while Caucasian conquests still further strengthened her in the south. Garrison towns and fortified lines were constructed to police the new territories and these provided a base for further colonisation and settlement.

Russia was now a mighty European power and its image in the west was greatly strengthened when, under Alexander I, the Russian armies and the Russian winter defeated Napoleon during his retreat from Moscow in 1812 and opened the way for his final overthrow.

The revolutionary movement

Participation in the Napoleonic Wars was one factor which stimulated the dissatisfaction felt by many of the intelligentsia and aristocracy of Russia. Many of the officers who went to fight in western Europe saw how much more advanced other countries were in economic and political affairs. The French and American revolutions had already encouraged discontent with the autocracy of the Tsars; now the demand for a constitution setting out the rights of the people grew apace. The matter flared up when Nicholas I came to the throne in December 1825 and a number of officers paraded their troops, demanding a constitution and an end to serfdom. They were crushed savagely and the leaders executed or exiled. Bitter repression was extended to everyone suspected of liberal ideas and Russia became a police state.

During the nineteenth century the name of Siberia became increasingly associated with exile and imprisonment. It should not be overlooked that to many of those who had gone there to settle in the past Siberia had been a land of promise, of freedom. The serfdom, which had been tightened up step by step throughout the Romanov period, bound the serfs rigidly to the landowners' estates and made them, in effect, the property of the landlords, but if they could escape to Siberia, where serfdom was not fully established, they could become free. From the sixteenth century, however, the state had used Siberia as a place of exile. This had the double advantage to the tsars of removing their troublesome subjects from the cities and helping to exploit the resources of the area. Siberia itself, however, benefited from the presence of some of the political exiles who spent much time experimenting with crops and trying to improve the living conditions of the people among whom they were cast.

Neither death nor exile could stamp out the reform movement. Liberal writers defied the censorship and intellectuals went to work among the peasants to try to ease their lot by taking to them some of the benefits of improving medical and

Petrodvorets, the summer palace of Peter the Great near Leningrad. The spacious gardens, now available to all, are noteworthy for their large number of varied fountains

technical knowledge. After the long-delayed and inadequate measures to emancipate the serfs, beginning in 1861, unrest became more violent because of disappointment at the smallness of the holdings created for the peasants and other terms of the emancipation and the general lack of progress. In 1881 the Tsar, Alexander II, was assassinated, ironically just as he was about to publish a constitution, which was promptly suppressed. Affairs went from bad to worse. The gospel of Marxism now spread to Russia. One of those who saw in it salvation for the masses was the embittered brother of a student executed for complicity in a plot in St. Petersburg. Under the assumed name of Lenin he was later to be the main figure in the foundation of the Soviet Union, but this was not until after the last of the Tsars, Nicholas II, had plunged the country into the horrors of the First World War and taken it to the brink of destruction.

The first half of the nineteenth century saw appreciable industrialisation (the first cotton mill in St. Petersburg was built in 1798) and corresponding improvement in communications. Canals were built, especially linking the north with the Volga, the St. Petersburg–Moscow highway was surfaced, and in 1837 Russia's first railway was opened. In 1851 St. Petersburg and Moscow were linked by rail and numerous lines were opened in the 1860s and 1870s. Lines from the Ukraine made possible large scale movement of grain to the growing industrial cities as well as for export, and also the movement of iron ore, fuel and finished iron goods, which enabled the Donbas to become Russia's principal iron-producing region before the end of the century.

The final phase of imperial expansion

After the debacle of the Crimean War (1853–6) Russian expansion in Europe was firmly blocked by the western powers. In the east, however, there was still largely a power vacuum between Russian settlement across the Siberian steppes and along the north Pacific coasts and, in the south, British power in India and the weak, but populous Chinese empire.

The seventeenth century exploration of Siberia and the Far East had been checked when the Russians, few in number, had come to the borders of the Chinese realm at the Amur river. In the Treaty of Nerchinsk, 1689, they had been forced to accept the Chinese claims to the Amur, which denied them access to this great river and the cultivable land much needed for the supply of their settlements so far from the western grainlands. During the Crimean war, Count N. N. Muraviev, the Governor-General of Siberia, decided to force a concession from the Chinese, and to show the strength of the Russians he took a flotilla down the Amur and was, in fact, just in time to repulse a British expedition against Kamchatka. The Chinese were duly impressed and by 1860 the Russians had taken possession of the Amur and Maritime Provinces, and founded Vladivostok as far south as possible. By no means all the Tsar's advisers were convinced about the desirability of extending the Far Eastern empire and Muraviev had to fight hard for his cause. Eventually the Russian position on the Asian mainland was consolidated but Russian interests in Alaska and California were sold to the United States. Pressure was successfully brought to bear on the Japanese to recognise Russian control over Sakhalin.

As a result of the efforts of another vigorous proponent of eastern development, Count Witte, the Trans-Siberian Railway was commenced in 1891. By treaty with the Chinese the final section was built across Manchuria to Vladivostok, thus avoiding the long and circuitous route of the Amur valley, which was developed later. Against the wishes of Witte, who believed in peaceful, commercial co-operation, other concessions were wrung from the Chinese as Russia belatedly sought to share in the scramble initiated by the western powers to exploit Chinese weakness by establishing ports under their own control. Russia obtained control over north Manchuria, the Liaotung Peninsula and Port Arthur. This brought them into conflict with the Japanese, at whose hands Russia suffered a humiliating defeat in 1904 and the loss of its Chinese gains, except the railway route.

During this period after 1860 the Russians also advanced into the central Asian deserts, which had long kept them at bay. The oases and montane valleys north and west of the Pamir, Tyan–Shan and adjoining mountain ranges were occupied by Turkic and largely Moslem peoples organised in khanates or sultanates, some with well developed agriculture, others with a nomadic way of life. This was the country of the old Silk Road to China, opened by Marco Polo in the fourteenth century. These ancient states, such as

Samarkand and Bukhara, not only attracted the Russians who were bent on military control and trade development, but also gave them excuses to intervene because of the raiders who periodically came out of the desert to attack Russian settlements in southern Siberia.

The Russians began to advance from the Lake Balkhash area in the middle of the nineteenth century, building the new town of Alma Ata in 1854. Tashkent was taken in 1865, and Samarkand in 1868. During the 1880s a railway was built from the Caspian Sea to Samarkand. Little by little the whole of the area right up to the borders of Afghanistan was brought under Russian control, leading to considerable tension as the British felt their interests challenged. Eventually, after the Russian defeat in the Far East by the Japanese, the respective spheres of interest of the Russians and British in the Near East were delineated and the borders stabilised.

Thus, as the rivalry of the great powers deepened in the early years of the twentieth century, a rivalry which in other spheres led inexorably to the catastrophe of the First World War, the Russian Empire spread over much the same proportion of the globe as the British, with no other comparable territorial domination. It was, however, a land empire, one consolidated block, dependent on land rather than sea communications. But compared with Britain and other western European powers, Russia was weak, economically, militarily and administratively. It was still an autocracy and a police state. After the attempted revolution of 1905, which had followed the humiliation of the Russo–Japanese war and growing discontent among intellectuals, proletariat and peasantry, Nicholas II had conceded the formation of a parliament, the Duma, but kept control of it by periodic dissolution and rigging of the electoral machinery. When Nicholas declared war on Germany on the side of the western allies, he thought he would thereby strengthen Russia, the citadel of autocracy and the divine right of kings. No ruler ever made a greater mistake.

The economy under the Tsars
The weakness of the Russian Empire in the war stemmed largely from lack of economic development and organisation in all sectors of industry—primary (agriculture, mining etc.,) secondary (manufacturing) and tertiary (including the transport network). It is true that industrialis-

ation had been proceeding fairly steadily, mostly in the Moscow, St. Petersburg, Ural and Ukraine areas. Whereas iron-working in the Urals had developed early in the eighteenth century, based on local iron ores, charcoal and water power, in the nineteenth century Donbas coal attracted heavy industry to the south. Cotton and flax mills were built in the towns around Moscow as well as in Moscow itself. Moscow and St. Petersburg became centres of engineering. These cities, those in the Baltic provinces and other northern towns manufactured clothing, shoes, chemicals, rubber and other goods.

By 1913 a substantial industrial base had been built up, but it was small in relation to the size and needs of the country, and did not compare with the industrialisation of Britain, Germany, France or the USA. Agriculture lagged perhaps even more strikingly compared with the western countries. Though large quantities of grain were exported, these shipments were achieved often at the expense of shortages within Russia. Serfdom had been abolished but poverty and indebtedness remained rife in the countryside. Land reform came late and had progressed but little before the catastrophe of war denuded the countryside of its able-bodied men and laid it open to the ravages of invasion. The weakness of the transport system meant that even when harvests were good, food supply in the towns was an uncertain business.

Russia, at the outbreak of the First World War, was a giant invested with some of the trappings of expanding capitalism, but lacking efficiency and the knowledge with which to realise its own potential.

THE CREATION OF THE SOVIET UNION

The war soon revealed the weakness of Russia. Soon the Tsar's armies were falling back on all fronts. Incredible suffering was borne by the Russian troops, not worse than that endured by those fighting in the horror-landscapes of Ypres, Verdun, Passchendaele and the Somme but on an even larger scale. The Russians, promised a quick victory, grew more disillusioned as the years went by. Successes were few and defeats many. The supply position grew worse. The railways became less and less reliable for want of maintenance and replacements, the farms, increasingly depleted of

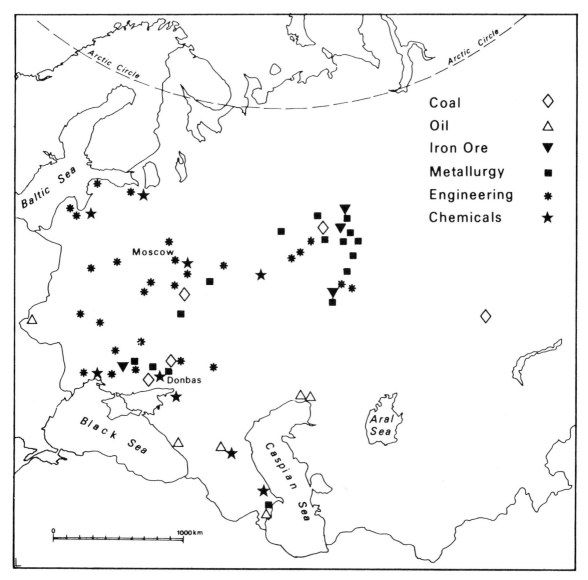

Fig. 1.3 The principal locations of heavy industry in Russia in 1913. The main industrial regions were the Donbas and Ural areas, while manufacturing industries were developed near Moscow and the capital, St. Petersburg (now Leningrad). Textile industries (not shown on the map) were particularly important near Moscow. Food processing industries were more widely scattered. There were also a few industrial plants in Siberia, mostly located beside the Trans-Siberian Railway

manpower, produced less, and it became more and more impossible to supply either the fighting troops or the armaments workers in the towns.

January 1917 was a terrible month, with temperatures well below freezing and the capital foodless. The leader of the Kadet Party, a liberal

group in the Duma, had already said that the government's incompetence bordered on high treason. February saw the consolidation of opposition, and in March the workers took things into their own hands. A wave of strikes began in Petrograd, as St. Petersburg had been renamed, and Moscow. Processions came out with banners carrying slogans such as 'Down with the war' and 'Give us bread'. Troops were ordered out, but after a few days, themselves mutinied. The Tsar made a last attempt to return from the front, where he had gone to take personal command from his generals, was prevented, and was forced to abdicate.

The 300-year rule of the Romanovs was ended and a Provisional Government was formed by the Duma. Its authority was challenged from the beginning by the Petrograd soviet (council) of workers and soldiers which formed an executive committee of leaders of the socialist parties. Lenin returned from abroad to take the leadership of the Bolshevik party, then in a minority position in the soviet. An abortive rising by the Bolsheviks in July was suppressed and Lenin and other prominent Bolsheviks went into hiding. The Provisional Government, however, failed to pursue its advantages and dissipated much of its energies in trying to continue the war. On the night of 7 November (25 October by the Julian calendar then in use in Russia) the Bolsheviks struck and took control

The new government set up by the Bolsheviks lost no time in seeking a peace treaty with Germany, though it resulted in great losses of territory. Opposition mounted within the country and the state of the economy worsened still further. The Germans occupied the Ukraine and the Donets region and anti-Bolshevik armies were created in the north Caucasus and Transcaucasus. Russia's former allies, dismayed at being left to continue the war without an eastern front to split the German effort and by the rise of communism, themselves invaded the country in a number of areas in the north and south while the Japanese established a puppet state in the Far East.

Counter-revolutionary 'White' armies gained control over large areas but the new Red Army gradually recovered most of the former tsarist empire and the civil war was virtually over by early 1921. Finally, the Japanese were expelled from Siberia in 1922.

Although the Bolsheviks had preached international revolution and communism, when they came to power in Russia they soon revealed strongly nationalistic tendencies. One demonstration of this was the movement of the capital from Petrograd back to Moscow. This symbolised a return to the traditional Russian centre as well as providing a base more secure from attack. One of the 'planks' in their election programme was recognition for national minorities and they soon set to work to give large minorities separate republics. Attempts by several, notably the Ukraine, to achieve complete independence were, however, resisted and all were finally incorporated into the Union of Soviet Socialist Republics, which was the name adopted for the new state in 1922.

The aim of the new government was the establishment of communism as early as possible. This involved nationalisation of all land, factories, transport facilities and other means of production, previously privately owned, so that they could be developed on behalf of the people as a whole. Much of this was not, however, to the liking of the people, especially the peasants who wanted to possess their own land and who had supported the revolution thinking they would thereby gain it when the landlords were overthrown. To recover their support and to stimulate production generally during the crisis period, Lenin proclaimed in March 1921 the New Economic Policy, which recognised the right to private production and profit. Later, when Stalin had secured power after Lenin's death, a policy of forced collectivisation of farmland was introduced, and by 1931 over half of the peasants had been brought into the collectives. Many of those who resisted suffered the fate of their forbears under the Tsars, being exiled to work in labour camps in Siberia and other regions.

The Second World War
Stalin distrusted the western powers, for which perhaps he had some justification in view of the allied intervention against the Bolsheviks after their withdrawal from the war. After negotiations for a non-aggression pact with the British and French he swung over in 1939 to concluding a pact with Hitler, with the result that Russia was again caught in an unprepared state when the Germans invaded the USSR in 1941. Once again the Russian armies had to fall back almost to Moscow, but they held on tenaciously to Leningrad which was besieged for two years. The Russians pursued a 'scorched earth' policy to hinder the German advance and evacuated people, factories and livestock to the eastern regions. This aided the development of the Ural, Siberian and Central Asian regions. The retreat in

Lenin, principal Bolshevik theorist and first head of the Soviet state, is commemorated in the USSR in every possible way. Here his portrait and name dominate a Moscow intersection

the south was finally halted at Stalingrad (now Volgograd) in 1942 where great battles through the winter finally forced the surrender there of the German Sixth Army on 30th January 1943. It was another 28 months before the Germans were finally defeated and the Soviet losses amounted to close on 30 million dead and missing, and countless wounded.

Emergence from the war on the victorious side enabled the Soviet Union to play a major part in redrawing the boundaries of eastern Europe. They now obtained a boundary with Poland nearer to the line suggested by Lord Curzon on ethnic grounds than they had previously been given at the settlement in 1920, while Poland received compensation by being given districts that had previously been held by Germany. The USSR reincorporated Estonia, Latvia and Lithuania and other small additions were made at the expense of the defeated countries. These included several important strategic areas such as

Kaliningrad, a former East Prussian naval port, and parts of the Carpathians. In the east, the Tuva area, near Lake Baykal, became part of the USSR after having been at one time occupied by Russia, when nominally Chinese. The southern part of Sakhalin, which had been held by the Japanese since 1915, and the Kurile Islands were added to Soviet territory.

Thus, by 1945 the Soviet Union had, with some minor differences which approximately cancelled each other out in terms of area, regained virtually all the territory of the former Russian Empire, except for Finland.

COMECON

Following the defeat of Germany the Soviet Union proceeded to protect itself and secure its frontiers against further possible attacks by establishing control over the eastern European states between itself and the western powers. Intense political support was given to the 'patriotic fronts' and communist parties of the countries or, as in the case of Germany, the part of the country which had been occupied by Soviet military forces, the political activity being backed up by the continued presence of Soviet forces. Bourgeois and capitalist political parties were forced out and Soviet-type economies imposed. When, in 1947, the US Secretary of State, George Marshall, put forward the European Recovery Programme, under which the USA would supply raw materials, goods and capital to the devastated countries, the USSR rejected this as an 'instrument of dollar imperialism'. The following year the Organisation for European Economic Co-operation began to distribute the ERP funds and in 1949, in opposition to the Marshall Plan and the OEEC, the Soviet Union drew the eastern European states together in the Council for Mutual Economic Assistance. (CMEA or COMECON)

Through CMEA, the USSR has attempted to co-ordinate the economies of Poland, the German Democratic Republic, Czechoslovakia, Hungary, Rumania and Bulgaria with a high degree of industrial planning for mutual interdependence, especially through exchange of Soviet minerals and other raw materials for manufactured goods from the other countries. Following the creation and enlargement of the European Economic Community in the west the Soviet Union has sought to strengthen COMECON in all possible ways. The organisation was not limited entirely to eastern Europe, Cuba having been admitted to membership in July 1972.

In addition Soviet influence is now widespread in south-east Asia, Africa and other parts of the world through trade and the granting of economic and military aid to countries of the Third World. After becoming General Secretary in 1985 (and President in 1987) Mikhail Gorbachev initiated a policy of seeking closer cooperation with western European countries and rapprochement with the USA and, at the time of writing Europe is facing a major realignment of political and economic forces and linkages.

BIBLIOGRAPHY

Armstrong, T. (1965), *Russian settlement in the north*, CUP, Cambridge.

Crisp, O. (1976), *Studies in the Russian economy before 1914*, Macmillan, London.

Dobb, M. (1966), *Soviet economic development since 1917*, Routledge and Kegan Paul, London.

Florinsky, M. T. (1953), *Russia: a history and an interpretation*, 2 vols., Macmillan, New York.

French, R. A. (1963), The making of the Russian landscape, *Advancement of Science*, **20**, 44–56.

Friedberg, M. and Isham, H. (eds) (1987), Soviet society under Gorbachev: current trends and the prospect for reform, Sharpe, New York.

Gilbert, M. (1972), *Soviet history atlas*, Routledge and Kegan Paul, London.

Hosking, G. (1985), A history of the Soviet Union, Fontana.

Kochan, L. and Abraham, R. (1983), The making of modern Russia, 2nd ed., Penguin

Lyashchenko, P. I. (1949), *History of the national economy of Russia to the 1917 revolution*, trans.

L. M. Herman, Octagon Books, New York, 1970.

Murzayev, E. M. (1988), 'The geographical thinking of N. I. Vavilov, *Soviet Geography*, **29**, 666–684

Nasonov, A. N. (1987), 'The "Russian Land" and Formation of the territory of the ancient Russian state,' *Soviet Geography*, **28**, pp. 571–576.

Parker, W. H. (1968), *An historical geography of Russia*, University of London Press, London.

Pethybridge, R. (1974), *The social prelude to Stalinism*, Macmillan, London.

Portal, R. (1962), *The Slavs*, Weidenfeld and Nicolson, London.

Rozman, G. (1976), *Urban networks in Russia, 1750–1800, and premodern periodisation*, Princeton University Press, Princeton, New Jersey.

Shaw, D. J. B. (1977), Urbanism and economic development in a pre-industrial context: the case of southern Russia, *Journal of Historical Geography*, **3**, 107–122.

Sumner, B. H. (1961), *Survey of Russian history*, Methuen, London.

2 Physiography

One of the most striking facts about the USSR is its very great size. With a total area of 22.4 million km², the Soviet Union is, by a large margin, the biggest country in the world. From the western frontier to the Bering Strait is a distance of more than 9000 km and the greatest latitudinal extent, from the Arctic Ocean to the borders of Afghanistan, is well over 4000 km. Within this vast territory there is an enormous variety of geographical conditions, and this applies to structure and relief as to other aspects. At the same time, however, the major physiographic units of which the country is composed are themselves very large, giving a general uniformity of physical conditions over wide areas within each of these major units.

Figure 2.1 divides the Soviet Union into major structural units, which serve as a starting point for more detailed discussion of landforms and relief. Fundamental to the whole arrangement of structural elements are two Pre-Cambrian 'continental platforms', one underlying most of the country to the west of the Urals, the other occupying central Siberia between the Yenisey and Lena rivers. Between these two platforms lies a broad area which, in Paleozoic times, developed as a geosyncline and received vast quantities of sediment derived from the denudation of the adjacent stable blocks. The sediments were affected by Paleozoic fold movements during both the Caledonian and Hercynian orogenic periods, but the resultant structures are visible only in two areas, the Ural mountains and the Kazakh uplands. In both the west Siberian and Caspian–Turanian lowlands, Paleozoic fold structures lie deeply buried beneath great thicknesses of younger sedimentary rocks. Caledonian and some Hercynian structures are also visible in the present-day landscape to the south of the Siberian platform in the Altay–Sayan and Baykalia regions, and there are Hercynian elements in the Central Asian mountain systems.

In this brief listing of the main structural elements, we should also note the presence of complex fold mountain ranges, of Mesozoic and Tertiary origin, which lie to the south and east of the Pre-Cambrian, Caledonian and Hercynian structural provinces. These ranges occupy a relatively narrow and discontinuous strip of territory along the southern edge of the Soviet Union as far east as Lake Baykal, but beyond the lake they swing round towards the north and north-east to cover virtually the whole area between the River Lena and the Pacific coast.

While a division of the Soviet Union into the units just described helps us to understand the basic structure of the country, it must be realised that, over very large areas, there is no direct or simple correlation between the present-day relief and the underlying structures. The surface configuration of the land is best described on the basis of the physiographic regions mapped in Fig. 2.2. The relation of these to the structural divisions in Fig. 2.1 is indicated in Table 2.1.

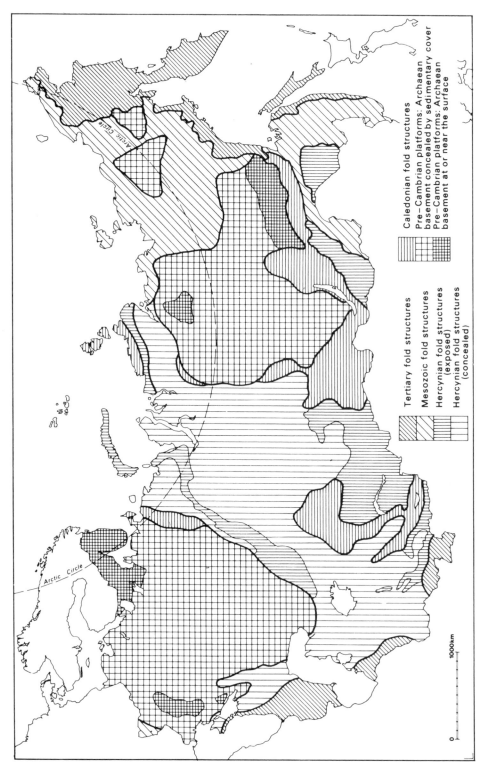

Fig. 2.1 Structure of the USSR

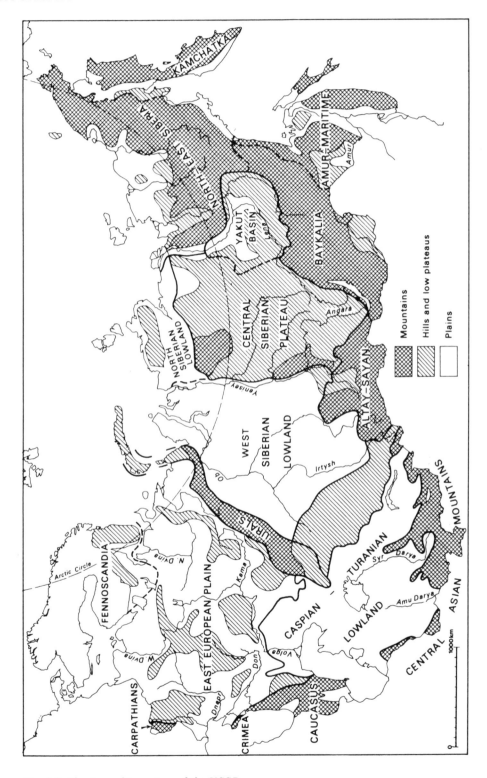

Fig. 2.2 Physiographic regions of the USSR

TABLE 2.1: PHYSIOGRAPHIC REGIONS AND THEIR RELATION TO STRUCTURE

Physiographic region	Structure
East European plain	Pre-Cambrian East European Platform, largely buried by sedimentaries of many different ages; platform exposed mainly in Fennoscandia and on a smaller scale in the Ukraine.
Ural mountains	Uplifted section of the Hercynian zone; Hercynian fold structures exposed
West Siberian lowland	Hercynian fold structures deeply buried by later sedimentaries
Central Siberian plateau	Siberian platform, largely buried by later sedimentaries
Kazakh upland	Caledonian and Hercynian fold structures, planated, re-elevated and exposed
Caspian–Turanian lowland	Caledonian and Hercynian fold structures, buried by younger sedimentaries
Mountains of the south and east	Varied structures
(a) Carpathians, Crimea and Caucasus	Tertiary fold mountains
(b) Central Asia	Caledonian, Hercynian and Tertiary fold structures
(c) Altay–Sayan	Caledonian and Hercynian fold structures
(d) Baykalia	Shield fragments, Caledonian, Hercynian and Mesozoic fold structures, much block faulting
(e) Amur-Maritime	Mesozoic and Tertiary fold structures with some Hercynian structures in basins
(f) North-east Siberia	Predominantly Mesozoic fold structures with older buried massifs
(g) Kamchatka	Part of the Pacific ring of Tertiary fold mountains.

PHYSIOGRAPHIC REGIONS: (A) THE PLAINS AND PLATEAUS

As the accompanying maps clearly demonstrate, plains together with hill lands and low plateaus, generally less than 1000 m above sea level, are the dominant elements. Such features occupy virtually the whole of the Soviet Union west of the Yenisey river with the exception of the narrow Ural ranges and the mountains of the southern frontier zone.

The East European Plain

The stable block of the east European platform occupies the whole of the USSR west of the Ural mountains and north of the Black and Caspian Seas, but the ancient crystalline materials of which it is composed outcrop only in two rather limited areas (see Fig. 2.4). The larger of these lies in the extreme north, in Karelia and the Kola peninsula, and is part of the Baltic or Fennoscandian shield. The smaller Ukrainian shield lies

some distance north of the Black Sea, crossing the Dnepr river between Dnepropetrovsk and Zaporozhye. Between these two areas, the ancient basement rocks are completely hidden beneath a cover of sedimentaries. These cover rocks are relatively undisturbed, having been protected from folding by the stability of the block on which they lie, but the latter has been subjected to a good deal of warping and faulting so that its surface is uneven and the thickness of the sedimentaries varies considerably. Between Kursk and Voronezh, for example, an upfaulted horst block brings the basement rocks very close to the surface, while between this block and the Ukrainian shield there is a deep structural trench in which a great thickness of sediments, including the Donbas coal measures, has been preserved.

The sedimentaries that underlie the bulk of the east European plain contain representatives of nearly every geological period and there is a general tendency for progressively younger rocks to appear at the surface in a transect from north-west to south-east, the oldest Paleozoics out-cropping along the Baltic coast and the youngest Quaternaries around the Caspian Sea.

It will be appreciated that, as a result of the situation just described, the structure of the east European platform has little or no direct effect on the detailed relief of the east European plain, which is much more influenced by the nature and arrangement of the sedimentary cover. In addition, more than half this physiographic region has been affected by the events of the Pleistocene glaciation, during which the Fennoscandian shield was a zone of ice dispersal and much of the remainder of the plain was a region of glacial deposition. There is ample evidence in contemporary landforms of at least two major ice advances, which are known to Russian scholars as the Dnepr–Don and Valday glaciations. These were probably contemporaneous with the Riss and Würm advances in central Europe*. The earlier of the two, which was also the more extensive, sent tongues of ice down the valleys of the Dnepr and Don to within 200 km of the Black Sea. The more recent Valday ice sheet advanced a shorter distance, reaching only to a line running near Minsk, Smolensk and Moscow. As a result, the east European plain carries a very complex assortment of glacial, fluvio-glacial and post-glacial deposits, which add greatly to the intricacy of the region's landforms. The plain as a whole is a very low-lying region. Only in a few places does the surface rise more than 300 m above sea level and at least half the area is below the 200 m contour. Under these circumstances, quite small differences in height assume great local importance, and the landforms of glacial deposition may appear as 'major' local relief features.

A clearer picture of the physiography of the east European plain requires a more detailed description of its component parts. The Russian section of the Baltic shield, occupying the Kola peninsula and the territory of Karelia, between Finland and the White Sea, is a barren land, where expanses of bare, ice-scraped rock alternate with shallow, drift-filled hollows, often occupied by lakes or marshes. Apart from one or two summits in the Khibin mountains (1)[†], which rise to nearly 1200 m, the greater part of this region is less than 300 m above sea level.

Beyond the faulted trough which marks the southern limit of the Baltic shield and contains the Gulf of Finland, Lake Ladoga and Lake Onega, the rocks of the shield disappear beneath the drift-covered sedimentaries of the plain. North of the latitude of Moscow, these sedimentary rocks dip south-east and eastwards towards basins developed between Moscow and the upper Kama, and denudation has given rise to a rudimentary scarp and vale topography. The main escarpments are developed on the Silurian limestone, which backs the southern coast of the Gulf of Finland, and the Carboniferous limestone forming the Valday Hills (2). The latter, which run in a north-east to south-west direction from the southern end of Lake Onega to the boundary of the Belorussian republic, are capped with one of the moraines of the Valday glaciation and reach 300–325 m in places. The relatively insignificant line of hills is one of the major watersheds of the European plain, separating the Volkhov, Western Dvina and other streams flowing to the Baltic from the southward-draining Volga system.

Another important morainic feature, probably marking the maximum extent of the final ice advance, is the Smolensk–Moscow ridge (3), which forms a traditional route from the western frontier to the capital. This northern section of the plain, between Moscow and the Baltic Sea, is one in which, since glaciation is relatively recent, the landforms of glacial deposition have been only partially removed by post-glacial erosion. The numerous moraines and other features are well-preserved and form well-marked relief features. The presence of moraines and other low hills provided ideal situations for the development of pro-glacial lakes during the Pleistocene period and many large, flat-floored, ill-drained depressions remain, some still carrying sizeable lakes such as Lake Ilmen (4) and Lake Beloye (5).

---

* Students of the Pleistocene Ice Age in central Europe conclude that it involved four major advances of the ice, separated by interglacial periods when the ice sheets withdrew from Europe. These four advances have been given the names of Gunz, Mindel, Riss and Würm, respectively. In the case of European USSR there is some doubt as to whether the area was affected by the first two advances. There is, however, strong evidence for at least two advances, the Dnepr–Don, contemporaneous with the third, and most extensive, Riss advance in central Europe, and the Valday, contemporaneous with the last, Würm advance.

† A number in brackets after the name of a relief feature refers to the relief map (Fig. 2.3).

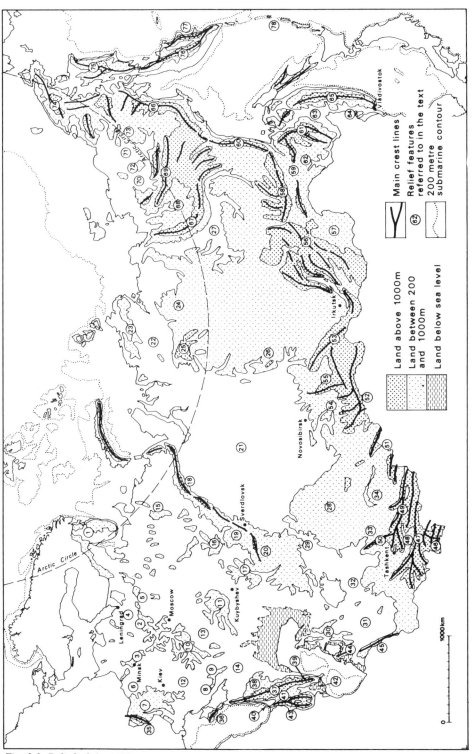

Fig. 2.3 Relief of the USSR

The zone between the limit of the Valday glaciation and that of the Dnepr–Don advance is somewhat different. The boulder clay from the latter is believed to have been rather thin and many of the features resulting from glacial deposition during this phase have either been removed by post-glacial action or covered by fluvio-glacial materials. The latter are most widespread along the upper valleys of the Dnepr and Don, where large depressions developed in which the glacial meltwaters accumulated. Parts are still very badly drained, particularly the area of the Pripyat (Pripet) marshes or Polesye (6). In the more southerly parts of this zone, where the ice sent tongues down the major valleys but did not succeed in over-running the interfluve areas, the pre-glacial relief is more clearly visible. This involves an alternation from west to east of low plateau uplands, often sharply dissected, and broad valley plains. Much of the western Ukraine is occupied by the Podolsk–Volyn uplands (7), which reflect the presence, at or close to the surface, of the Ukrainian shield. The highest point in this area is 472 m and the general level declines towards the Dnepr but rises again to the east of the river in the small upland of the pre-Azov heights (8), which reach 325 m. To the north-east of the latter are the Donets heights (9), rising in places to about 350 m, which represent the sediments of the Donets trench folded against the edge of the Ukrainian massif.

The south-western edge of the Moscow basin is marked by the beginning of the central Russian elevation (10), which runs southward between the Dnepr and the Don, terminating at the northern edge of the Donets valley. It is separated by the Oka–Don lowlands from a third upland, the pre-Volga heights (11). All three uplands carry an extensive loess cover, while the plains of the Dnepr (12) and Don (13) are characterised by broad terraces at various heights connected with post-glacial changes in the level of the Black Sea. Where the Dnepr crosses the Ukrainian massif, however, its valley is sharply incised and there were rapids before these were drowned by the construction of the Dneproges hydro-electric barrage at Zaporozhye.

South of the Ukrainian massif, broad, gently-sloping plains, developed on Tertiary sediments with a patchy loess cover, occupy the southern Ukraine and the northern Crimea. In the Azov–Caspian depression (14), Hercynian structures are concealed beneath a sedimentary cover with Quaternaries at the surface.

The north-eastern section of the east European plain has so far received little attention and can be dismissed quite briefly. The terrain in this region is a good deal more monotonous than in the areas already discussed. Uplands are confined to the much-eroded Timan range (15), which runs north-westwards from the northern Urals to the Arctic Ocean, and the Perm (16) and Ufa (17)

Erosion scars on the sides of valleys cut into the loess of the southern Russian steppes near Rostov on Don

plateaus on the west flank of the Urals. The remainder of this part of the plain is composed of near-horizontal sedimentaries, among which the Permian is the most widespread outcrop. Glacial and fluvio-glacial deposition adds some variety to the landscape.

The Ural Mountains

While the fold structures of the Ural ranges are of Hercynian origin, the present height of the area is due to much more recent uplift. The original Hercynian mountain ranges, produced by orogenic movements at the end of the Paleozoic era, occupied the whole area between the European and Siberian platforms. This extensive mountain system was subjected to a long period of sub-aerial denudation, which reduced the area to a peneplane by the beginning of the Tertiary era. Late Tertiary uplift raised the eroded stumps of the old Hercynian ranges in the area now forming the Urals, thus initiating a new cycle of erosion which has carved out the present relief. As a result, the Ural mountains consist of a central belt of metamorphic and intrusive materials, which are very rich in minerals, flanked on either side by belts of tightly-folded Paleozoic sedimentaries.

Relief in the Urals consists of low, parallel, north-south ridges, broken in many places by cross-faulting. The bulk of the region is between 200 and 1000 m above sea level and the highest summits reach only 1500–1900 m. Despite their traditional role as the boundary between Europe and Asia, the Urals offer no really serious barrier to movement between Europe and Siberia. The region may be divided into three rather different parts. The northern Urals (18) extend approximately as far south as latitude 61°N and contain the highest peak in the whole system: Mount Narodnaya (1984 m). The mountains are here at their narrowest, consisting of two broken parallel ridges which coalesce into one north of the Arctic Circle, curving first north-west and then north-east through the tundra wastes to the Yugorskiy peninsula and thence into Novaya Zemlya. The central Urals (19), between latitudes 61° and 55°N, are wider but a good deal lower, the highly resistant rocks which make up the main range in the north being poorly represented here. This central section, with its subdued relief, lies mainly below 500 m, and the more important transport routes, including the Trans-Siberian Railway, cross the Urals in this section, which is also the site of the

region's largest cities. Southward from Chelyabinsk, in the southern Urals (20) the ranges fan out and the system reaches its maximum width of 150–200 km. Heights of 1500 m or more are reached in several of these ranges which, in the extreme south of the region, open out to form a broad, dissected plateau, not unlike that of the adjacent Kazakh upland.

The West Siberian Lowland

Beyond the Urals and stretching more than 1500 km to the Yenisey and western edge of the Siberian platform, is the west Siberian lowland (21), quite the most striking single relief feature in the whole of the Soviet Union. Throughout this vast area of some $2\frac{1}{2}$ million km², the land is nowhere more than 200 m above sea level, and at least half the area is below the 100 m contour. The Paleozoic basement, with its Hercynian structures, lies buried beneath 1500 m or more of sedimentary rocks. Surface outcrops are entirely of Tertiary and Quaternary materials, the latter being more widespread.

In the extreme north, the surface deposits are the product of a post-glacial marine transgression, while south of these, to about 60°N, they are the result of glacial deposition and include both boulder clay and outwash materials. Beyond the limit of the maximum ice advance is a zone some 300 km wide where fluvio-glacial deposits predominate, south of which there is a belt of territory in which loess is widespread and overlies both Tertiary and Quaternary sediments. The rather drier and slightly more elevated zone in the south contains the bulk of the settled area and the region's farmland.

This enormous lowland is drained mainly by the Ob and its tributaries the Irtysh and Tobol; the Yenisey, which runs along the eastern edge, has a much smaller catchment area. These great rivers, which are among the biggest in the Soviet Union, have extremely gentle gradients and thus flow very slowly, carving out huge flood plains which may be as much as 100 km across and yet may lie only 10 or 20 m below the interfluve surfaces. Under these circumstances, flooding is a common occurrence, particularly during the spring thaw when it is intensified by the fact that the upper reaches of the rivers melt first while the lower parts are still frozen and water thus spreads out over great areas. Most of the region is very poorly drained and contains some of the world's

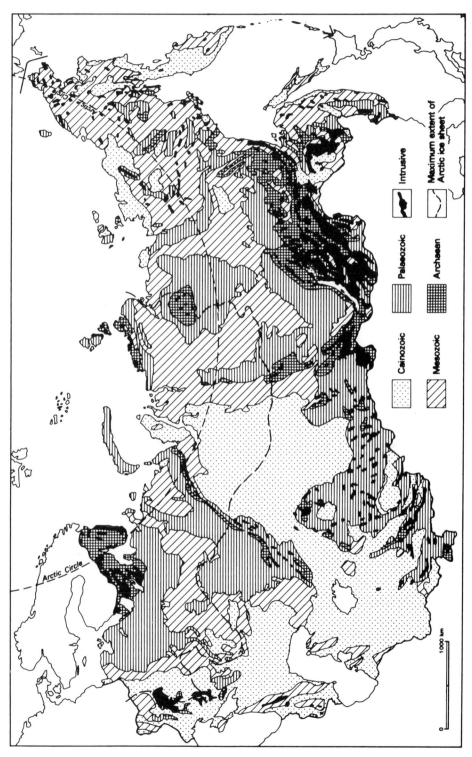

Fig. 2.4 Geology of the USSR

largest swamps, notably the Vasyuganye swamp, between the Ob and Irtysh rivers, which is roughly the size of England.

Features similar to those of the west Siberian lowland are carried eastward around the northern edge of the central Siberian plateau in the north Siberian lowland or Khatanga depression (22), in which Hercynian structures are buried at depth and the surface is composed of Quaternary marine sediments. Hercynian and Caledonian fold structures are exposed in the Byrranga mountains (23) of the Taymyr peninsula, which forms a dissected plateau between 300 and 500 m above sea level, with occasional peaks rising to 1200 m.

The Central Siberian Plateau
The territory between the Yenisey and Lena rivers is occupied by the central Siberian plateau, a region which covers the greater part of the Siberian platform. Within the area of the plateau, the ancient basement rocks are exposed only in the relatively small northern section known as the Anabar shield (24). Elsewhere they are covered by sedimentary rocks, mainly of Paleozoic and Mesozoic age. These are thinnest in a central zone running south towards Lake Baykal, in which only the Paleozoics are preserved. To the east and west lie the Lena and Tunguska structural basins, in which the sedimentary rocks are very thick and include large quantities of Carboniferous material, containing huge coal reserves, though the surface outcrops are mainly of Mesozoic age.

The relief of the plateau, however, bears little relation to this structural arrangement and consists of a series of dissected erosion surfaces cut indiscriminately across Archaean, Paleozoic and Mesozoic rocks alike at heights varying between 300 and 800 m. In a few places, particularly resistant materials give uplands which stand out above the general summit level, as in the Putoran mountains (25) in the north-west, which reach 1700 m, or the Yenisey range (26) (1104 m) in the south-west. The plateau is a region of considerable relief since it has been vigorously dissected by tributaries of the Yenisey such as the Angara, Stony Tunguska and Lower Tunguska, and also, to a lesser extent, by those of the Lena. The Lena itself, however, and the lower reaches of its main tributaries, the Aldan and Vilyuy, flow in a broad structural trough between the eastern edge of the Siberian platform and the Mesozoic fold ranges. The sedimentary rocks which fill this trough form

a triangular lowland, some 800 km across, known as the Yakut basin (27).

A comparison of the map of physiographic regions (Fig. 2.2) with that showing structure (Fig. 2.1) will indicate that the central Siberian plateau occupies only about two-thirds of the Siberian continental platform and that the latter extends further south to include most of the territory between the upper Lena and the Manchurian border. In this region, labelled Baykalia in Fig. 2.2 the Pre-Cambrian structures are exposed but have been broken by a series of roughly parallel faults to give a mountainous terrain. Consequently the southern part of the Siberian platform is described later, along with other southern mountain areas to which it is physiographically, though not structurally, more akin.

The Kazakh Upland
The Kazakh upland (28) borders the west Siberian lowland on its southern side and separates the latter from the Caspian–Turanian plains of Soviet Central Asia. The region was affected by fold movements in both Caledonian and Hercynian orogenic periods, but the resultant complex structures, like those of the Urals, have been planated and re-elevated in more recent times. Consequently the upland now consists of an alternation of plateaus and shallow depressions, and for the most part stands between 500 and 1000 m above sea level, with a maximum elevation of 1565 m roughly in the centre of the region. In several places the depressions preserve Mesozoic and Tertiary sedimentaries, but the greater part of the surface is composed of Paleozoic rocks together with Archaean metamorphic and igneous materials which are very rich in minerals. The upland is a region of dry steppe and semi-desert with arid landforms and permanent streams are confined to the northern half of the region.

The Caspian–Turanian Lowland
Like the west Siberian lowland, to which it is connected by the narrow, Quaternary-floored corridor of the Turgay Gate (29), the Caspian–Turanian lowland is part of the Hercynian structural province in which the Paleozoic structures are deeply buried and the surface rocks are entirely of Tertiary and Quaternary age.

Around the Caspian Sea is an area now some 30 m below world sea level. Quaternary plains marking the former extent of the Caspian stretch nearly 500 km northwards from the present shoreline. Over the Turanian lowland as a whole, the general arrangement is one in which low plateaus with steep, scarped edges, developed on Tertiary rocks, overlook the plains developed on Quaternary sediments. The most prominent of these plateaus is the Ustyurt (30), which rises to heights of 150–200 m between the Aral and Caspian Seas.

The Caspian–Turanian lowlands are, of course, an area of inland drainage. Apart from relatively small sections in the west and east, which drain towards the Caspian and Lake Balkhash respectively, drainage is towards the Aral Sea. Since this is a desert region, the only permanent watercourses are those which originate in the high mountains to the south and only two of these, the Amu Darya and Syr Darya, reach the Aral Sea; the remainder, including some sizeable rivers such as the Chu and Sarysu, as well as a number of smaller ones from the north, peter out in the desert. The arid landforms which are the characteristic of the region include large expanses of sand desert: the Karakum (31) between the Caspian and the Amu Darya, the Kyzylkum (32) between the latter and the Syr Darya, the Muyunkum (33) to the east of the Syr Darya and the Taukum (34) on the southern side of Lake Balkhash. In contrast, there are large stretches of clay desert and many ribbons of alluvium derived from ancient and modern rivers. The large number of abandoned river courses suggests that many of the landforms of this region are derived from a wetter period during the Pleistocene. To-day most of the region is arid and, over large areas, agriculture is impossible without the aid of irrigation. The most productive soils are those of the piedmont zone along the southern edge of the desert, where loess has accumulated.

The regions so far discussed occupy approximately three-quarters of the land surface of the Soviet Union. Throughout that entire area there are only a handful of places at which the land rises above 1500 m and at least half is below the 200 m contour. These plains and low plateaus, which are dominant from the western boundary to the Lena river, take the form of an amphitheatre facing north to the Arctic Ocean and bounded on its eastern, southern and south-western sides by high mountain ramparts. Only in the west, through the 500 km-wide corridor between the Carpathians

and the Baltic, and in the extreme south-east, where her territory overlaps the mountain rim and has a frontier with China along the Amur and Ussuri rivers, is the Soviet Union in easy overland contact with her neighbours. Elsewhere, her frontiers are the mountains or the sea.

PHYSIOGRAPHIC REGIONS (B) THE MOUNTAINS OF THE SOUTH AND EAST

The remaining quarter of the Soviet Union is occupied by a complex series of mountain systems which, although in a broad sense forming a single major physiographic unit, in fact have a great variety of structural origins and include Pre-Cambrian, Caledonian, Hercynian, Mesozoic and Tertiary (Alpine) elements. The oldest fold structures, those of Pre-Cambrian age, now form the European and Siberian platforms, around which are arranged the later orogenic belts. Caledonian structures occur mainly around the southern edge of the Siberian platform, in Baykalia, but are also present in the Altay–Sayan district and in parts of Central Asia. Traces of Hercynian folding are to be found in Central Asia and the district between the Siberian platform and the Amur. The most widely developed mountain systems, however, are those associated with more recent folding in Mesozoic and Tertiary times, these young fold ranges occurring in the Carpathians, Crimea, Caucasus and parts of Central Asia as well as throughout the Soviet Far East.

There are fairly important contrasts between the Mesozoic and Alpine fold ranges on the one hand and those formed in Caledonian and Hercynian times on the other. The latter have, in most cases, undergone long periods of sub-aerial denudation and owe their present height mainly to recent block uplift, often taking the form of a series of basins and ranges. Mineral-rich metamorphic and intrusive core zones are often exposed in these ranges. The young fold mountains, however, particularly those produced by the most recent, Alpine, orogeny, have not been denuded to such a degree, though they have often been deeply dissected by fluvial and glacial erosion so that steep slopes and sharp crest-lines abound.

For the sake of clarity, the mountain zone will be discussed under the headings provided by the regional nomenclature in Fig. 2.2.

Carpathians

In the extreme west, the Soviet Union now contains a small section of the Carpathian mountains (35) and a small foothold in the plains of the Danube. This situation, which has considerable strategic significance, results from the annexation in 1945 of the former Czechoslovak province of Ruthenia, now known as the Sub-Carpathian Ukraine. In this district, the Carpathians take the form of a series of parallel ridges some 100 km wide with peaks between 1000 and 1800 m above sea level.

Crimea

The bulk of the Crimean peninsula is a continuation of the Tertiary plains of the southern Ukraine, but in the extreme south there is a narrow zone of Alpine folding about 30 km wide. There are three parallel ridges (36), of which the southernmost is the highest and rises a little above 1500 m. The shelter from northerly winds pro-

Limestone ranges of the Crimea, a vineyard in the foreground

vided by these mountains gives a moderate climate along the Crimean coast, reflected in its specialised agriculture and its popularity as a holiday area.

Caucasus

This is a much larger and more complex mountain system, also of Tertiary origin, which occupies the isthmus between the Black and Caspian Seas. With a length approaching 1000 km, the Caucasus are equivalent in size, height and structural complexity to the Alps of central Europe and Soviet geographers distinguish a large number of physiographic sub-divisions. We will consider this region in three sections: the Greater Caucasus, the Transcaucasian depression and the Lesser Caucasus.

The Greater Caucasus, or main Caucasian range (37) is a major anticlinal feature which stretches right across the isthmus from Novorossiysk on the Black Sea to the Apsheron peninsula on the Caspian, and is structurally continuous both with the Crimean ranges and the mountains of Soviet Central Asia. In the west, the anticline has been breached to expose the igneous and metamorphic rocks formed at depth during the mountain-building process and this core zone is flanked on either side by zones of tightly-folded Mesozoic sedimentaries. The main crest-line exceeds 3500 m over much of its length and there are several peaks above 5000 m, the highest of all being Mt. Elbrus (5642 m), one of a number of volcanoes which were active here in late Tertiary times. On the northern flank of this western section of the main Caucasian range, a zone of Tertiary sediments extends northwards between the Quaternary plains of the Terek and Kuban. This is the Stavropol plateau (38), a dissected platform which in places reaches 800 m above sea level but lies for the most part between 200 and 300 m, and is particularly important as a source of natural gas.

The eastern half of the main Caucasian anticline has not been breached so that Mesozoic rocks, among which Jurassic limestones are the most widespread, extend right across it. Karstic landforms are especially widespread in Dagestan (39). The crest-line in this eastern section is somewhat lower than in the west, but a number of peaks rise above 4000 m. The whole of the main Caucasian range has, of course, been vigorously dissected by normal erosion, especially in the west, where rainfall is higher. The upper slopes have been affected by ice action and still carry a number of small glaciers. The barrier presented by these mountains to northerly climatic and human influences does much to account for the distinctive character, both physical and human, of the Transcaucasian region.

The Transcaucasian depression itself falls into three parts. In the west is the small Kolkhida or Colchis lowland (40), developed along the lower reaches of the Rioni river. This triangular plain extends about 100 km inland from the Black Sea and is floored with Tertiary and Quaternary sediments. Heavy deposition of material eroded from the Caucasus ranges has given a flat alluvial plain, much of which has required artificial drainage. The lowland is terminated on its eastern side by the granitic Suram massif (41), through which the only route is the Suram Pass, a narrow defile with a summit at 850 m. Beyond this obstacle is the Kura basin (42), which stretches some 500 km eastwards from Tbilisi to the Caspian. Like the Rioni basin, that of the Kura is infilled with Tertiary and Quaternary sediments. The former occur as low hills and plateaus at heights of 200–600 m around the edge of the basin. The Quaternary deposits are ill-drained in their lower parts and the large Kura delta has much swampland despite its near-desert climate.

The Lesser Caucasus is an area of complex structure and falls into two main parts. Northeast of a line running north-west to south-east through Lake Sevan, the rocks are sedimentaries of Mesozoic age, compressed into folds by the Alpine earth movements, and there are several peaks above 3000 m. The western part of the lesser Caucasus, however, consists mainly of the lava plateaus of Armenia (43), which stand at heights of 1000–2000 m. Within these plateaus are several down-faulted troughs, such as the one containing Lake Sevan and that which carries the Araks river along the Turkish frontier. Above the level of the plateaus rise numerous volcanic peaks, of which the highest is Mt. Aragats (4090 m). The lava plateaus continue into eastern Turkey, where Mt. Ararat (5165 m) overlooks the Araks valley from the south-western side.

The Mountains of Soviet Central Asia

West of the Amu Darya river, in the Turkmen republic, the system of mountain ranges is quite simple and only the Alpine orogeny is represen-

ted. The anticline of the main Caucasian range is continued across the Caspian into the Bolshoy Balkhan mountains (44), which reach 1880 m, and thence south-eastwards to the Kopet Dag (45) range. The latter, which rises to a maximum height of 2942 m near Ashkhabad, forms the northern edge of the Iranian plateau and marks the boundary between Iran and the USSR.

To the east of the Amu Darya, a much more varied and complex series of mountain systems is to be found, involving structures from the Alpine, Hercynian and Caledonian orogenic periods. The youngest, and also the highest of these fold ranges are the Pamir mountains (46), which occupy the eastern half of the Tadzhik republic. These ranges constitute the 'Pamir knot', a focal zone in the Eurasian alpine fold system whence chains of young fold mountains fan out to the south-west (Hindu Kush), south-east (Himalayas) and northeast (southern Tyan Shan). The Pamirs contain the highest peaks in the Soviet Union, notably Lenin Peak (7134 m) and Mt. Communism (7485 m). Below the main crest lines, which carry glaciers and extensive permanent snowfields, are large plateau areas at heights around 4000 m which are deeply dissected by narrow river gorges.

The northern limit of the Tertiary fold system of the Pamir mountains is marked by the deep-set valley of the Surkhob river. To the north of this valley, the Alay ranges (47), a series of re-elevated

A tanker lorry with trailer climbs on a snowbound road in the mountains of east Siberia

Hercynian blocks, run eastwards from Samarkand to the Chinese border. The main crest-lines exceed 3000 m over most of their length and carry several active glaciers. Subordinate ranges run off towards the south in western Tadzhikistan, separating the deep valleys of the Amu Darya's north-bank tributaries, while north of Samarkand low ridges finger out into the Kyzylkum desert.

The Alay ranges in turn give way northwards to the Fergana basin (48), the most impressive of several down-faulted Hercynian basins in this region. 300 km long and 160 km across, the Fergana basin is surrounded on all sides by mountain ranges save in the west, where there is a gap some 10 km wide, through which the Syr Darya makes its exit. The centre of the basin is a flat desert plain, while around the edges are a series of terraces and alluvial fans, backed by low, loess-covered hills composed of Tertiary and Quaternary sediments. The Fergana basin thus displays, in a relatively compact area, all the landforms which are associated with the Central Asian hill-foot zone, between the great desert and the southern mountain ranges.

North of the Fergana basin, the succession of east-west trending mountain blocks and troughs continues. The Tyan Shan system (49) is a broad complex of parallel ranges which includes both Caledonian and Hercynian structural elements and extends eastwards from Tashkent to and beyond the Chinese border. Crest-lines rise well above 3000 m and several peaks in the east top the 5000 m contour. In the north-west, several low ridges, such as the Karatau range, (50) finger out into the desert. Over most of its length, the northern face of the Tyan Shan system is very abrupt, dropping along fault lines to a loess-covered pediment zone at heights between 500 and 1000 m. Beyond this lie the desert basins of the Chu river and Lake Balkhash, which separate the fold mountain systems from the Kazakh upland. The Balkhash basin narrows eastwards but is joined to the Sinkiang province of China by the narrow Dzhungarian gate (51), which has for centuries been a routeway between Turkestan and the Asian interior.

The Altay–Sayan Ranges

The basin of the upper Irtysh river marks the north-eastern limit of the Kazakh upland, and beyond this a complex zone of Hercynian and Caledonian mountain ranges extends as far as

Lake Baykal. These fall into two main groups, the Altay (52) and the Sayan (53), the division between them coinciding with the valley of the upper Yenisey.

Of the two systems, the Altay is the more complex with an intricate pattern of ranges which have a predominantly east-west trend in the south and important north-south elements in the northern part. In particular, the northward-protruding Salair and Kuznetsk Alatau ranges enclose the Kuznetsk basin, or Kuzbas (54), with its deep cover of Carboniferous rocks and its valuable coal measures. The Minusinsk basin (55) on the upper Yenisey is similarly enclosed between the Kuznetsk Alatau and Sayan ranges. The Altay system as a whole has undergone a good deal of block faulting so that high, dissected plateau uplands alternate with enclosed basins and valley troughs. A number of peaks close to the Mongolian border exceed 4000 m.

The Sayan consists of western and eastern ranges which trend south-west to north-east and north-west to south-east respectively, both with crest-lines above 2500 m, and these enclose the high basin interior of the Tuvinian ASSR.

Baykalia

Around and beyond Lake Baykal, in the region marked 'Baykalia' on Fig. 2.2, there is a very great variety of structural elements including Caledonian and Hercynian folds and a large exposed section of the Siberian platform, known as the Aldan shield. In the area between Lake Baykal and the Yablonovyy range (56), which forms the watershed between the Arctic and Pacific drainage basins, the landscape displays a distinct south-west to north-east grain imparted by large-scale block faulting. This is *horst* and *graben* country on a massive scale, with a great altitudinal range between the summits of the uplifted mountain blocks, which are often between 1500 and 2000 m above sea level, and the floors of the down-faulted troughs which are often below 800 m. One such trough is occupied by Lake Baykal, a water body some 640 km long and 45 km across. While the mountains on either side reach 2000 m, the bottom of the lake is 1300 m below sea level, a total height range of 3300 m. Baykal is the world's deepest lake, with a maximum depth of 1752 m.

Between the Yablonovyy ranges and the Manchurian border is the high-level basin of Dauria

Lake Baykal, which occupies a major tectonic depression in southern Siberia, is the deepest lake in the world

(57), in which Hercynian structures are buried beneath Mesozoic sedimentaries, and where low hill ranges alternate with broad, open plains.

The Aldan shield section of Baykalia is a massive upland, much less broken by faulting than the area closer to Lake Baykal, and has extensive summit plateaus above 1500 m. The watershed between the Pacific and the Arctic is carried eastwards across this zone by the Stanovoy range (58), the rounded summits of which exceed 2500 m. To the south of the Stanovoy and parallel to it, the Tukuringra and Dzhagdy (59) ranges lead off into the Amur-Maritime region. At its eastern end, the Stanovoy range runs into the coastal Dzhugdzhur mountains (60), composed mainly of Mesozoic volcanics, which mark the limit of the north-east Siberian region.

The Amur-Maritime Region
This region occupies the entire area south of the Stanovoy mountains and between the Manchurian border and the Pacific, and consists of an alternation of fold mountain ranges, Hercynian in the west, Mesozoic and Alpine in the east, and broad, open lowlands. The Dzhagdy range approaches within 100 km of the Pacific, where it

merges with the Bureya mountains (61). The latter have a north-east to south-west trend and fall away towards the Amur, beyond which they are continued by the Hsiao Khingan of Manchuria. Within the enclosing arc of the Dzhagdy and Bureya ranges is the large Zeya–Bureya plain (62). Here, Hercynian structures are partially concealed beneath younger sedimentaries, which have been carved into terraces at various heights by the Amur, Zeya and Bureya rivers.

On the south-eastern side of the Bureya range is the lower Amur plain (63), which extends from the Amur–Ussuri confluence to the sea. The floor of the plain is mostly about 50 m above sea level and a great deal of it is poorly drained. Broad, open sections with well-developed terraces alternate with narrow stretches where the river has cut through the ridges which cross its course. The latter, composed mainly of Mesozoic rocks, rise to heights varying between 400 and 1000 m. Running southwards from the Amur–Ussuri confluence to the sea near Vladivostok is the Khanka–Ussuri plain (64). This is a fault-bounded depression, the Quaternary floor of which is some 50 m above sea level, and in Quaternary times was a sea strait separating the Sikhote Alin from the mainland. The Amur and Ussuri low-lands are of particular importance in that they contain the bulk of the settlement and agricultural land of the Far Eastern region.

The Sikhote Alin ranges (65) are among the youngest fold mountain systems in the USSR and include both Mesozoic and Tertiary structures. There are seven or eight parallel mountain chains, the peaks of which vary in height from about 1800 m in the centre to 1300 m at the northern and southern ends. Two more parallel ridges of Tertiary age, separated by a down-faulted, flat-floored depression, make up the island of Sakhalin.

North-East Siberia
North-east Siberia is one of the most remote, thinly settled and little explored parts of the Soviet Union, and knowledge of its geology and structure is still incomplete. The region comprises a complex system of Mesozoic fold ranges, between which there are large median masses. Some of these have been uplifted to form high plateaus, while others are represented by low-lying basins with a cover of Quaternary materials facing out to the Arctic Ocean.

The Dzhugdzhur range (60) backs the Pacific coast as far as Okhotsk, beyond which a more broken coastal upland leads eventually to the Gydan (or Kolyma) range (66). From these coastal ranges, a series of mountain chains run off north and north-westwards to the Arctic Ocean. Of these, the most clearly marked is the Verkhoyansk range (67), which forms a continuous barrier more than 1500 m high, with numerous peaks above 2000 m, immediately east of the Lena lowlands. The Verkhoyansk range is separated by the much-dissected Yano–Oymyakon plateau (68) from the Cherskiy (69) mountain system. The latter consists of a series of rather short, broken ranges, the summits of which rise to heights of 2000–2500 m. Between the Cherskiy and Gydan ranges, large median masses underlie the Indigirka (70) and Kolyma (71) lowlands and the Alazeya (72) and Yukagir (73) plateaus. In the extreme north-east, the Chukotka range (74) forms the backbone of the Anadyr peninsula.

Kamchatka

The Mesozoic ranges described in the last few paragraphs are separated by a narrow, broken lowland corridor, floored with Tertiary and Quaternary sediments, from the Tertiary folds of the Koryak-Kamchatka-Kurile arc, part of the Pacific zone of alpine folding. The Koryak range (75) consists of a series of roughly parallel ridges rising to 2000 m with numerous extinct volcanoes and extensive lava flows. This fold system is continued into the Sredinyy range (76), with several peaks above 2500 m, running down the centre of the Kamchatka peninsula. To the west of this central mountain system is a broad, rather poorly drained coastal plain, while to the east, beyond the valley lowland of the Kamchatka river, lies a highly distinctive volcanic zone (77). This consists of lava plateaus between 500 and 1000 m above sea level. High above these plateaus rise 20 or 30 volcanic cones of which about a dozen are still active. Klyuchevskiy Peak, the most impressive of all, reaches a height of 4750 m. This volcanic zone is continued southwards through the Kurile Islands (78) to northern Japan.

CONCLUSION

Within the space available, it has been possible to give only a very general picture of the great

variety of structure, rock-type, relief and landforms to be found within the vast territory of the USSR. In particular it has been possible to identify only the major physiographic regions and to name only a few of the most important relief features. It will no doubt be realised that, within each region discussed, and indeed within each 'feature' named, there is a great deal of variety and each could be further subdivided. It is an inevitable result of the great size of the Soviet Union that any first-stage regional break-down of the country gives regions which are themselves often very much larger than, say, the individual countries of western Europe.

The USSR, then, is a land of enormous contrasts in the nature of its physical geography and, in this respect, as in its sheer size, is more akin to a continent. At the same time, however, the individual physiographic units which go to make up the Soviet Union are in many cases so large that they give rise to monotonous uniformity of relief over wide areas. The east European plain, the west Siberian lowland or the Caspian–Turanian lowland, for example, are crossed by road or rail in journeys involving days rather than hours, and the traveller would notice little change in the relief or landforms over distances measured in hundreds of miles. In such areas it is aspects of the physical environment other than relief, notably climate, soils and vegetation, which are the most important influences on the whole complex of human geography. Paradoxically, however, on a local scale, quite minor relief features, involving very small changes in altitude, may be of great significance. A difference of a few metres in part of the east European plain, for example, may make all the difference between virtually useless swamp and moderately productive farmland or, in Central Asia, between irrigable flood plain and waterless sand desert.

Our hypothetical traveller across one of the great lowlands of the USSR would, therefore, be struck not only by the monotony of the landscape but also, if he were sufficiently observant, by the fact that this monotony was the product of the constant repetition of a small number of physiographic elements rather than the result of complete uniformity. A further point is that, were he to travel from north to south across one of these plains, he would also perceive a slow change in the nature of the landscape and in agriculture, resulting from changes in climatic conditions and associated variations in soil and vegetation types, factors which, in the lowland regions, are much more important than relief alone.

Quite different is the situation in the mountains of the south and east. Here, and only here, do we find those great contrasts in relief and landforms over short distances and the associated variety of human response which are commonplace in central and western Europe, and this serves to emphasise still further the contrast between these regions and the bulk of the USSR.

BIBLIOGRAPHY

Berg, L. S. (1950), *Natural regions of the USSR*, Macmillan, London and New York.

Bogdanoff, A. (1957), 'Traits fondamentaux de la tectonique de l'URSS,' *Revue Géog; Physique et Géol. Dynamique*, 2me. série **1** (3), pp. 134–65.

Fiziko-geograficheskiy atlas mira (1964), Moscow. A key in English to this atlas appears in *Soviet Geography, Review and Translation*, **6** (5–6), 1965, pp. 1–403.

Krasny, L. I. (ed.), *Structure géologique de l'URSS*, CNRS, Paris.

Markov, K. K. and Popov, A. J. (eds.), (1959), *Lednikovyy period na territorii evropeyskoy chasti SSSR i Sibiri*, Moscow.

Nalivkin, D. V. (1960), *The geology of the USSR*, Pergamon, Oxford.

Nalivkin, D. V. (1973), *Geology of the USSR*, Oliver and Boyd, Edinburgh.

Parker, W. H. (1969), *The world's landscapes: 3, The Soviet Union*, Longman, London.

Rikhter, G., Preobrezhenskiy, V. and Nefedyeva, V. (1976), The Soviet land revealed, *Geographical Magazine*, **48**, (5), 266–272.

Suslov, S. P. (1961), *Physical geography of Asiatic Russia*, Freeman, San Francisco and London.

Tushinskiy, G. K. and Davydova, M. I. (1976), *Fizicheskaya geografiya SSSR*, Moscow.

Velitchko, A. A. (1979), 'Soviet glaciers were late developers,' *Geographical Magazine*, **51**, (7), 472–478.

3 Climate

No subject is of greater importance to mankind today than climate. During the 1980s the world began to realise that human activity was profoundly modifying climatic conditions in many ways. Scientists had long realised that our era might be but an interglacial period and that the ice sheets that had retreated some thousands of years ago might at any time begin to return, and that man might have to learn to live in periglacial conditions in what are now temperate regions. Such a deterioration in the natural pattern of climates, however, seemed far enough away to be virtually ignorable. Suddenly, however, danger to mankind appeared from an unsuspected source. Where hitherto warming of climates from artificial heat generation particularly in cities, had seemed relatively harmless, the discovery by scientists that the world's protective ozone layer was being destroyed by the use of CFCs (chlorofluorocarbons) used as propellants in aerosols and in refrigerator coolants, alerted governments to the need for action to reduce these threats. At the same time recognition of the extent of atmospheric pollution by power stations and the exhausts of internal combustion engines strengthened awareness of the vulnerability and inter-dependence of climates around the world.

Because it has been less developed economically than many western countries, the Soviet Union has been less guilty of influencing the climate of the world than have some countries, but its record in terms of air pollution has been locally poor because of the lack of design and maintenance for control of vehicle emissions and suspect power station technology. Disaster at the Chernobyl nuclear power complex in the Ukraine in 1986 resulted in a radioactive cloud which polluted air and, through rainfall, the soil over areas from Lapland to Great Britain.

Recognition of the global nature of climatic problems led the Soviet Union to join international co-operative schemes to avoid future hazards of this kind and to join research with other countries into the threat to the ozone layer.

In this chapter, however, we are concerned primarily with the climate as experienced by the Soviet people, for whom it has always been a major factor in daily life and economic planning on a scale quite unknown in, for example, Britain or the USA.

The Soviet Union is the most thoroughly continental as well as the largest of the nation-states, being virtually surrounded by other lands or by partly frozen seas. Nearly half of the Soviet Union to the north of its broadly settled belt is fundamentally unsuited to permanent human occupation owing to lack of heat, and into this agricultural wasteland, Great Britain could be fitted 40 times. In addition, to the south of the generally settled belt in the Soviet Union is an area of desert and inland drainage at least as large as Australia's.

Thus human settlement has become more or less channelled, over the centuries of expansion of the Russian Empire, into an elongated triangular area, with its base on the European frontier and its apex around Lake Baykal, forming a wedge between the cold northern forests and the deserts of Central Asia (Fig. 10.3).

Although the climate seems to exercise a decisive influence on the general pattern of population distribution in the Soviet Union, we must beware of ascribing to it an absolutely 'determining' power over the activities of man and even his psychology and degree of civilisation. The control of man's activities by climate is to be looked at through the medium of costs in the broadest sense. For instance, there are risky and expensive climates for farming but very few impossible ones; with enough financial subsidy all sorts of fruits and vegetables can be grown on the Arctic coast, but probably will not be if they can more cheaply be grown elsewhere, and transported to the north. In other words, the climate is a persuasive but not a compelling director of human affairs. However in these days of increasing competitive regional specialisation and more efficient transport, it seems likely that natural factors are becoming more, rather than less, influential in modern farming operations.

On the other hand, urbanisation has profoundly modified the significance of climate in the life of the Soviet people. Before the 1930s, the great majority of Russians were peasants and were effectively immobilised for long periods during the severe winter when there was little or no work, followed by a hectic pace of work in the short summer. However, today the average Russian is an industrial urbanite working indoors throughout the year, and although he may gravitate to cities with a somewhat better climate, amenities in the broad sense, like housing and education, are usually more important to him.

SALIENT FEATURES OF THE SOVIET CLIMATE

Together with its extreme continentality, already noted, the high latitudes in which most of the Soviet Union lies should be kept in mind. Transferred to the southern hemisphere at the same latitudes, most of the Soviet Union would lie well out in the Southern Ocean and into Antarctica. But the most meaningful comparison is with North America, and here it appears that some four-fifths of the Soviet area, and about nine-tenths of that part of the country which receives adequate moisture, are in the latitudes of Canada rather than of the United States. Moscow itself is

further north than Edmonton, Alberta (the most northerly city of any real importance in Canada) while Leningrad, the second city and former capital, is in the latitudes of southern Alaska. Whereas no part of the coastline of the continental USA ever becomes frozen, almost all the Soviet coastline is subject to varying periods of freezing in winter.

A strong sense of climatic uniformity over large areas of the Soviet Union in any one season is gained from considering a few facts. In January it would be possible to skate on the rivers and canals from the Arctic to the Caspian Sea or sledge from the Polish border to the Pacific, while in July it is comfortably warm almost everywhere. Of course there are exceptions, such as the cool Arctic coast in summer and some warm sheltered parts of the Black Sea coast in winter.

A feature of the Soviet climate often overlooked but no less significant is its dryness. All of Britain receives at least 500 mm of precipitation in the year and well over half the United States is so favoured, but less than a quarter of the Soviet area comes up to this level. Even allowing for differences in evaporation, the USSR is more similar to Australia in this respect than to any other important country although Australia is, taken all round, much more arid.

The Pressure Systems

Because of the very large mass of land at relatively high latitudes, marked seasonal differences in pressure conditions occur, resulting largely from the profound and rapid changes in the prevailing temperatures.

During the frigid winter there develops over Siberia and Central Asia a high pressure system of an intensity unique in the inhabited world at any season. In the windless centre of this system radiation leads to the accumulation of dense air which drains into certain enclosed valleys, especially in north-east Siberia, as noted below, bringing the coldest temperatures recorded on earth (outside Antarctica). This outflowing air from the Siberian centre of high pressure extends also, less intensely, over European Russia and occasionally crosses the North Sea to paralyse eastern England, but throughout this western peripheral region its dominance is challenged by the inflowing air from the Atlantic.

Spring is a fleeting season in most of the Soviet Union, bringing rapidly rising temperatures, and spreading from across the country in a generally north-easterly direction from the south-west Ukraine, about March, to the mouth of the Lena in late June. The heating of the land and the consequent ascending air soon leads to the replacement of the winter high pressure system by one of low pressure, which opens the way for the inflow of moist oceanic air. The rise in temperature also promotes convectional activity with locally heavy thunderstorms in summer. In the Soviet Far East, the tail end of the Asian monsoon is experienced in August and September, though with far less intensity than further south. With the abruptly falling temperatures in October, the Siberian high pressure sytem builds up again and the seasonal cycle recommences.

The Burden of Winter

When considering the negative features of the Soviet climate, the length and severity of winter occupies a dominant position. Over half the area of the country experiences more or less continuously freezing temperatures and snow cover for over half the year.

Compared with North America, which lies open to tropical air masses from the Gulf of Mexico, the Soviet Union is cut off from the air from the Indian Ocean by an almost continuous mountain rim. The general westerly airstream makes the Atlantic, not the Pacific, the tempering influence, and both continentality (i.e. annual range of temperature) and winter severity and length are intensified from south-west to north-east.

Thus, almost all of Siberia has average winter temperatures below −18°C, i.e. colder than any part of the United States, let alone Britain, and almost all of European Russia has average temperatures below freezing point. The coldest places in winter, averaging −51°C in January, are in north-east Siberia (Verkhoyansk and Oymyakon), relatively close to the Pacific coast. Some idea of the intensity of the winter in the eastern half of the country may be gained from the fact that most of the ground is permanently frozen ('permafrost' or *merzlota*)—a phenomenon which acts as a severe deterrent to all kinds of human activities.

The only part of the Soviet Union which has a January mean temperature more than a few degrees above freezing point is the south-eastern corner of the Black Sea (the Kolkhid lowlands)

but even this area suffers from occasional killing frosts.

The Heat and Water Budget

Annual precipitation figures and mean temperature values are only moderately significant for estimating the effect of climatic elements on man. As far as permanent settlement, fundamentally agricultural, is concerned, two values have been of overwhelming significance. These are 'accumulated heat', measured by 'day-degrees' above the threshold at which plant-growth begins, about 10°C, and 'effective moisture', that is, precipitation minus evaporation. In general, because of the sharp seasonality of the temperature and rainfall regimes (most rain comes in the summer) the most valuable comparisons to be made are again with North America.

Unfortunately the outstanding reservoir of accumulated heat (Fig. 3.1) in the Soviet Union is the Central Asian desert and semi-desert, which have heat and moisture balances comparable to most of the southern parts of the USA, but where economic farming depends on the availability of water for irrigation. The traditionally most intensively farmed land in the country, in the northern Ukraine, as well as the vast area of virgin lands in western Siberia and Kazakhstan, ploughed up in the late 1950s, have a heat budget similar to that of the prairie lands of the Dakotas and Minnesota. The region around Moscow, which was all the Russians were able to occupy for centuries, has a heat budget similar to that of the poorer parts of the Canadian prairies or north-west Europe. But even in the northern half of the country (mostly covered by coniferous forests) which has a shortage of heat, there are a few sheltered valleys—including the coldest in the world in January—which accumulate just enough heat for some passably successful agriculture in the short summer. Average figures for accumulated temperatures for different areas are given in the regional descriptions of agriculture in Chapter 7.

The moisture balance is described in greater detail in Chapter 5. Here it may simply be noted that for the distribution of 'effective moisture' (Fig. 3.2), the most important fact is that most of the lands which have a 'surplus' are at least marginally deficient in heat. Of the regions of superabundant moisture, the only area where it is accompanied by ample heat is the very small basin

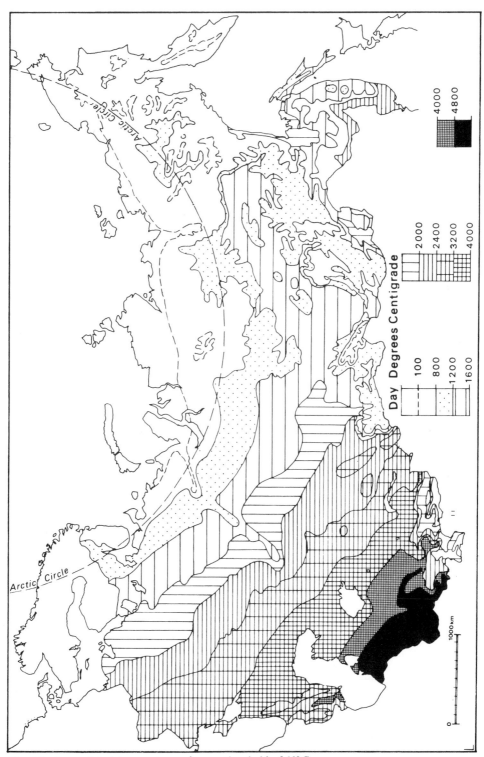

Fig. 3.1 Accumulated temperatures above a threshold of 10°C

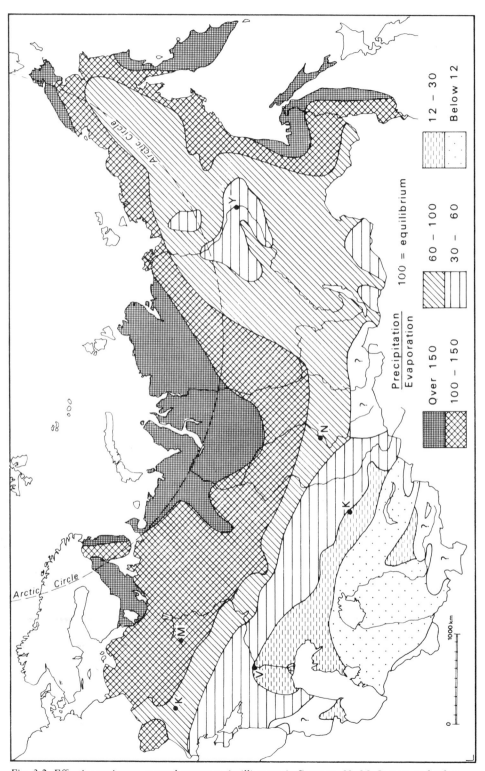

Fig. 3.2 *Effective moisture, annual averages (millimetres). Sources: N. M. Ivanov and others*

of the western Transcaucasus, facing the Black Sea. Somewhat drier regions, but where the moisture is everywhere adequate, comparable to Britain or USA east of the Mississippi, include European Russia north of Kiev and Kuybyshev and the Far Eastern monsoonal area of the Amur Valley. The large dry zone of inland eastern Siberia is to some extent ameliorated by the fact that the permanently frozen subsoil acts to conserve the limited moisture in the topsoil, preventing normal percolation.

In summary, it may be said that south of a line from Odessa to Lake Baykal (excepting the tiny anomaly of the western Transcaucasus), shortage of moisture is something of an obstacle to agricultural progress. The lands along this line may be compared to the western extensive wheat-growing margins of the North American Great Plains or Australia, which similarly fade into semi-desert ranching country and eventually desert.

Thus it can be seen that a major tragedy for the Soviet Union lies in the lack of broad geographical coincidence of its endowments of heat and moisture. There is nothing quite comparable to the great mass of the eastern United States where adequate supplies of both necessities coincide geographically.

THE IMPACT OF THE SOVIET CLIMATE

The Creation of 'Natural Zones'
Apart from the various direct influences exercised by climate on human activities—and on simply living—we must not forget its most basic indirect action in playing the dominant role in the creation of the prominent system of 'natural zones' in the Soviet Union. The Russian landscape is in a very real sense stamped in the image of its climate. The distinctions in the landscape which meant most to man as he strove to colonise these lands were not the mountains and the valleys, but the forests and the grasslands in their various forms. (See maps in Chapter 4.)

In the geologically recent formative period following the pleistocene period, the low and subdued nature of the relief over the vast areas of the Russo-Siberian plains enabled climate to use a 'broader brush' in the creation of continuous, fairly homogeneous latitudinal 'natural zones' of

soil and vegetation than anywhere else in the world. These and the soil-forming processes involved will be described in detail in Chapter 4. It will therefore be enough to say here that heat and water are the active elements in creating a soil and allowing a certain vegetation and animal life to live in it and give fertility and character to it. Thus the relative amounts of heat and water, considered above, are the key elements in creating and maintaining a distinctive soil-vegetation character and therefore the broad 'natural zones' become, to a significant degree, faithful reflectors of their climatic conditions. To the north of the wooded steppe zone there is adequate moisture but inadequate heat, the soils (mainly podzols) are leached and acidic, and trees dominate the scene roughly as far as the Arctic Circle. To the south where, except for some Black Sea lowlands, there is enough heat but not enough moisture, the soils (most notably the black earth or chernozem) are rich in humus in the more humid areas and in salts in the more arid, with grasses and drought-resisting shrubs (not trees) as the characteristic vegetation.

The climatic emphasis is reflected in the fact that the Soviet Union has a much greater proportion of tayga (coniferous forest) and tundra than has North America, slightly more grassland and desert and very much less well watered and heated forest land. This pattern has had incalculable effects on the history and character of colonisation and the wealth of the nations concerned, and indicates the fundamental significance of the climatic factor.

Climate and Farming
The influence of climate on agriculture seems obvious, but can easily be over-simplified if the historical, economic and technological circumstances are not kept in mind throughout. For instance, although over two-thirds of the Soviet arable area is now on black-earth soil (chernozem), the Russians, throughout most of their agricultural history, until the eighteenth century, were largely prevented from using it by non-Russian people who did not themselves till it.

It is true that the present distribution of population, urban as well as rural, has come, over the centuries, to coincide more or less with the distribution of arable land, and that there has been a natural tendency for farmers to move from poorer to better land. However, as in other industrialised countries, accessibility to the urban markets often transcends the normal disincentives of a poor agricultural climate, e.g. around Moscow or Leningrad, and influences the type of farming practised. Location, traditions, comparative costs and general government policy may be just as important as any 'pure' climatic factor in deciding the distribution and specialisation of farming.

All the same, an important effect of the development of a national market and improved communications in large countries like the Soviet Union or the United States is, or should be, to make climatic factors more rather than less important in regional farm specialisation. In practice this process is frequently delayed in the Soviet Union by the Government's ambivalent attitude to regional self-sufficiency in agriculture, or overloaded transport lines or compulsory purchase orders for crops which neglect regional disadvantages.

There is some evidence that the Soviet government has recently come to realise that attempts to extend agriculture towards the climatic margins, instead of intensifying investment in optimum areas, while not by any means physically impossible, can be very expensive. Consideration of the heat and water budget leads to the conclusion that the fairly narrow zone of overlap of adequate heat and moisture in the European black-earth wooded steppe region and neighbouring areas is likely to yield the best return on investment for most of the widespread crops (and animals) in the country. Even here the crops are frequently devastated by a scorching wind (*sukhovey*) which causes rapid wilting of plants through excessive transpiration. The strain of coping with a short growing season or unreliable rainfall on the margins, say in Siberia or the virgin lands, not only means uncertainty about harvests, but ties up large quantities of equipment and labour which are only in use for a small part of the year. Even in the *relatively* well-favoured sub-tropical Caucasus regions, which receive only occasional killing frosts, the question arises whether it would not be cheaper in the long run to import the nation's needs of citrus fruits from Italy, Cuba or even Australia or the South Pacific.

There is not really much difference between the total area of climatically usable agricultural land in the United States and the Soviet Union. However, the significant difference lies in the fact that much more of the Soviet land is climatically marginal and therefore an unduly risky investment. Recently there have been moves toward a thorough appraisal of the essential quality of Soviet farmland with the aim of intelligently maximising specific long-term environmental assets. Although there is no doubt that the virgin lands extension which took place in western Siberia and Kazakhstan in the 1950s has effected an indispensable, massive and permanent shift of the centre of gravity of grain production to the east, the yields are low and vulnerable to climatic misfortunes. The realistic conclusion may well have been reached that intensive investment in proven good quality farmland is necessitated by the costly climatic marginality of much of the rest.

Climate and the Transport System

From what has been written already it will not be surprising that heavy climatic burdens are placed on the movement of people and goods during

The Krestovy Pass (2437 m), summit of the Georgian Military Highway, built in the 19th century, is now an important commercial link, here seen in May with snow banks still lining the route.

much of the year. From more temperate countries, where occasional natural hazards such as slips on the road, flash-floods or high winds at sea form the chief obstacles to movement, it is difficult to imagine the range of prolonged and paralysing problems imposed by the Russian winter.

Until a century or so ago, most of the Russian freight was handled by the rivers, almost all of which are obstructed by ice for several months, followed by destructive floods and finally by dangerously low water in late summer. Transport on the large north-flowing rivers of Siberia is particularly limited because their lower reaches are still frozen while the upper reaches are in their spring flood. Almost all the seaports are troubled by ice and apart from the inactivity, considerable damage is done to bridges, port facilities and the like, on rivers, lakes and seas. Further, the presence of permafrost under nearly half the area of the country makes it difficult and expensive to build permanent railways or roads.

On land, the roads were (and in many cases still are) more impassable in the quagmire conditions of the spring thaw and early summer than in the winter, when sledging was relatively easy. The very cold weather sometimes causes metals to become brittle and makes lubrication and motor operation very difficult, apart from snow blockage and damage to electrical machinery, including transmission lines. For further details see Chapter 11.

New technology, whether on the railways, in the air, or in building the much overdue highway network, is gradually reducing the extent of the chaos which regularly prevailed and making it more feasible to plan in terms of regular, large-scale interregional trade. But the climate still adds high overheads to the operation of the Soviet transport system, compared with that of most modern countries, quite apart from the human discomfort, not to say hardship, involved.

Climate and everyday living
In addition to the almost crippling climatic disabilities under which certain specific essential economic activities labour, there remains the pervasive general effect upon everyday life, migration and even the state of mind of the Soviet people. The Soviet climate, in general, is something hostile to be wrestled with, or at least taken very seriously.

Throughout their history, the Russians, as distinct from the Georgians, Tadzhiks or other southerly-dwelling peoples who were conquered by them in the nineteenth century, have lived in a climate which must be considered rather miserable by most standards. Since the majority of them were, until quite recently, rural dwellers (the urban segment became a majority of the Soviet population only in 1962), they have lived close to nature. The forced winter inactivity, where peasants, if they could, frequently passed the time in a lethargic, and sometimes drunken state, has sometimes been blamed for supposed ingrained and unfortunate habits of mind in Russia. Ellsworth Huntington thought that the Russian peasants should be shipped south to sunnier Mesopotamia for the winter, for the good of their souls. Clearly, since the present Soviet population is largely an urban and an indoor, industrially-oriented one, the climate is being placed at further remove for more and more people. On the other hand, problems of winter in the rural areas are not rendered less by reduction of numbers; food still has to be produced, and moreover winter in the city also has its hardships.

A wistfulness for 'southern climes' has long been present among the Russians, as it has among other 'northern' peoples. A passage from Peter Kropotkin, the nineteenth century Russian philosopher-geographer, on his first journey to western Europe in mid-April, may be indicative:

'And the contrast of climate! Two days before, I had left St. Petersburg thickly covered with snow, and now, in middle Germany, I walked without an overcoat along the railway platform in warm sunshine, admiring the budding flowers. Then came the Rhine, and further on Switzerland bathed in the rays of a bright sun, with its small, clean hotels, where breakfast was served out of doors, in view of the snow-clad mountains. I never before had realised so vividly what Russia's northern position meant, and how the history of the Russian nation had been influenced by the fact that the main centres of its life had to develop in high latitudes. Only then I fully understood the uncontrollable attraction which southern lands have exercised on the Russians, the colossal efforts which they have made to reach the Black Sea, and the steady pressure of the Siberian colonists southward, further into Manchuria.'

There were, of course, many more primary reasons than the search for the sun for the persistent southerly movement of Russians in the last two centuries or so, notably the promise of improving farmland and colonial enrichment. However, now that these aims have been achieved and consolidated and the Soviet population has become more urban, mobile and better educated, the question of climate as an 'amenity' is beginning to make itself felt and modify significantly the population map.

The 1959 Census recorded a massive displacement of the Soviet population towards the harsher climates of the east, especially to the Urals, Siberia and Kazakhstan. Much of this can be attributed to the Second World War and Stalin's enforcement of labour movement to new factories in the eastern regions, but much also to Khrushchev's grandiose virgin lands campaign— probably the last major extension of the Russian 'pioneer fringe'.

But since the late 1950s, as discussed in Chapter 6, there has been clear evidence of a substantial net outflow from Siberia, where the climate is indeed very severe, in spite of a disproportionately high government level of investment and new discoveries of industrial resources. These 'losses' are officially admitted, but commentators are at pains to stress the importance of the relative lack of other amenities, like good housing and education, in addition to the climate. On the other hand, the more southerly regions, notably Soviet Central Asia, the north Caucasus and the southern Ukraine (including Crimea) have recorded a considerable degree of net in-migration from the north. Although articles in the Soviet press frequently deplore the movement to these areas, where there is already a labour surplus (Central Asia and Caucasia have a natural increase which is much higher than the national average) and the exodus from Siberia and north European Russia, they still continue. This indicates a considerable degree of personal freedom of migration decision-making compared with the recent past, but it is difficult to avoid the conclusion that one important component in the recent southerly drift is that of climatic attraction. There is some evidence that the Soviet government will accept these trends, locating 'foot-loose', labour-intensive industries in these new reception areas, while encouraging mainly resource-oriented, automated industry in areas which cannot retain labour, thus obviating the need for costly subsidies to unattractive areas.

The southern areas attract holidaymakers in large numbers not only in summer but also for winter sports as the Caucasus ranges provide better climatic conditions than the other ranges and are also relatively easy of access from the main centres of population.

CLIMATE AND MAN IN PARTICULAR PLACES

Up to now, we have examined broad patterns and general connections between the climate of the Soviet Union and various activities and needs of its people. But since geography does not entirely come into its own until it focuses on the integration of man and his environment in particular places, we will attempt to bring out some significant features of the climate-man complex at selected points across the Soviet map. These points are necessarily cities, but they will be assumed to include the surrounding rural districts (Fig. 3.3).

Kiev

The first Russian city of any wealth and consequence was Kiev, in the wooded-steppe region of what is now the northern Ukraine, and although this success was mainly due to trade in furs, a sizeable population of farmers developed in the vicinity. Although by west European standards the winter in this region is severe (about three months with freezing temperatures) the growing season is long enough and warm enough for the safe cultivation of most of the European grains (including maize), vegetables and animals. Kiev is in the narrow region of overlap of adequate moisture (with a summer maximum) and adequate heat and sunshine, with a quite fertile type of black or grey earth and a fairly easily cleared mixture of open grassland and deciduous trees producing mild humus. By comparison with most of the USSR today, it is a mild and well-watered region and is one of the most intensively farmed as well as being relatively well regarded as a living-place for urban workers. The apparent regional trends in both agriculture and industry in the Soviet Union seem to favour the long-term growth of the Kiev area (quite apart from its importance as the capital of the Ukraine and a historico-cultural centre). From many viewpoints, the climatic factor in the future of Kiev is increasingly favourable in the Soviet context.

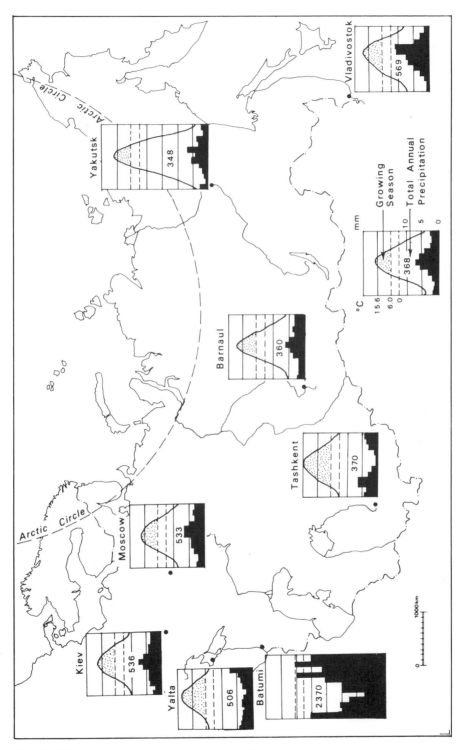

Fig. 3.3 Climatic characteristics of cities studied

Snow cover in the south of the USSR is of short duration on low ground but attracts large numbers of skiers to the mountains. Here a chair lift gives access from the Dombai valley in the Caucasus to the ski-field

Moscow

When the city-state of Kiev was sacked by Batu's Mongol Hordes in 1240, the Slavic remnants had to regroup in the watery, forested region between the Upper Volga and Oka rivers, centred on Moscow. Although on the broad scale of the Soviet Union the two areas have a not very dissimilar climate, the marginal differences are, in fact, significant enough to produce differences in the total environment which must have been painfully noticeable to the early escapees from Kiev. Moscow's climate is generally damper, cooler and more cloudy than Kiev's with twice as much snow, lying for five months instead of three, and appreciably less accumulated heat, sunshine and, thus, evaporation. This leads to a more waterlogged landscape, a predominance of coniferous rather than deciduous trees and a more podzolic, less humus-rich soil. Costs of items like snow clearance are heavy.

The traditional crops were the hardy rye, oats, flax, potatoes, cabbages and other vegetables and the competitive position of the area's grain farming has been suffering ever since the infiltration again of the better lands south of the Oka by Russian peasants several centuries ago. The current presence in its midst of the largest urban market in the country has promoted dairying and market gardening and saved the physically somewhat marginal farmland from continued decline. As far as the large urban population is concerned, the unique metropolitan amenities undoubtedly make up for the somewhat dismal climate.

Barnaul

Barnaul lies in the Altay Steppe, near the eastern end of the continuous zone of wooded steppe, black-earth and optimum overlap of heat and water, just before it is brought to an end by the mountains and permafrost which characterise eastern Siberia. During the first incursions of peasants into Siberia in the eighteenth century, the Altay Steppe was perhaps the most coveted, in spite of being the most distant, of the good lands and still today, for Siberia, it is one of the most reliable and productive farming areas.

However, a climatic comparison with Kiev, on the western end of this long strip of wooded steppe, may be illuminating. Barnaul is distinguished by much colder winters, a significantly shorter growing season, and less accumulated heat, although the mid-summer temperatures are much the same. Although Barnaul has less precipitation in the year, it is more concentrated in the spring and summer when it is needed and evaporation is less, so that the average supply of effective moisture is much the same in both cases.

The severity of the winter means that crops like wheat have to be sown in the spring, rather than the autumn as in the Kiev region, and this places a greater strain on equipment and labour and makes the farming routine more hectic and rather more limited in range of crops and livestock. In common with the rest of Siberia, the length and severity of the winter, when coupled with shortcomings of other amenities, makes retention of people increasingly difficult. However, the climatic conditions for permanent agriculture in this region are better than one can find anywhere in that half of the Soviet Union which lies to the east of it. In fact, it is likely to continue to be one of the most densely populated agricultural communities of any size in Siberia.

Yakutsk

Yakutsk, on the bend of the Lena river in northeast Siberia, has much the most severe continental climate of any of these chosen case-studies and is the only one situated on permanently frozen ground. The mean January temperature is about −40°C compared with a mere −18°C at Barnaul and, of course, much lower actual temperatures are recorded. Since little snow falls in this centre of the winter high pressure system, permafrost several hundred feet thick is formed and maintained, and the danger of frostbite is acute during even a short walk in the town. However, although snow lies for seven months and night frosts may occur in any month except July, circumstances have accorded this region a warmer summer (warmer than Moscow in July) and a longer growing season—about three months—than in normal for this latitude. Therefore, although relatively unimportant nationally, this region is the only notable outlier of agriculture in the great wilderness of Siberia, north of the closely settled belt. Frost-resistant strains of wheat and other crops are grown for local use. Rainfall is light, and although the great majority of it falls in summer, and the permafrost conserves the surface water, irrigation is usually needed. But the permafrost, which is potentially unstable at its upper edge,

Collapse of part of an apartment house in Yakutsk following subsidence caused by melting permafrost

presents constant problems for building, transport and urban life in general.

This is an extremely severe environment by any standards and, were it not for the fact that it is the centre of a huge area where there are valuable mineral workings and also the focal point of a vigorous native group, the Yakuts, the hardships would be too difficult, or too costly, to bear.

Vladivostok

The southernmost Soviet port on the Pacific, Vladivostok has a severe winter for its latitude and position on the sea, since it is under the influence of strong outflowing Siberian winds. It has freezing temperatures for about five months and its excellent harbour has to be kept open by ice-breakers, while new, virtually ice-free ports, Nakhodka and Sovetskaya Gavan, on the open sea, have been developed. Although the growing season is long enough for normal European or north Chinese agriculture, the rainfall maximum comes in late summer, at the tail end of the Asian monsoon, when ripening crops may be damaged, while irrigation is often needed in the early summer. The humidity in late summer can be uncomfortable and clouds of biting insects add to the discomfort. Thus an unfavourable climatic 'comfort index', added to isolation and a limited amount of second-class farmland, conform to the common Siberian pattern as far as human attractiveness is concerned.

Some Southern Exceptions

All of the stations discussed so far have much more severe winters than are ever experienced in England or Australasia and a more or less restricted growing season. They represent about four-fifths of the total area and the total farmland of the Soviet Union, and therefore can be said to typify the most common range of climatic problems. On the other hand, the next three to be considered, although anomalous in Soviet terms, are basically much more akin to stations in New Zealand and the southern parts of Australia or the United States, in whose latitudes they lie, with their warmer summers and winters and a growing season which lasts at least three-quarters of the year.

Tashkent

In the centre of the Asian landmass, Tashkent has long hot, and dry summers, with accumulated heat values similar in amount to those of Arizona, inland California or inland southern Australia. As in those places, irrigation is necessary for intensive farming of cotton, fruit and vegetables, while away from the irrigated areas only extensive dry farming techniques and ranching are possible. Actually Tashkent has a higher rainfall, concentrated in the winter, than much of the semi-desert to its west, owing to the fact that it is at the foot of the mountains which intercept the rain-bearing depressions. The fertile loess soil, common to this piedmont strip, is easily blown about so that the summers are hot, dry and dusty. Provided that an adequate supply of irrigation water continues to be forthcoming, the hot, long growing season and the increasing demand for cotton and fruits in the USSR will ensure the continued importance of the area, but the supply of water, as well as fertile soil, is strictly limited. The summer heat does not seem to deter the immigration of Russians into Tashkent, any more than it does Americans to Arizona. In addition, the native people of the region have a much higher rate of natural increase than the country as a whole, and do not tend to leave the area. Thus, the water limitation will probably be the crucial long-term problem in the man-land equation here.

Batumi

On the shores of the Black Sea, Batumi, with the Kolkhid lowlands to the north of it, are climatically unique in the Soviet Union in several respects. It is the only part of the country where the January mean temperature is over 6°C, and correspondingly higher day temperatures, giving a year-round growing season. However, occasional killing frosts do occur and threaten the citrus fruits. Secondly, it is the only part of the Soviet Union which has abundant rainfall as well as adequate year-round heat. In fact, draining of excessive water has been the major prerequisite to developing the land for tea and other plants which need abundant water but also good drainage. The swamps were also breeding-grounds for malaria-carrying mosquitoes until recently. Thirdly, it is the only Soviet area with a very heavy total rainfall and a winter-autumn maximum—although it receives a good deal in the summer as well.

All of this gives the area a luxuriant sub-tropical vegetation, with tree-ferns, eucalyptus and plants from sub-tropical China, Japan and India—altogether a most untypical situation in the Soviet Union. Since this is the only practicable area of the country for the growth of frost-

Seaside crowds at Yalta in the Crimea, vessels for tourist trips in the foreground

sensitive crops like citrus fruits and tea, it has received special attention from the government. Because of topographical and local frost difficulties the area which is really well suited to commercial growth of these exotic crops is quite limited and, were it not for the autarchic traditions of Soviet-agriculture, might well be better employed growing less exotic crops like maize or rice.

Yalta

Yalta is the centre of a string of tourist resorts along the very narrow southern coastal strip of the Crimean peninsula, backed by steep mountains protecting the coast from the cold northern winter winds. After the Russians had conquered the Crimea in the late eighteenth century and found this sheltered coast to have no regular freezing period and abundant winter sunshine, the nobility soon began to build palaces there as an escape from Moscow or St. Petersburg—they are mostly converted to sanatoria today. The January temperatures—averaging about 3°C—are appreciably cooler than at Batumi and there is a brief plant resting period. The rainfall regime is a partial Mediterranean one, with a winter maximum but much less heavy than at Batumi and with more sunshine. The summers are relatively dry and sunny, again much less humid than Batumi. On a year-round basis this is probably the pleasantest climate in the Soviet Union which fact, combined with its natural beauty, explains the concentration of tourist and medical establishments. Most of the very limited area not taken up by resorts is given over to growing vines and other fruits. Altogether it presents a climatic scene not unlike many coastal places in the northern Mediterranean, southern Australia, or California, but one which is quite unique in the Soviet Union.

CLIMATIC COSTS, BENEFITS AND ADAPTATIONS

The Soviet climate has been analysed in terms of how it appears to the inhabitants who have to live in it, how it affects their activities, and how this compares with the climatic lot of people in other parts of the world, notably North America, Britain and Australasia. By way of summary, a balance sheet of the costs, benefits and trends will now be presented.

Costs

(*a*) The length and severity of the winter effectively rules out at least half the country's area, in the north and east, for normal agriculturally-based settlement. It incidentally causes most of this area to be permanently frozen underground and to have poor vegetation and soil conditions. Transport operations are severely restricted and made expensive, and so are all economic activities, like mining, which have to be carried on in the non-agricultural areas. Labour is also expensive and difficult to attract and retain in these winter-bound areas.

(*b*) The generally drought-ridden character of parts of the country which are not severely crippled by the winter places a serious extra cost-burden on the national economy. Irrigation development in Central Asia in the very dry areas as well as heavy investment in equipment and men, coupled with a high risk of drought failures in the drier parts of the black-earth belt, add to the costs of farming enterprises.

(*c*) The lack of a broad overlapping zone of the large areas which have *either* enough moisture *or* enough heat is a major tragedy compared, say, with the situation of the United States. The fact that the relatively small overlap zone is in relatively high latitudes compared with that of the USA or much of Europe is an additional misfortune and cost.

Benefits

It is difficult to discover really positive, large-scale benefits accorded to the Soviet people by their climate, but certain silver linings to the clouds may be pointed out. In other words, things could have been worse!

(i) In spite of the narrowness of the zone of overlap between heat and water in relation to the country's total area, the vast size of the Soviet Union ensures that, in absolute terms, the generally productive land is extensive enough, given good management and enough investment, to feed the population adequately in the foreseeable future.

(ii) The low rainfall totals are significantly ameliorated by the fact that most of the year's precipitation is received in the growing season and most of the rest is kept in cold storage until the spring.

(iii) At least there are some areas of sub-tropical climate, however small, which constitute just as much psychological and recreational assets as strictly economic ones.

Adaptations

It has often been implied in Soviet government pronouncements since the time of Lenin that general policy was to develop all regions of the country regardless of comparative costs. Indeed there has been a considerable extension of the cultivated area, culminating in the massive virgin lands scheme of the 1950s. However, in recent years, the evidence seems to indicate a new recognition of the climatic limitations and the cost of failing to pay close attention to them—largely, but by no means entirely, with reference to farming.

In general, the geographical directions of change seem to be towards the west and south rather than to the east and north as in previous decades. This results from a recognition of (*a*) the fact that an increase in food production can be obtained more economically from the already more intensively farmed and climatically well-endowed parts of the European black-earth zone than from further extension at the climatic margins, and (*b*) that, with a more mobile, educated and affluent population, a significant migration is under way from the areas generally considered climatically 'difficult' to live in, towards those that are easier—given the presence of other amenities and employment.

BIBLIOGRAPHY

Berg, L. S. (1950), *Natural regions of the USSR,* Macmillan, London and New York.

Borisov, A. A. (1965), *Climates of the USSR,* Oliver and Boyd.

Budyko, M. I. (1974), *Climate and life,* English edition, D. H. Miller (ed.), Academic Press, New York and London.

Fiziko-geograficheskiy atlas mira (1964), Moscow. A key in English to this atlas appears in *Soviet Geography, Review and Translation,* **6** (5–6), 1965, pp. 1–403.

Gibson, J. R. (1969), *Feeding the Russian fur trade.* University of Wisconsin Press.

Hooson, D. J. M. (1966), *The Soviet Union.* University of London Press, London.

Lyakhov, M. Ye. (1987), 'Tornadoes in the midland belt of Russia,' *Soviet Geography,* **28,** 562–570

Lydolph, P. E. (1978), *Climates of the Soviet Union, World survey of climatology,* Vol. 7, Elsevier.

Lydolph, P. E. (1959), 'Federov's complex method in climatology.' *Annals Association American Geographers,* **49,** pp. 120–144.

Lvovich, M. I. (1963), *Chelovek i vody,* Moscow.

Parker, W. H. (1968), *Historical geography of Russia.* University of London Press, London.

Rogers, J. A. (ed.) (1962), P. A. Kropotkin *Memoirs of a Revolutionist.* Doubleday, N. Y., p. 178.

Shashko, D. I. (1962), 'Climate resources of Soviet agriculture', in Akademiya Nauk SSSR (1962), *Pochvenno-geograficheskoye rayonirovaniye SSSR,* trans. by A. Gourevitch, *Soil-geographical zoning of the USSR (in relation to the agricultural usage of lands),* IPST, Jerusalem, 1963.

Suslov, S. P. (1961), *Physical geography of Asiatic Russia,* Freeman, San Francisco and London.

4 Biogeography—the Vegetation, Soils and Animal Life

The vegetation and soil provide two of the most essential natural renewable resources of the USSR, as they form the basis of her agriculture and of the supply of many kinds of organic raw materials. Green plants convert radiant solar energy, atmospheric carbon dioxide and the soil's water into organic food material by photosynthesis. This *Primary Production* is consumed by grazing animals or herbivores which constitute the *Primary Consumers*; the latter are the food sources of the carnivores, or *Secondary* and *Tertiary Consumers*. Each of these groups contributes dead or waste organic material to the soil where it is converted by bacteria, fungi and other organisms of decomposition into nutrient chemicals which become absorbed in solution by plant roots.

Thus a forest or a grassland may be considered a macro-environment, a definable geographic unit where all its plant life, fauna, micro-organisms and soil, together with the atmospheric elements and the geological substratum, function together in a condition of balanced interaction and inter-relationships. Such units are *ecosystems* or *biomes,* or as defined in the USSR by Sukachev, *biogeocenoses.* Changes in any one of the components can thus produce important and sometimes unpredictable consequences upon the others, as exemplified by mankind's ability to transform ecosystems by replacing natural vegetation by agricultural systems and therefore creating transformations in soil character and in the animal life.

The distribution of two of the most important of such components in the USSR, namely the natural vegetation and the soil cover are illustrated by Figs 4.1 and 4.2. These have developed during the ten millenia of the post-glacial climatic periods, following the Quaternary glaciations. During the latter almost all of the former Tertiary warm-temperate ecosystems were destroyed. With the disappearance of the Pleistocene ice-sheets and the subsequent gradual warming of the climate, a succession of communities of plants came to occupy the area which is now the USSR. Such communities consisted of species of plants whose habit or life-form was most closely adapted to the environment. Those occupying most space, absorbing most nutrients and having the most bulk or biomass are the dominants, such as the forest trees. Much smaller plants of different habit are the subordinates, such as shrubs, herbs, lichens or mosses. Each of these major plant communities or formations became associated with dependent and interdependent animal groups ranging in size from soil micro-organisms to large vertebrates so that each represents a major life-zone or biogeocenosis.

Within these large natural units, the soil is also a living component. In tsarist Russia and since the 1917 revolution, soil in the USSR has been the object of intensive research, conducted by such pioneers as Dokuchayev and Glinka and continued by modern work such as that of Lobova and Gerasimov. Their work was directed not only to the soil's agricultural potential but also to its functional role in natural biological communities. They stressed that soil character is determined by several factors operating over the wide areas of the Russian plains, namely the geological parent

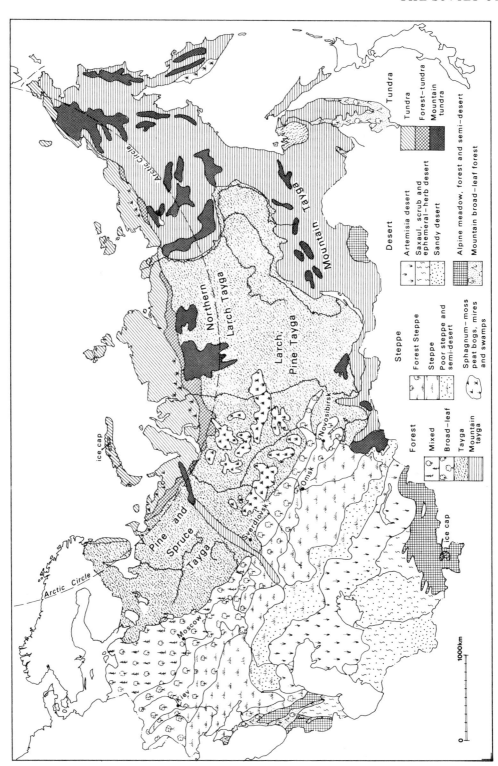

Fig. 4.1 Vegetation zones of the USSR

material, the relief, the climate and the living organisms, both plant and animal. The interaction of these factors gives rise in a mature soil to several distinct layers or horizons. Taken together, the horizons ideally contain the four essential constituents of soil which are mineral matter, organic matter, air and water.

Such soils and the plant formations they support have been evolved by natural selection on the lowland areas of the USSR where a substantial uniformity of topographic, geological and geomorphological features is found. They have developed in a climatic environment which consists of a continuous gradation from north to south, comprising temperature, rainfall, irradiation and potential evaporation. This has resulted in a series of broad latitudinal zones trending from west to east, each with its own characteristic plants, soils and animal life, and identified by the dominant type of plant life in each. Human exploitation of the zones has resulted in varying degrees of changes in their character, although the basic relationships between the environment and the biotic communities are still functionally significant.

In the far north is the treeless Arctic tundra zone of cold deserts; farther south, the higher temperatures and longer vegetative period have produced various kinds of forest vegetation, the latter still forming the most characteristic landscape of many rural areas of the USSR. The forests are replaced southwards by open steppes or temperate grasslands which are the vegetational response to the sub-humid conditions unsuitable for tree growth in these areas. To the south and east the steppes grade, as aridity increases, to the semi-deserts and deserts. This regular gradation becomes modified where mountains produce a more rapid climatic zonation and hence a vertical zonation of biotic communities, and it is much less evident in eastern Siberia and the Soviet Far East. Sub-tropical conditions occur in relatively limited areas in the far south, such as in Georgia, and these are zones where highland barriers exclude the deep chill of continental winters.

A system of nature reserves is being established in the USSR, in which the plants and animals characteristic of the above zones are protected. Scientific study of their ecology and interdependence is organised in each reserve combined with observations of the components of their habitat such as temperature, rainfall and atmospheric quality. Such measurements and subsequent modelling provide evidence of the manner in which natural and semi-natural ecosystems are being affected by human intervention.

Although the creation of a system of nature reserves was begun in the USSR in 1921, the global programme of nature conservation instituted by the United Nations and operated from 1971 by UNESCO, has provided the initiative and has emphasised the urgent need for expansion in the number of reserves. In 1988 there were 7 biosphere reserves but a total of 24 future reserves were envisaged, each being representative of the most widely distributed ecosystems in the USSR. Each is required to provide scientific data for comparison with analogous ecosystems in other countries. Protection of rare and endangered species is also provided by the reserves, many of which are now listed in the Soviet 'Red Data Books' published in each Republic in 1978 and 1985.

The Soviet social system and planned management of the economy apparently embodies great potential for the rational use of natural renewable resources embodied in various Conservation Laws. These are administered by Councils of Ministers in the Central Government and in each of the Union Republics. Various State Committees for Nature Conservation together with trade unions and educational institutions also play an important role in the promotion of the concepts of conservation of nature.

Despite the existence of such authorities it is apparent that abuse of the environment is at least as widespread in the USSR as it is in the West; this appears to be a consequence of the inadequate implementation of the Conservation Laws which are allowed to be ignored with almost complete impunity in many parts of the country and by all members of society. In addition, the lack of coordination between and among these authorities allows decisions of the central government and its agencies to plan and promote enterprises which they consider to be socially or economically beneficial but which are inimical to ecological principles.

In the following sections the biogeographical zones are discussed mainly in order of latitude, beginning with the tundra. The plant zones and their associated soil types in the main climatic regions are shown in Table 4.1. Beneath the conventional name of the soils are the names given to soil units on the maps published by FAO/ UNESCO (1972). The map (Fig. 4.2.) illustrates the distribution of soils; this should be compared with the map showing the vegetation zones (Fig. 4.1).

The Tundra

The *Arctic tundra* covers the most northerly areas of the USSR, north of the limit of forest vegetation; a closely related type of vegetation, the *alpine tundra* extends far southwards into mountainous areas where low temperatures exclude tree growth, such as the Urals or the various highland areas of Siberia and the Soviet Far East.

West of the Urals the Soviet tundras occupy a relatively narrow strip of the coastland of the White Sea and the Barents Sea, but they become much broader in northern Siberia where there are

large extensions of land northwards into the Arctic Ocean.

Tundra plant life is adapted to a rigorous climate, characterised by a very short, cool summer of two or three months when temperatures rarely exceed 10°C (50°F), and prolonged very cold winters with sub-freezing temperatures and low levels of precipitation. In the west, Atlantic marine influences provide slightly milder winters and marginally higher precipitation than in the Siberian sector, where winter temperatures may fall as low as −50°C (−58°F).

However the plant life is assisted by the long period of daylight in summer, allowing active photosynthesis even at 'midnight', contrasting with total dormancy in the dark winter. Tundra herbs and shrubs are able to take advantage of the short growing period by very rapid development immediately conditions allow in spring; mid-

TABLE 4.1 VEGETATION, SOIL TYPE AND CLIMATIC RELATIONSHIPS IN THE USSR

Vegetation	Soil type	Climate
Tundra	Tundra soils (Gleysols)	Arctic
Boreal Forest	Podzols (Podzols)	Sub-arctic
Mixed forest	Brown podzolic (Cambisols)	Cool temperate continental
Broad-leaf forest and wooded steppe	Grey forest soil (Orthic luvisols)	
Steppe	Black earths (Chernozems)	Continental semi-arid
Poor steppe	Chestnut soils (Kastanozems)	
Desert and semi-desert	Grey desert soils (Xerosols and Yermosols)	Arid continental
Mountain	Mountain soils	Mountain climates
Warm-temperate forests	Red podzolic soils (Acrisols)	Humid sub-tropical.

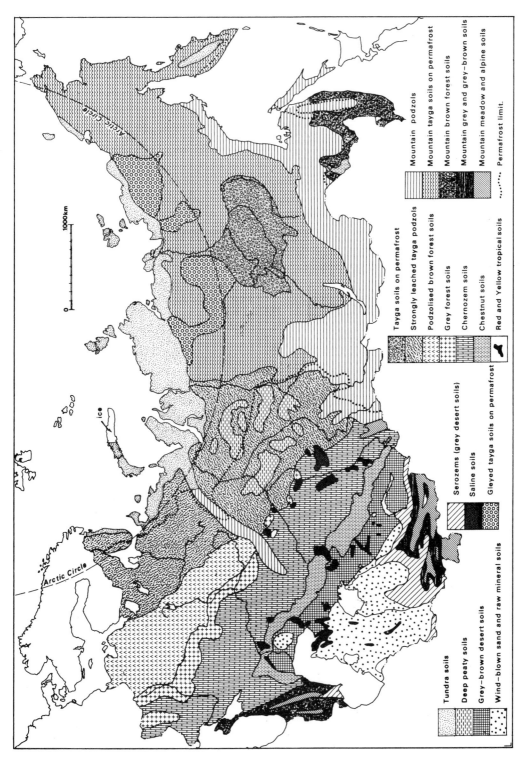

Fig. 4.2 Soil zones of the USSR

summer growth is followed by the formation of flowers and seeds within just a few days. Food reserves are built up and stored in the very extensive root systems and rhizomes in preparation for rapid resumption of growth in the following spring. Because seedling plants have little chance of survival, most species reproduce vegetatively and are long-lived. Most of the bulk or biomass of each plant is below the soil surface, but above ground the plants appear dwarfed or stunted, often with recumbent foliage trailing close to the soil surface; many species have a tufted habit, or form a dense, compact low-growing cushion. Such adaptations afford some protection from the chill and the dessicating effect and also the abrasive action of strong winds laden with ice particles. Low-growing lichens are particularly resistant to intense cold and prolonged drought and are also important for nitrogen fixation, apart from their value as forage for herbivores, notably reindeer.

Several latitudinal subdivisions of the Soviet tundra are recognised, based upon vegetational habit and distribution which respond to the gradual climatic amelioration from north to south. The Arctic tundra is a desolate north coastal strip where marine and glacial deposits have become colonised by scattered patches of algae, mosses and lichens; these are flowerless and rootless plants, extremely tolerant of tundra conditions, and found also on highland areas far to the south as alpine tundras. With decreasing latitude the longer growing season produces the 'typical' tundra, much richer in species, particularly of perennial herbs of grassy or graminoid habit such as sedges, rushes and grasses; the bog-moss sphagnum and other mosses of waterlogged ground form mires and swamps which are extensive in this area, but well-drained sites support clusters of flowering herbs on southerly exposures. Perennial woody dwarf shrubs, though they also occur, become most abundant farther south in the next sub-zone, the shrubby tundra. The latter contains many of the species of the more northerly divisions but is particularly characterised by a denser plant cover composed of dwarf-shrubs such as Arctic crowberry and bearberry, with dwarf birch and dwarf willow. To the south of the shrub tundra sub-zone is the forest-tundra, a transitional zone where the pioneer representatives of tayga forest trees appear among communities of tundra plants. The trees are commonly stunted or deformed larch, birch, alder and spruce species occupying more sheltered habitats whereas higher and more exposed situations carry the typical or arctic tundra vegetation.

Tundra plant landscapes are strongly influenced by the permanently frozen substratum, the *permafrost* which everywhere underlies their soils. The plants are rooted in a shallow surface layer of soil above the permafrost termed the *'active layer'* usually less than one metre in thickness, which freezes in winter and thaws in summer. Permafrost is conducive to very highly disturbed conditions in the active layer; being impervious to water it gives rise to waterlogging of the lower part of the active layer when the latter thaws in spring, and in this state solifluction of the soil takes place on sloping sites. The displaced soil forms lobes, bulges and hummocks. On flat sites the seasonal freezing and thawing of the active layer forces its stony basement material up to the surface where it forms defined patterns, often polygonal or circular in shape and called 'patterned ground,' with mosses or herbs growing in a fringe around bare spots in the centre of the patterns. Such movements create 'spotty' tundra landscapes; similar upheavals give rise to small mounds termed 'hummocky' tundra, each hummock with its plant cover of shrubs and mosses, and separated from others by swampy depressions.

Soils that are saturated with water in their lower layers are thus widespread in the tundra, arising from the accumulation of water from snow-melt and from precipitation. The lower horizon becomes deficient in oxygen and develops the 'gleyed' condition consisting of blue/grey layers of ferrous iron clay or stony loam which is a poor rooting medium for plant life. The lack of oxygen in the active layer retards the decomposition of the soil organic matter by bacteria and fungi; the decaying material thus accumulates near the surface as peat, and the soil, lacking the activity of micro-organisms becomes deficient in nutrients.

As a result of the very brief warmer season, the low soil temperatures and the widespread gleyed condition, the total yearly organic output or 'production' of the tundras is small. This primary production is also relatively under-utilised by herbivorous animal life so that much of the undecomposed output remains near the surface as peaty material.

The mammal and bird populations of the tundra biome exhibit two contrasting types of adaptation to this harsh environment. Those which visit the tundras during the summer breeding season, and those which remain throughout the year. Many bird species of geese and duck and mammals such as reindeer or caribou, are migrants, spending spring and summer in the tundra where there is abundant food and absence of human disturbance.

The other group are permanent residents, fewer in number of species but having a wide circumpolar distribution over both the Eurasian and North American tundras. They are adapted to endure the rigours of the eight-month winter, having an insulating coat of thick fur or feathers, exemplified respectively by the Arctic fox and the ptarmigan or the snowy owl. Most numerous are the herbivorous lemmings, rodents which pass the winter in burrows beneath the blanket of snow, not in hibernation but actively feeding upon roots of plants. Wide fluctuations in the numbers of herbivores and their dependent carnivores, occur. Lemmings have a well-expressed three-four year cycle of abundance and scarcity, associated with an equivalent cycle in numbers of their predators, the fox and the snowy owl. These are often conditioned by climatic or biotic factors, affecting the quantity and nutritional quality of the plants which provide the main part of the animals' food. The ecosystem is in a state of delicate balance, exposed to a change in the condition of any of its components that would affect all the others.

Soviet northlands of the tundra and the northern tayga zones are in urgent need of protection from the various impacts of increasing human presence; their ecosystems are fragile and vulnerable to damage which may produce irreversible changes in their structure.

The northern margin of the tayga is receding southwards, accompanied by an advance of the tundra subzones in the same direction. This is the combined result of many types of human influence, particularly economic development which includes industrial settlements with associated road and railway building, airfield construction and geological prospecting. Additional pressure on the natural vegetation and soils is produced by grazing of large reindeer herds during their seasonal migrations across the tundra and northern tayga, together with fires which originate in the camp fires of the herdsmen, fuelled by shrubs and

A notice at a conservation park (zapovednik) in the Caucasus mountains, organised in 1936 to protect the 'unique natural complex'

small trees.

Long-lasting damage to tundra vegetation and soil occurs when heavy, all-purpose vehicles produce deep furrows with their wheels or tracks, crushing the fragile cover of vegetation and thus exposing the underlying permafrost to rapid thawing. Subsidence of the permafrost results in the formation of thermokarst cavities and depressions which widen into ravines or become pools or swamps. Surface pipelines cut across and obstruct migration routes of tundra animals and overhead wires form a hazard to birdlife.

Some of the most intensive forms of ecosystem degradation have been recorded in the Bolshezemelskaya tundra of north-western USSR, associated with the mining and industrial operations of the Pechora coalbasin. Around this area's main Vorkuta industrial complex there is serious atmospheric pollution by gases such as sulphur dioxide and nitrogen oxide which are destructive to plant life, causing the disappearance of lichen growth and otherwise altering the species composition of the moss and herb cover. Waste water discharged from the mines, together with sewage from the settlements causes the loss of fish, invertebrate populations and plant life of streams and lakes. The density of the bird life has also been reduced within a zone 10 km wide around the Vorkuta complex.

Northern ecosystems and such influences exerted upon them by human activities are being studied at nature reserves situated at three widely differing situated sites: in the west is the Kandalaksha Bay system of reserves adjacent to the industrial areas of the Kola peninsula; farther east, in northern Siberia is the new Taymyr reserve; and in the southern Arctic Ocean is the offshore Wrangel Island reserve. The Taymyr reserve and the Kandalaksha Bay reserves are within the influence of large urban and industrial complexes where mining and metal smelting and thermal electricity generation are producing emissions injurious to atmospheric quality and to tundra and tayga ecosystems. As an example, the Taymyr reserve is situated north-east of the city of Norilsk with its copper, nickel and cobalt mining and smelting operations; discharges from these industries have created serious pollution of the city and its environs. The gases such as sulphur dioxide when emitted and exposed to moist air, fall as acid precipitation, with harmful effects on the vegetation, animal and human life. The smoke-stacks of Norilsk also eject solid particles into the air each day, reducing incident radiation and thus affecting plant photosynthesis. A similar situation exists in the Kola peninsula where the smelting of non-ferrous metals at the centre of Monchegorsk produces emissions damaging to trees, mosses and lichens of the area.

The Taymyr reserve, the largest in the USSR and occupying 13 483 km² contains each of the three main sub-zones of Soviet tundra ecosystem, the Arctic, the typical, and the shrub tundra. This situation allows it to monitor and to protect a vast range of tundra plant life and its mammal and bird fauna, the latter including rare species of geese and duck. The mammals now include the herbivorous musk-ox, a native resident of the tundra which became extinct in USSR but has been recently reintroduced experimentally in small herds imported from the North American tundras.

In such reserves the migratory species of geese and duck which visit the tundras each spring for summer breeding will be protected against the immoderate hunting that has catastrophically reduced their numbers in recent years.

Domesticated reindeer herding forms an important part of the economy of the Soviet northlands; controlled systems of management of the grazings are applied by the herdsmen to ensure optimal pasture utilisation without destruction of the plant cover. When the system is not applied, pasture impoverishment with destruction of the lichens occurs, many years being required for lichen regeneration in the tundra climate. In the natural conditions of the ecosystem predation by wolves would control the numbers of reindeer so that overgrazing is avoided. Protection of the herds and hence of an important source of meat is ensured by a campaign of wolf extermination.

The Forested Zone

The extent of the forested zone may be seen from Fig. 4.1 which shows that it occupies more of the surface area of the USSR than any other type of vegetation, although huge clearances for economic and urban growth that have occurred, particularly in the west, are not shown on the map. As there are large variations in soil and climate in this vast area, three forest divisions are conventional. The main area is the *tayga* or boreal coniferous forest, covering much of the north-west in Europe and extending across Siberia for 6000 miles (10 000 km) to the Pacific Ocean. To the south-west the tayga grades into the much smaller *mixed forest*, composed of both coniferous and broad-leaved deciduous trees; and in the Soviet Far East the tayga is replaced by a species-rich *mixed deciduous forest*. Additionally, very small areas of *warm-temperate woodland* occur in the extreme south of the USSR, in the Black Sea area.

THE TAYGA

This occupies nearly 40% of the USSR and is the earth's largest single forest formation, providing both the greatest reserves of softwood timber and the largest volume of timber cut for industrial purposes. The dominant trees are spruce, pine, fir and larch, with different species of these genera appearing according to site conditions, although very large tracts of forest are commonly dominated by only one or two species, providing a general character of great uniformity. The conifers are associated widely with species of deciduous hardwood trees of the genera birch, alder, poplar and willow which invade the zonal coniferous forest when the latter is disturbed by clearance or by fire.

All the conifers, however, are perfectly adapted in life-form or habit to the harsh continental environment. All, except the larch, are evergreen, retaining their foliage for several years, thus taking full advantage of the short summer when photosynthesis can be resumed immediately after winter inactivity; their narrow needle-leaves with their waxy thickened coat or cuticle, are both frost-resistant and drought-resistant, the latter an adaptation to reduce dessication during the winter, when moisture lost by the plant cannot be replaced by the roots as the soil water is frozen. The short lateral branches, providing the spire-like profile, assist stability in strong winds and allow the shedding of snow. The dense canopy of foliage reduces light on the forest floor, particularly in the case of spruce, but in suitable light conditions there is a ground vegetation of dwarfed berry-bearing shrubs. These include plants of the heath family such as blueberry, cranberry, bilberry, cowberry and Labrador tea, yielding edible berries which are a valuable source of food for animals and humans. Lichens are also present, but in low light conditions feather mosses may form the sole ground cover.

The composition of the forest is dependent upon influences such as climate, topography, soil, fire and human action. The western part of the tayga, extending from northern Europe eastwards into the west Siberian lowland is more oceanic than the part east of the Yenisey river where the east Siberian conditions of winter cold and dryness are at their most extreme. In this western sector, slight and gradual temperature increase south of the tundra zones results in a series of west to east forest sub-zones, each with individual combinations of species: they consist broadly of a northern sub-zone adjacent to the tundra, a central or typical tayga, and a southern tayga.

A more detailed subdivision has been described for the west Siberian tayga on the basis of its plant landscapes and their ecological features: in the north is the sparsely wooded tayga nearest to the tundra; next is the northern sub-zone; south of the latter is the central or typical tayga, and in the south is the southern tayga. A final sub-zone, adjacent to the wooded steppe is the sub-tayga.

The main features of each of these sub-zones may be summarised as follows.

The *sparsely wooded tayga* is an ecotone, or transitional zone, in which the environment becomes, marginally suitable for growth of the hardiest trees. Larch and birch form this open northern fringe of the tayga. Both leafless in winter they grow first as single species, often of dwarfed or deformed habit, commonly separated by mires, bogs or marshes with a tundra dwarf-shrub moss and lichen type of surface vegetation. Southwards, the trees gradually form clusters occupying depressions or similar sheltered sites, but tundra shrubs and herbs occupy the more exposed higher ground.

The *northern tayga* has a much denser forest cover; larch and birch are continuously present,

A riverside clearing in the Siberian tayga

but are subordinate to species of the typical 'dark coniferous forest' namely spruce, fir, and pine, whose dominance results from the longer period of growth in the climate of this sub-zone. The densest cover of these forest trees is found in river valleys, but on the interfluvial areas between valleys various lake, bog or swamp complexes occupy very extensive tracts, all of acid or nutrient-poor (oligotrophic) character. Their surface is an alternation of low ridges and shallow pools formed of a vegetation of hydrophytic mosses, particularly species of *Sphagnum* accompanied by sedges, shrubs and lichens. The drier areas with shrub willow and lichens provide good reindeer pastures.

The *central tayga* is the sub-zone of some of the densest stands of timber, and here the trees are able to grow very tall in a climate which is less extreme. The characteristic dark coniferous forest species of mixed fir, spruce, and silver fir occupy more moist, heavier soils of river valleys, but pines are most abundant on glacial sands or loams and occupy large areas where these soils occur to the east of the Ob river. As in the northern tayga sub-zone, the interfluvial areas are the habitat of the vegetation of mires, bogs and swamps, many of which are continuous over thousands of hectares. Large tracts of birchwood occur often mixed with pine where forest fires or forest clearances have removed the original woodland cover.

The *southern tayga* is of great economic importance; its lower latitude provides both higher summer temperatures and a longer growing season resulting in more vigorous growth, high quality timber and enhanced primary productivity. The most typical dense stands of fir, silver fir and spruce are concentrated in the eastern part of this sub-zone, but in the western areas, between the Ob and the Irtysh rivers, the forests are mainly composed of pine. Pine dominance is here favoured by relatively dry conditions produced by the rain-shadow of the Ural mountains together with well-drained sandy soils derived from sandstone rock formations. But an increasing area of arable and pasture land is characteristic of this sub-zone, on sites where secondary birchwood scrub forests have been cleared, separated by stands of coniferous forest. Large tracts of swamp, bog and marsh occur on the poorly-drained flatter areas above the river valleys.

The southern tayga grades southwards into the *sub-tayga*. The conifers of the dark spruce-fir tayga become scarcer, although pine remains, growing in association with birch and aspen woodland, the latter on scattered sites separated by arable land and areas of damp meadowland which have replaced forest cleared by fire or by felling. The virtual absence of typical dense conifer forest together with the warmer and longer summer promote rapid evaporation of soil moisture with a tendency for soils to become saline, as salts derived from saline bedrock are drawn upwards towards the surface. Bog and mire complexes are present in the sub-tayga with their characteristic vegetation of sphagnum mosses, sedges and dwarf-shrubs, but they are much richer in nutrient salts than the acid bogs of the northern sub-zones and are classified as mesotrophic or eutrophic. They grade southwards into the meadow-steppe-saltmarsh associations of the wooded steppe and steppe zones.

East of the Yenisey the tayga contrasts with that of the west Siberian lowland as a consequence of the varied relief of eastern Siberia, the extreme dry continentality of the climate, and the widespread permafrost. The east Siberian tayga also occupies a much greater extent than that of west Siberia, where the wide west-east areas of steppe and wooded-steppe replace the tayga vegetation in the south. By contrast, the tayga is virtually continuous in east Siberia, apart from scattered areas of steppe and wooded steppe which replace it in exceptionally dry situations.

The east Siberian tayga lacks the latitudinal subdivisions of forest vegetation characteristic of west Siberia; it lacks also the vast development of swamp, mire and marsh vegetation of the latter, because of its better drainage and its semi-arid climate.

The tayga is distinguished by a widespread dominance of the Dahurian larch, a forest tree perfectly adapted to the prolonged, severe and dry winter, a cold-hardy plant able to root satisfactorily in the thin soil above the permafrost. However the species of fir and spruce of the typical dark coniferous tayga of western USSR are much less prominent, although pine accompanies larch to an increasing extent towards south and south-east Siberia. But forest fires are frequent, with burnt areas colonised by scrub birch and pine. The middle Lena lowland of Yakutia has a sparse larch tayga interspersed by areas of steppe or meadow-steppe associated with black soils or chernozems resembling those of west Siberia. Above the tree-line in many mountainous regions of east Siberia, the sparse, open montane larch tayga is replaced by *goltsi* or bare highland summits with their thin, sparse cover of moss, lichen, herb and shrub vegetation.

The soil cover of the tayga is a product mainly of climate and vegetation, operating over a very long period of time, but the relief and the type of soil parent material are also important. With the exception of some areas of eastern Siberia, tayga soils are derived from Quaternary deposits of the continental ice-sheets such as morainic clays, fluvioglacial sands and alluvial silts and loams.

Nearest to the soil surface is a layer of dead conifer needles and cones which fall at intervals, together with other organic debris. As this litter decomposes, the residues become the soil's humus, a slow process due to the toughness of the waxy and resinous needles and the low temperatures. The latter tend to retard bacterial action and consequent nutrient release, a process in which abundant soil fungi play an essential role in decomposition. Various acids produced in the humus increase the solvent action of percolating rain and snow-melt, hence calcium and other nutrient bases are leached from the underlying parent mineral layer so that it becomes the eluviated grey podzolised layer or 'A' horizon. Below it is the more compact illuviated 'B' horizon where oxides, humus and mineral colloids are deposited. This is dark-brown or rust-coloured from its content of iron-oxide and often contains an indurated layer or 'pan' of iron which inhibits plant root development and impedes soil drainage.

However, the heterogeneous nature of glacial drift and the uneven topography result in the formation of a mosaic of podzol types and their associated vegetation. An open type of forest with a rich herbaceous ground vegetation produces a richer type of humus; this is the 'turfy' podzol. Permeable and hence freely-drained parent materials such as sands favour podzolisation but on the more impermeable glacial clays, or valley sites with poor drainage, the profile becomes seasonally saturated and hence lacking in oxygen, so that a gleyed horizon is formed. Such gleyed podzols are common as are the peaty-gleyed-podzols with their surface layer of peaty material, the residues of swamp or mire vegetation. The extreme case of permanent soil saturation creates the peat bog or mire (muskeg) which has none of the mineral horizons of the podzol but consists of partially decayed fragments of the mire vegetation such as mosses and sedges.

All such soils are generally of low agricultural value and productive capacity and are incapable of sustained cropping unless treated with liming to counteract their acidity and mineral fertilisers to improve their nutrient status. In the case of the peaty soils, such improvements require to be preceded by soil drainage techniques.

In east Siberia, the soils beneath the tayga are affected by the widespread permafrost. The usual genetic soil horizons may be absent when the permafrost is close to the surface and the low soil temperatures inhibit the processes of soil formation. In addition, horizon development is often prevented by periodic churning of the active layer above the frozen ground as a consequence of upheavals generated by seasonal alternation of freezing and thawing. The soils are commonly super-saturated by moisture accumulating above the impermeable surface of the permafrost. However in the extremely dry climate of east Siberia plant roots are able to utilise the supply of moisture released from the permafrost when its surface layers thaw slightly in spring, so that agriculture derives some benefit.

The tayga ecosystem, or biome, includes a wide variety of mammal, bird and insect life, despite the limitations imposed by the severe winter and the relatively small number of plant species. Insects, which are abundant and become active in the warmer seasons, include those living in the soil and taking part in the processes of decomposition of organic material and recycling of nutrients. Insect larvae, notably the sawfly species, feed on and often severely damage the conifer foliage; and the watery sap within the needles is food for aphids. Conifer seeds nourish red squirreis, crossbills, tits, capercaillie, and also the nutcracker crow, which like the squirrel and the chipmunk, hoards this food. Seeds that fall add to the diet of berries from the abundant bilberry shrubs and other plant food for the animals living on or in the forest floor, such as hares, voles and mice. But the occasional year of poor seed production means starvation or migration for many such creatures. Though some species hibernate, many small animals avoid winter cold by living under the protective blanket of snow.

The largest of the herbivorous mammals is the elk or moose, the largest of the deer species. Usually solitary animals, they may travel for long distances in the forest, seeking food from clearings where the young shoots of the deciduous trees such as willow, birch or aspen are found. In summer they wade out into lakes to graze on the littoral or aquatic vegetation of the shallower water.

These mammals and birds sustain a number of predators; long prized for their valuable fur are the sable and the pine marten. The latter inhabits the western part of the tayga but the former's main range is in east Siberia. Both are agile members of the weasel family whose prey are squirrels, chipmunks and other rodents. Elk are the natural prey of wolves, and the large brown bears of the tayga, though they may attack elk or boars, depend largely on a variety of food such as berries, plant roots, fish or carrion.

A principal component of the tayga ecosystem is mankind, whose role is that of a top predator or tertiary consumer. He exercises great and increasing influence on the distribution and the number of mammals and birds. Large tracts of the forest are allocated to the hunting collectives in which wild animals such as ermine, sables, squirrels foxes, bears and elk are trapped or shot. The state takes its quota of the furs, and controls the number of animals that are culled to protect the species and to provide a satisfactory yield in the future. However populations continue to be threatened by overhunting and by poaching.

Large tracts of the forest are being set aside for conservation, including nature reserves and national parks; as an example, the Barguzin Nature Reserves provides refuge for the varied flora and fauna of the tayga east of Lake Baykal, particularly the sable, a valuable animal whose survival has been threatened by overhunting. Careful conservation of the few remaining animals in this reserve has allowed the population to increase so that small numbers are exported to other Soviet reserves as the basis of new colonies.

The Soviet tayga is one of the largest remaining plant formations on earth; it is also perhaps the state's most impressive natural and characteristic renewable resource. Apart from its commercial importance, it plays an essential role in environmental protection, shielding the soil from the direct impact of wind and heavy rain and increasing the soil's absorptive capacity, so that runoff is regulated. Riparian forests lessen the turbulence of rivers and the losses of water by evaporation, and in or near towns trees are of cultural, recreational and aesthetic significance, apart from their beneficial effect on atmospheric quality.

As regards exploitation, a vast amount of timber is harvested; each year it averages some 400 million m^3. Heavy losses of timber result from fires and the ravages of insect pests, but even greater are those due to wastage during extraction and processing—estimated at 160 million m^3 each year. Only about half of the original production reaches the consumer, due to non-utilisation and needless burning of material wastefully regarded as scrap. Yet the size of the forested area is expanding.

The heaviest harvesting of timber has taken place in the middle and southern zones of the European and Uralian tayga. These areas are favoured by higher net primary production, resulting from climatic and soil influences and they have greater accessibility to the needs of western markets than have the forests of Siberia. The increasing consumption of pulpwood by industry and the fuller use of pine for resin has given rise to increased felling of spruce, exceeding the amount of allowable cutting. Management is more intensive, giving a sustained yield through better use of thinnings and an improvement in felling operations. But in this European/Ural area the reserves of mature timber have been reduced, and the insufficient replacement of conifers in cleared areas has resulted in the invasion of fast-growing deciduous species of birch, alder and aspen which are of less economic value.

The most extensive reserves of mature timber now lie in Siberia, but the excessive proportion of larch eastwards in major reserve areas and their relative inaccessibility and distance from western markets, reduce the prospect of their widespread harvesting. Roundwood exports to Japan and China from eastern Siberia and the Far East have been small, less than 2% of the total Soviet timber extraction. Along the route of the Baykal–Amur railway (BAM) the predominantly larch tayga is of sparse montane species of low growth rates, and much of the logging is expected to be around the two terminal areas in the upper Lena and the Komsomolsk districts where there are more favourable economic and physical conditions. The establishment of proposed wood processing industries would require to be integrated with the actual logging and accompanied by infrastructure development in order to avoid overloading on the transport system and the depletion of local timber reserves. At the present time the emphasis on cost-effectiveness would be likely to preclude, or at least limit, such developments.

THE EUROPEAN MIXED FOREST ZONE

West of the Ural mountains, the tayga changes into a forest zone consisting of a mixture of coniferous and broad-leaved deciduous tree species. Here, the winters are cold, yet are not so severe as those of the tayga, and the longer and warmer summer becomes suitable for the dominance on the more fertile soils of oak, ash, lime and other deciduous species; they discard their leaves in autumn when sunlight becomes insufficient for photosynthesis and to avoid injury by dessication in winter when sufficient water uptake is prevented by low soil temperature or freezing. However, on the areas of sandy, nutrient-poor soils, the hardier conifers of the tayga such as spruce and pine maintain their local dominance. Elsewhere, many tracts of undrained marsh and fen occur, as in the Polesye, with characteristic vegetation of alder, birch and willow and peat-forming moss and sedge plant communities in wetlands.

Long-continued human exploitation and economic development in this historic heart of Russia have severely reduced the original extent of the mixed forest zone and have modified the floral character of the remaining areas. They contrast with the tayga in the complexity and diversity of species of both plants and animals. Below the canopy formed by the crowns of the dominant trees, as exemplified by an oak forest, several distinct layers of subordinate strata are usually present consisting of shrub, herb, moss and lichen associations, the species of each stratum being adjusted to their particular micro-environment. Each association is accompanied by characteristic mammal, bird and insect populations, for which a much greater variety and abundance of food is available than in the tayga.

This relative richness of plant life is reflected in the formation of soil of greater nutrient capacity. Beneath the deciduous trees, the annual leaf fall provides a natural supply of fertiliser, its energy-rich organic material being exploited by a vast array of soil animals and microorganisms whose activity produces a deep, dark-brown upper humus horizon. Leaching of the soil occurs, produced by snow-melt and spring and summer rains resulting in the formation of a lighter-coloured eluviated 'A' layer, below which is a darker-brown illuviated 'B' horizon of accumulation of clay minerals and colloids. The distinction of horizons is much less apparent as material from each horizon is mixed with that of the others by the soil earthworms and other numerous soil fauna. These are largely absent from the acid humus-iron podzols of the tayga zones. Such soils of the mixed forests are the brown or grey-brown podzolic soils. A particularly rich herbaceous ground vegetation produces an additional supply of humus to the surface layers, hence the name 'sod-podzolic' applied to these soils in the USSR. A significant proportion of the production of forage crops and their attendant beef and dairying farms is supported by these 'non-chernozem' soils.

Locally the soils are varied as a consequence of the effects of the Pleistocene glaciations of eastern Europe. The parent materials of the soils range from freely-drained sands and gravels of glacier-retreat outwash to the heavier drift deposits or boulder clays associated with ice advances. Soil quality varies according to such origins and also to the local relief and micro-relief. Modern soil research in the USSR has shown that a complex mosaic of soil types or associations is found on the Russian plain in accordance with several interacting factors such as gradient, ground-water character and level, solid geology and vegetation. These are recognised to be of great importance to the character of the agriculture and land use of the collective and the state farms of the region.

The mixed forest zone ecosystem or biome is exemplified in the several nature reserves that have been established in western USSR. One of the most important of these is in a large tract of the near-primaeval forest situated across the border between the USSR and Poland, the 'Old Forest' (Urwald) of Bialowieza. It lies within a large area of lowland, forming part of the basin of the Pripyat river where a variety of soils has developed from the Quaternary deposits, ranging from nutrient-rich eutrophic lacustrine peats to podzolised moderately acid brown or grey-brown forest soils and to poorer acid podzols. This sequence is thus an adequate representation of the typical soils of the mixed forest zone and they support a great diversity of plant and animal communities.

Many of these are found in the great area covered by the Bialowieza National Park (670 km²). In the northern part are mainly acid podzols formed upon sandy parent materials and these have a cover of pines and spruces with their typical dwarf-shrub ground vegetation of bilberry, and their societies of herbs, mosses and lichens. Farther south are the richer, heavier soils which carry dense and luxuriant forests of oak, lime, ash, hornbeam, maple and aspen; occasional conifers are also present in this part of the forest.

The animal life consists of many species of herbivores of which the largest are the elk and the European bison; the latter with roe deer once grazed in large numbers throughout the forest zone until its populations were reduced to the point of extinction by hunting and by the destruction of its forest habitat. Protection of the few surviving animals within this fragment of their natural habitat has allowed their population to be increased to herds of large numbers. The predators, however, also enjoy protection and include the timber wolf, the lynx and the brown bear.

THE MIXED FORESTS OF THE SOVIET FAR EAST

In the Soviet Far East broad-leaved deciduous forests form the characteristic plant landscapes of the lowlands of the lower Amur river and the Ussuri river. The monsoon climate of the area with warm, abundantly rainy summers, favours the dominance of these tree species on lower elevations. With increasing altitude and latitude, species of the Siberian tayga, predominantly larch, become mixed with the broad-leaved trees and ultimately replace the latter entirely in highland areas such as the Sikhote Alin. Forests are, however, absent from wide areas of flat river bottomland near the lower courses of rivers such as the Zeya and the Bureya as they enter the Amur. Frequently flooded, such tracts are occupied by marshes, lakes and swamps or by rich grassland meadows on better-drained alluvial land.

The soils associated with the broad-leaved forests are analogous to those of their European counterparts: they are brown or grey podzolic soils, rich in humus, but with the leached horizon resulting from the effects of the summer precipitation. Near the flood-plains of the rivers, soils derived from alluvium or lacustrine parent materials are gleyed or peaty-gleyed clays or silts

Giant plants on a state farm in Sakhalin. Their growth is encouraged by the climate and the rich volcanic soils

with a black upper layer of humus formed by the vegetation of wet grassland.

The flora and fauna of the areas consist of a unique mixture of Siberian boreal elements and sub-tropical or tropical Chinese elements, together with relict species of the former Tertiary tropical forests which survived the recurring rigours of the cold climates associated with the Pleistocene glaciations. Forests of Mongolian oak with hornbeam, lime and maple form a lowest tier of primary production; at higher altitudes there are forests of mixtures of species of pine, walnut, cork-oak and ash with an equally rich undergrowth of shrubs, herbs, mosses and lianas such as grape-vine. The vegetation supports a great abundance of herbivores, many of which, like the carnivores, are endemic to this region. This fauna, like the plant life, is a mixture of tayga and tropical species, exemplified by the top predators, the Siberian brown bear and wolverine of the tayga, the tiger and leopard of the southern Amur forest, both the latter being protected in nature reserves.

The greatest natural floral and faunal wealth of this area is found in the basin of the Ussuri river. Here there is a variety of highland and lowland relief and climate leading to a proliferation of the species. Several reserves have therefore been established, exemplified by the Biosphere Nature Reserve of Sikhote Alin, where many unique species of birds, reptiles, amphibians and other fauna are given sanctuary.

The Wooded Steppe and Steppe

To the south of the southern tayga sub-zone in eastern Europe and Siberia, the lighter and variable precipitation, together with the higher summer temperatures and increased evapotranspiration produce a reduced tree cover where scattered copses of deciduous trees are interspersed with tracts of open grassland. This zone, the wooded steppe, is one of transition. Beyond it to the south, trees disappear, except in valleys, and grasses, sedges and other herbs become dominant, forming open steppe or prairie. Due to their fertile soils, the two zones provide the major part of the USSR's cereal production so that their original character has been almost totally changed by agriculture.

The wooded steppe west of the Ural mountains consists of isolated groves of oak, lime and maple, but in Siberia where the continental climate is more severe, this zone is formed by the hardier birch, aspen and willow species. Separating the stands of trees throughout the wooded steppe zone are areas of meadow-steppe rich in species, where the typical steppe feather and fescue grasses are associated with many species of flowering herbs, all providing excellent pastures.

The soils of both wooded steppe and open steppe contrast with those of the dark coniferous tayga; the steppe soils are famed for their upper layer of black humus, up to one metre in thickness, giving the name black earth or *chernozem* to this soil. This horizon is the residual product of the decay of the dead root systems and foliage of the herbaceous vegetation. Within it, the activity of the abundant soil fauna, bacteria and fungi releases plant nutrients which accumulate in the upper soil layers, providing immense reserves of fertility. Although leaching of the profile occurs when water from precipitation and snow-melt moves down through it, the process is largely counteracted by the evapotranspiration of water created by summer heat and the strong dry winds or *sukhovey* of the open steppe. The soil's fertility is further enhanced by its parent material of lime-rich silt or loam of a loess-like texture and composition; it thus has a neutral or slightly alkaline reaction, unlike the acidity and marked podzolisation of many tayga soils. Some degree of podzolisation has, however, occurred in the case of the soils of the wooded steppe, resulting from a former phase of increased precipitation which allowed the trees to advance into the meadow-steppe. This is the origin of the grey forest soils of the zone, or leached chernozem soils. Such soils are present in valleys or other depressions of the plateau-like steppe landscape which bear oak forests, whereas the typical chernozems now under cultivation are developed on the flatter, freely-drained interfluves.

The typical chernozems become modified southwards in response to increasing aridity; the southern or dry steppes carry a sparser cover of drought-resistant feather grasses, associated with shrubs and mosses. The soils are less fertile, having a shallower and poorer humus layer, being derived from a more open plant cover. The upper layers are dark-brown in colour, hence the name dark-chestnut soils.

These dark-chestnut soils, like the chernozems, have a wide distribution, extending across the southern part of the east European plain eastward into Siberia and Kazakhstan, forming a transition between the chernozems and the semi-desert soils of the southern areas of Kazakhstan. The fertility of both types of soil is reduced by the frequency of droughts and by strong dessicating winds; they are also affected by wind or water erosion unless an adequate protective plant cover is maintained. Such adverse influences have given rise to the policy of planting drought-resistant trees to act as windbreaks or shelterbelts.

Large areas of both soil types are also affected by the processes of salinisation and alkalisation, which refer to the deposit of salts within the soil profile, accompanied by loss of fertility. Strong evaporation of soil moisture during the hot summers, together with dry winds lead to a concentration of white alkali in the *solonchak* soils or of black alkali in the *solonets* soils, although these processes often result from the ascent of saline ground water derived from bedrock formations. In such steppe areas as the Kulunda steppe of the Ob plateau, such soils are widespread. Valleys and other depressions of the plateau surface have become inland drainage areas, with large saline lakes and marshes, the latter with characteristic salt-marsh or halophytic vegetation. Of all saline soils those with soda salinity present most problems for farming as soda is less easily removed by leaching and is more poisonous for plant life; hence research activity is being directed towards their amelioration.

For many centuries the steppes were grazed by vast herds of herbivores, deer, antelope, wild horses and wild asses. Their mobility allowed migration of the herds during winter or at other times of scarcity of pasturage. Their grazing effected continued plant productivity by dispersing, trampling and burying seeds and fertilising the soils by their manure. Sharing the steppe with them were innumerable colonies of burrowing rodents such as lemmings, bobac marmots and ground squirrels, whose deep tunnels in the soft ground gave protection from both winter cold and predation by the steppe wolves, foxes, polecats and eagles. Grass-eating insects provided food for many bird species, in nests among the tufts of vegetation. The steppes were also the vast grazing and hunting territory of nomadic tribes of herdsmen.

The extension of agriculture and other forms of economic development over the steppes has removed almost all components of this former ecosystem as in North America and in other areas of mid-latitude grassland. Some of the mammals and birds have been able to survive by retreat to habitats not yet disturbed by man, although insects and rodents are still the most numerous animals. Many plants and animals are today protected in nature reserves such as Askaniya-Nova of the Ukrainian steppe. They include species once abundant but now scarce and endangered—birds such as the bustard, and the crane; mammals such as wild horses; and even the once universally abundant species of feather-grass and other herbs. But the saiga antelope, which was reduced to the point of near-extinction by over-hunting, has been saved by its exceptional capacity for rapid reproduction, and by the timely realisation of its unusual efficiency in the conversion of grass pasture into much-relished meat. Large managed and protected herds of this steppe animal are now established and many thousands of the animals are culled annually to supplement the food supply of the USSR.

Another characteristic mammal of the steppe ecosystem is the bobac marmot, whose existence, at one time endangered, is now protected in the Lugansk State Reserve in the Ukraine. Licensed hunting elsewhere may be permitted only if damage to crops by the marmot has been demonstrated.

The steppe reserves of the USSR have been shown to be in urgent need of greater protection from human encroachment in an area where human population densities are high and intensive agriculture takes place. Adequate protection of the reserves from public recreational activities which create hazards for the protected flora and fauna does not always appear to be provided by local government authorities.

The Semi-desert and Desert

The semi-desert zone extends eastward from the lower Volga to the Altay mountain foothills, crossing the central area of Kazakhstan. It is a transitional zone to that of the desert, indicated by a vegetation combining species of both steppe and desert; steppe feather grasses are mixed with drought-resistant dwarf shrubs of the *Artemisia* genus, but after sufficient rain many ephemeral plants appear briefly, such as tulip and crowfoot. Pasturing of herds, however, reduces the herb cover and favours the less palatable artemisia shrubs. The sparse, open vegetation is supported by light-chestnut and brown semi-desert soils, with shallow profiles, poor in humus, with a lime-rich horizon below the organic horizon. Large areas of solonchak and solonets soils also appear in depressions, with a typical halophytic vegetation of succulent saltworts and glassworts.

To the south is the true desert, the arid Turanian lowland, a large basin of inland drainage which extends from the Caspian Sea to the highland areas of middle Asia in the east, and to the Kopet Dag mountains in the south. It is occupied by the southern part of Kazakhstan and by the Central Asian republics of Turkestan, Uzbekistan, Tadzhikistan and Kirgizia.

The plant and animal communities of the desert have developed in an environment of prolonged, hot summers, cold winters and a meagre rainfall, highly variable in amount and distribution and averaging less than 250 mm. Many plant species endemic to the region have adapted to the moisture deficit by various forms of xeromorphy, including reduction in leaf area, leaf-shedding, development of hairy, thickened or waxy leaf-surfaces and deep root systems. Drought-evading ephemeral plants are also widespread during periods of spring rains and yield valuable grazings for livestock. The vast sandy desert areas of Karakum and Kyzlkum are the habitat of the numerous plants adapted to colonise unstable sands. These include psammophytes, such as the sand sedges and marram grass. Tall, woody perennial plants of the genus *Haloxylon,* or saxauls, are also a feature of the sand deserts, often growing near the base of sand dunes where their long root systems are able to reach ground water. A characteristic sequence of psammophytes occurs in the stabilisation of the barchans, or mobile dunes of the Karakum, with shrubby sand acacias forming the pioneers of the shifting sand of the summits, and sand sedges with the black saxaul occupying the more stable habitats, together with shrub species of *Calligonum,* an entirely leafless xerophyte.

The desert soils are developed from various parent materials. Much of the southern part of the Turan lowland is Quaternary alluvium, spread by ancestors of the modern Syr Darya and Amu Darya rivers. Soils from this source include the sands and loams of the sand deserts. Soils are also derived from the weathering of Tertiary and older rocks of the area. The zonal soil is grey-brown desert soil, a sandy loam or sandy clay, relatively poor in organic matter and with a surface crust of carbonates. Extensive patches of highly saline soils are ubiquitous, and occur where saline ground water reaches the surface, forming solonchaks, and also 'takyr' soils which develop a highly saline surface crust, resulting from evaporation of temporary lakes or surplus irrigation water. The most fertile of the desert soils are the grey desert soils or *serozems*. They are derived from loess, an extensive sheet of which extends over the foothills and piedmont areas of the Central Asian mountains. Serozems support a dense annual cover of ephemeral herbs which release humus and abundant carbonates, endowing this soil with its fertility when irrigated.

The animal life of the desert ecosystem is, like the flora, characterised by specialised adaptations to the arduous conditions and also includes many endemic species. Most, like the reptiles, are nocturnal, and find shelter from the heat in burrows, often those made by the desert rodents, such as the jerboas. The innumerable insect species are the prey of lizards which in turn are the food of the viper and the cobra. But the hoofed mammals, the Kulan or wild ass, and the goitred gazelle are now rare, as are the species of predatory cats, the largest of the latter being the Asian cheetah, now kept in semi-captivity for future reintroduction into the wild.

Although irrigated agriculture, urban development and the introduction of domestic herds and hunting have been inimical to the complete survival of the desert ecosystem, it is still represented within the many desert nature reserves that have been established in this area. Each of the separate desert biocenoses is protected by a large reserve. Some, like the Repetek Reserve of the Karakum, have existed for over fifty years; others such as the Ustyurt, a large reserve of 2230 km², sited on the plateau of that name, are of very recent introduction.

The Highland Zones

The biogeographical features of highland areas are the product of numerous interacting factors: they reflect the local variations of slope, type of rock, exposure, temperature and precipitation, and human influence. The nature of the various vertical zones of vegetation and soils which accompany the climatic changes produced by altitude depends on the latitude of the highland areas and on their regional climates.

Mountain soils are affected by gravitational movements such as soil creep, favoured by alternating conditions of freeze and thaw. Thus soil horizon development is often modified or disturbed and may contain coarser rock fragments which, combined with a steep slope, endow the soil with rapid drainage. The soils are also subject to sheet wash and damage by the formation of erosional gullying, particularly after deforestation or other disturbances of the plant cover.

As exemplified by the Caucasus system, these glaciated alpine-type ranges rise to well over 5000 m. They form a climatic divide, separating steppe or semi-desert in the north from the humid semi-tropical conditions to the south. The lower slopes of the western part of the Great Caucasus have a natural vegetation of deciduous forests of oak and beech, lime and maple, with a species-rich shrub and herb undergrowth. This plant life is associated with a brown or grey forest soil, analogous to that of the lowland European forests farther north. Beech extends upwards to above 1500 m but is replaced by dark coniferous forests of larch and fir by 1800 m. These higher forests grow in cool humid conditions and their soils are podzolic, resembling those of the northern tayga. Dense thickets of rhododendron appear above the tree-line, forming a sub-alpine belt which, with its cover of sedges, grasses and many other flowering herbs, offers rich summer grazing for sheep, cattle and goats. Above the sub-alpine zone the true alpine environment with the typical dwarfed perennial herbs, lichens and mosses, rises to the summit zone of snowfields and glaciers.

Farther east, on the wetter westernmost ridges of the Tyan Shan mountain system, there are similar vegetation and soil zones to those of the Caucasus. A lower, semi-arid wormwood shrub-steppe belt is succeeded upslope by a broad-leaf forest of aspen, maple and a species-rich undergrowth, rising to 1200 m. Higher still is the zone of conifers, in this case dominated by spruce, and beyond the tree-line at about 1800 m is the sub-alpine belt of dwarf juniper extending to 3000 m. Above it, the alpine zone is particularly rich in beautiful mountain flowers, many of which are rare and confined to small and remote habitats. Many form dense and compact clusters or cushions, a growth-form providing tolerance of the strong and dry mountain winds.

Comparable zones occur in other mountain systems farther north and east as in the Altay ranges of southern Siberia. In north and north-east Siberia, the tree-line is much lower and in the colder and drier conditions the rich sub-alpine and alpine zones above the conifer belt are absent and are replaced by open, sparse and stunted moss, lichen and grass-herb plant communities of the *goltsi* or bald mountain summits.

A characteristic zonation of animal life is associated with the belts of vegetation. The high alpine tundras and sub-alpine meadows are the habitat of ptarmigan, dotterel, mountain thrush and snowcock. The rich herb flora attracts large and colourful populations of butterflies. Grazing mammals are represented by various species of wild goats and wild sheep and the vegetation also supports colonies of burrowing marmots. Predators include golden eagles and vultures, but the snow leopard, whose favoured prey is deer, lives in remote habitats and is rarely seen. Seasonal movement is characteristic of mountain fauna, many species retreating to lower and less exposed areas in winter.

As a result of the destruction of habitats by deforestation and intrusion of domestic grazing animals, many faunal and floral species of Soviet highland environments have become scarce or endangered, particularly where mountains such as the Caucasus are accessible to densely populated lowland areas. However, the biota of the Caucasus appears to be well protected by the largest concentration of nature reserves in the Soviet Union which total more than thirty. As an example, the Caucasus Biosphere Nature Reserve covers 2635 km² of the western part of the range and includes all the altitude zones from the lowland broadleaved forests of beech and the higher conifer forests, to the subalpine and alpine meadows and glaciers of the summits over 3000 m. Man's modification of the plants and animals has been minimal so there is a rich and endemic flora and fauna. The reserve makes a special study of mountain ecosystems and the management of mammals such as the chamois and the European bison. The latter, absent from these mountains since the last were killed in the 1920s have now been bred and successfully returned to this southern part of their former range.

The Sub-tropical (warm-temperate) Zone
The sub-tropical environment occurs in two relatively small regions situated in the extreme south

where shelter from the rigours of the continental climate is provided by highland barriers, and where there is also a maritime influence from the Black Sea. The larger of the regions, the Kolkhid lowland, is part of Georgia in the Transcaucasian depression. Here a hot summer and a mild winter are accompanied by an abundant and well-distributed rainfall. The much smaller environment is in the southern coastland of the Crimean peninsula, where the very dry, hot summer alternates with a mild and wet winter.

The Kolkhid lowland has a natural vegetation of evergreen hardwood sub-tropical forest rich in species of oriental beech, alder, tree-ferns and climbing epiphytes such as vine and ivy, developed on red and yellow clay loams developed from alluvium and from weathered igneous rocks. Little of this original ecosystem now remains: the cleared forest soils now support crops of tea and tobacco and much of the former marshland is now drained. Cultivation extends to the lower slopes of the Caucasus facing the Black Sea where there are the remains of the former rich broad-leaved forests of oak, beech, chestnut and hornbeam. The brown podzolised soils of these forests are terraced for plantation agriculture of tea and citrus fruits.

With the exception of one small area, the Lenkoran lowland, the eastern part of the Trans-caucasian depression contrasts sharply with the Kolkhid lowland. This occupies much of Azer-baydzhan and is also mainly lowland, partly below sea-level, drained by the Kura and Araks rivers. Here the climate, vegetation and soils resemble those of poor steppe and semi-desert. Rainfall is scanty so that crops require irrigation of the grey serozem soils after clearance of the vegetation of salt-bush and ephemeral desert herbs. Intense evaporation in the dry, hot summers has produced large areas of saline soils.

In the Crimea, the USSR's very limited area of sub-tropical, or, more correctly, Mediterranean type climate, extends along the southern coast of the peninsula, protected from continental influences by the Crimean highlands. This is associated with the characteristic sclerophyllous type of plant life, many species of which are adapted to the summer heat and drought by their xerophytic small, stiff, leathery leaves, and their evergreen habit. The zonal vegetation was an open mainly evergreen forest dominated by the evergreen oak and the two conifers, Aleppo pine and a juniper of

arborescent life-form. Forests of beech, maple and hornbeam formed a higher belt, associated with a brown podzolic soil, but the forests are now fragmented by clearances for grazing, fires or agricultural development. Many of the typical Mediterranean shrubs remain amidst the vineyards, such as laurel, rockrose and myrtle.

Soil erosion

Soils damaged by erosion are widespread in the USSR (Fig. 4.3), having been initiated by misuse of land by human agency after removal of the protective cover of natural vegetation. Agriculture, itself the principal cause of erosion, is seriously affected but other components of the environment are usually also damaged. The increased run-off from deforested valley slopes produces floods, and silt deposition reduces the depth of lakes and reservoirs. Turbid rivers, their waters choked with suspended soil particles, become unsuitable habitats for aquatic plants and animals.

Running water and wind are the chief agents of soil erosion; they may operate in combination or singly. Three phenomena of erosion produced by running water are known as sheet, rill, and gully erosion. Initially, water moving as a sheet or film downslope becomes concentrated into small channels or rills which deepen and widen into gullies and ravines. The gullies develop branching systems, eventually merging together so that they cover the entire slope, removing the whole soil cover down to the underlying rock surface.

Wind erosion or deflation is common in the largely treeless steppe, semi-desert and desert area, and also hilly and mountainous areas where the soils are shallow and light-textured, such as sandy loams, sands and silts. Strong winds, acting on exposed sites carry away the upper soil horizons, killing crops by uprooting them or by burying them under dust. The dust accumulates on fields, roads or on sources of freshwater or settlements.

Study of gullies, common in the steppe and wooded steppe areas of the USSR, has shown that they originated with sheets of rainwater moving over bare, friable soils, the cohesive properties of which were weakened by unsuitable agricultural methods. Gullies also developed from the deep tracks made by farm vehicles. Large areas of some of the most fertile chernozem soils have been lost since the early colonisation of the steppes

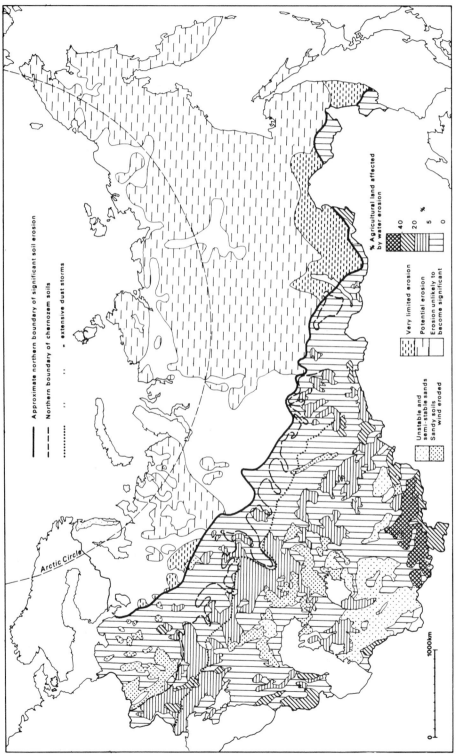

Fig. 4.3 Soil erosion   Sources: Atlas sel'skogo khozyaystvo, Silvestrov (1971) and others

of southern Russia 300 years ago, eventually covering the Ukraine and in the present century spreading into western Siberia. The Novosibirsk region of the upper Ob shows the development of such erosion processes. This area's dissected plateau terrain shows the influence of both an earlier phase of normal river erosion initiated in the Quaternary period, and a much more recent stage of soil erosion which has augmented the gully development of the earlier period. In this later phase, man's improper agricultural methods have allowed a combination of rapid sheet erosion, gullying and wind erosion to take hold on sloping land, resulting in the loss of valuable loamy soils. Control of these processes is thus an important and urgent task in this area.

Deflation or wind erosion has been particularly active both in this Siberian steppe and wooded steppe, and in the western or European steppe. The extension of agriculture on the west Siberian and northern Kazakhstan steppes in the 1950s was accompanied by deflation and dust storms affecting very large areas of former grazing land. As an example, the Pavlodar and the Omsk oblasts suffered wind erosion affecting more than 20% of the sown grain. Farther west, dust storms destroyed tens of thousands of hectares in Bashkiriya and the Kuban.

Recent research on the distribution of soil erosion in the USSR has indicated that the podzolised soils of the forested zones of European areas are as vulnerable to erosion damage after deforestation and cultivation as those of the steppe and wooded steppe zones farther south. Although the forest regions have undulating relief, moderate slope angles and low ridges of higher land such as the Central Russian upland, large tracts are subject to active erosion processes of sheetwash, gullying and river channel development.

The extent and intensity of the erosion is greater in the south than in the north, in accordance with the degree of deforestation and agricultural use. In the north, erosion in the tayga zones is produced mainly by meltwater runoff but, in the south, erosion in the areas of mixed and mainly deciduous forest is largely runoff from heavy warm-season rainfall. Soil erosion in the tayga sub-zones appears to be much more extensive than had been hitherto supposed and is present in the far north in such areas as the Kola peninsula, Karelia and in the Vologda and Archangelsk oblasts. In the southern tayga it has resulted from the more intensified agriculture in that zone where the better-drained land on steeper valley sides has offered sites superior to the flatter interfluves for farming. The presence in

Accelerated erosion in the Caucasus mountains resulting from deforestation and overgrazing

the zone of considerable remaining forest cover, together with perennial grassland, has, however, reduced the scale of erosion hazards which are much more extensive farther south, affecting the brown or grey podzolised forest soil complexes.

The southern part of this area of non-chernozem soils thus contains the greatest proportion of land most seriously affected by man-induced soil erosion, and it becomes particularly pronounced in the area of transition from the forest zone to the wooded steppe where cultivation has long been established.

Soil removal, affecting the most productive nutrient-rich colloidal horizons has been calculated as amounting to less than two tonnes per hectare per year in the southern tayga subzone, but increasing southwards to a maximum of more than fifteen tonnes per hectare in the richer non-chernozem soils farther south.

Eroded soil material in the runoff enters the drainage systems of the area as it is carried to the seas; data on the turbidity of river water confirm the calculations on annual sediment runoff. It is at its lowest in the far north but it increases in the south, particularly in the rivers draining the hillier parts of the forest zone such as the Valday hills or the Smolensk-Moscow ridge, where 70 to 80% of the land is subject to erosion.

The areas of the most severe soil erosion include the foothill and mountain regions of the southern republics of the USSR, (Fig 4.3) such as the humid 'subtropics' of the Black Sea coasts of the Caucasus and the irrigated desert areas of Central Asia. Irrigated areas are subject to sheet erosion which removes fine soil from the fertile ploughed land under the effect of the scouring and entraining force of irrigation water. Even quite gentle slopes in foothill areas may be eroded when excess water is applied by means of sprinkler equipment, giving rise to sheetwash, followed by further soil loss when surplus water becomes concentrated along irrigation furrows.

Conservation measures applied to the problem of soil erosion include the use of protective crops, particularly grasses; contour ploughing and sowing; autumn ploughing, which improves the filtration capacity of the soil thus reducing runoff; control of grazing pressure by farm livestock; improved methods for sowing crops and soil preparation; and the use of tree shelter-belts to check runoff, to act as windbreaks and to improve moisture intake by controlling snow-drifting.

BIBLIOGRAPHY

Averyanov, S. F., Minayeva, E. N. and Timoshkina, V. A. (1971), *Increasing agricultural productivity through irrigation and drainage*, in Gerasimov et al. (eds.), 1971, pp. 131–160.

Borzhonov, B. B., Borozdin, E. K., Dyachenko, N. O. and Zabrodin, V. A. (1976), 'The domestic reindeer industry, influence on the flora and fauna of the tundra of the USSR', in symposium, *Geography of polar countries*, XXIII International Geographical Congress, Leningrad, pp. 134–135.

Conservation Law of the Russian Republic, *Pravda* 28, Oct. 1960, p. 2, translated in *Current Digest of the Soviet Press*, Vol. XII, No. 44, 30 Nov., 1960 and subsequent legislation.

Doncheva, A. V. and Kalutskov, V. N. (1977), 'Prediction of the environmental impact of mining and metallurgical production in the tayga zone,' *Soviet Geography*, **18**, pp. 223–229.

Durrell, G. & L. (1986), *Durrell in Russia*, London.

FAO/UNESCO (1972), *Soil maps of the world*, UNSECO, Paris.

Gerasimov, I. P. (1983), *Geography and ecology*, Moscow.

Gerasimov, I. P. and Glazovskaya, M. A. (1965), *Fundamentals of soil science and soil geography*, translation IPST, Jerusalem.

Gerasimov, I. P., Armand, D. L. and Yefron, K. M. (eds.) (1971), *Natural resources of the Soviet Union*, their use and renewal, translation ed. W. A. D. Jackson, Freeman, San Francisco.

Knystautas, A. (1987), *The natural history of the USSR*, London.

Kolomenskaya, L., Lulikova, L., and Checheneva, L. (1980), *C prirodoy ryadom*, (Next to nature), Moscow.

Komarov, B. (1978), *The destruction of nature in the Soviet Union*, London.

Kosov, B. F. *et al.* (1977), 'The gullying hazard in the Midland region of the USSR in conjuction with economic development,' *Soviet Geography*, **18**, pp. 172–178.

Kovalev, R. V. (ed.) (1969), *Genesis of the soils of western Siberia*, Jerusalem.

Kryuchkov, V. V. (1976), The change of the northern environment as a result of its use, in symposium, *Geography of polar countries*, XXIII International Geographical Congress, Leningrad, pp. 129–131.

Lobova, E. V, (1967), *Soils of the desert zone of the USSR*, Jerusalem.

Medvedkova, E. A. and Malykh, G. I. (1973), 'Cartographic evaluation of the use of timber resources in the Irkutsk oblast'. *Soviet Geography*, **14**, pp. 184–194.

Mekayev, Yu. A. (1981), 'Nature reserves in the southern European USSR; their condition and future prospects', *Soviet Geography*, **22**, pp. 523–528.

Neishtadt, M. I. (1977), 'The world's largest peat basin, its commercial potentialities and protection', *Bulletin*, International Peat Society, No. 8, pp. 37–43.

Osakov, Yu. A., Kirikov, S. V. and Formozov, A. N. (1971), 'Land game', in Gerasimov *et al.* 1971, pp. 251–292.

Ponomanev, G. V. (1973), 'Changes in the wildlife population in the Sos'va section of the Ob basin as a result of human activity in the tayga', *Soviet Geography*, **14**, pp. 356–362.

Pryde, P. R. (1984), 'Biosphere reserves in the Soviet Union', *Soviet Geography*, **25**, pp. 398–408.

Rosswall, T., and Heal, O. W. (eds.) (1974), *Structure and function of tundra ecosystems*, Abisko, Sweden.

Shelkunova, R. P. (1976), 'The lichen cover change caused by the human activity at the north of the Yenisey basin,' in symposium, *Geography of polar countries*, XXIII International Geographical Congress, Leningrad, pp. 136–137.

Silvestrov, S. I. (1971), 'Efforts to combat the processes of erosion and deflation of agricultural land,' in Gerasimov *et al*, 1971, pp. 161–183.

Sochava, V. B. (1980), *Geograficheskiy aspektoy Sibirskoy taygi*, Irkutsk.

Sokolov, V. (1981), 'The biosphere concept in the USSR', *Ambio*, **10**, 2–3, pp. 97–101.

Sokolowski, A. W. (1983), 'Restoring the bison's habitat in Bialowieza', *Ambio*, **12**, pp. 197–202.

Sukachev, V. and Dylis, N. (1964), *Fundamentals of forest biogeocoenology*, Edinburgh.

Suslov, S. P. (1961), *Physical geography of Asiatic Russia*, London.

Uspensky, S. M. *et al*, (1976), 'Protection of natural complexes of the Arctic and Sub-Arctic,' in symposium, *Geography of polar countries*, XXIII International Geographical Congress, Leningrad, pp. 126–129.

Vasilyev, P. V. (1971), 'Forest resources and forest economy,' in Gerasimov *et al*, 1971, pp. 187–215.

Walter, H. (1973), *Vegetation of the earth*, London.

Walter, H. and Box, E. O. (1983), Middle Asian Deserts, in *Temperate deserts and semi-deserts*, Ecosystems of the world, 1983, (ed.) N. E. West, London.

Zorina, Ye. F., Kosov, B. F. and Prokhorova, S. D. (1977), 'The role of the human factor in the development of gullying in the steppe and wooded steppe of the European USSR,' *Soviet Geography*, **18**, pp. 48–55.

5 Water Resources

The vast area of the USSR allows her to claim a large share of the global supply of moisture termed the hydrosphere. Her water resources include the maritime water of coasts, estuaries and seas, the freshwater of rivers, streams and lakes, the moisture supply of the soil and the ground water contained in the strata of rock formations. But it is the freshwater supply, derived mainly from rain and snow, that forms the most precious natural asset, for unlike most other resources, freshwater has no substitute. It therefore occupies a unique place among the nation's renewable resources. Yet the demands upon it are constantly increasing, for adequate water supplies are essential for almost all kinds of economic and social development.

The Soviet government acknowledged these aspects of the water resource in the law of the Conservation of Natural Resources in the RSFSR (1960); all sources of water are affected by it and emphasis was laid upon the need for planning and conservation of water supplies for the expansion of agriculture, industry and forestry.

The Water Balance
The relationship between the supply of freshwater from precipitation and the subsequent losses of water from the land is shown in Table 5.1. Two fifths of the precipitation in the USSR (3340 km³) are carried back to the sea by gravity and surface runoff (S). The most important part of the water balance is the volume of water absorbed by the land, the total soil water supply (W). This water either remains in the upper soil layers where it

TABLE 5.1 ANNUAL WATER BALANCE OF THE USSR

	km³
Precipitation (P)	8 480
Total runoff	*4 220*
Underground (stable) runoff (U)	880
Surface runoff (S)	3 340
Total soil water supply (W)	5 140
Evaporation (incl. transpiration) (E)	4 260

Source: M. I. Lvovich 1977.

may evaporate or it is utilised by vegetation in growth and in transpiration from leaf surfaces (E). A proportion of soil water, however, moves downward beyond the reach of plant root systems where it becomes the ground water; here it may migrate through water-bearing rocks and into rivers as underground runoff (U), or it is retained in the rocks as a valuable subterranean reservoir. Hence the equation: $W = P - S = E + U$

A map of the river systems of the Soviet Union appears to indicate a country very well endowed with water resources. In fact, although the Soviet land area occupies 15% of the global land mass, its annual runoff equals only 11.2% of the earth's total river runoff. Table 5.2 shows features of river distribution and individual characteristics. By far the greater proportion of river flow passes through sparsely peopled regions, exemplified by the Ob, the Lena and other great Siberian rivers which flow northwards to the Arctic Ocean; however, the greater part of the Soviet population and the economic activity are clustered in the European area of the country. In contrast to the 155 000 rivers in Siberia there are only 45 000 in the European part, and in terms of quantity of water the European inhabitants, who form about 70% of the population, have to share only about 18% of the water.

To this unsatisfactory geographical distribution of water we must add also the fact that there is an uneven seasonal distribution; 60% of the rainfall comes in spring and summer, so that spring flooding of the rivers is a constant and widespread annual event, followed by low-water periods later in the year, and then a long period of freezing when the surface flow ceases completely.

There are also considerable annual and areal fluctuations in precipitation which affect the water supplies: cycles of alternating high and low water years occur in different regions; an increase in river flow in the European area and in western Siberia frequently coincides with a decrease of flow in eastern Siberia; but in the 1930s the opposite variation favoured eastern Siberia, contrasting with a sharp decrease in river flow in European USSR, the Kazakh steppe and in western Siberia.

In addition to the rivers, the Soviet Union contains also an immense number of lakes and reservoirs; there are 513 of these with a surface area of 20–1000 km² and 24 with a surface area of more than 1000 km² each.

The distribution of water resources

A more detailed discussion of the distribution of the water resources has to take into account not only the circumstances noted above, but also such factors as topography, vegetation and soil. In European Russia the rivers flow radially from upland watersheds situated centrally in the Russian plain, but the dominant directions of flow are either northwards to Arctic waters or southwards to the Black Sea or the Caspian Sea. In spring, the thawing of the deep snow cover produces high flood levels in these rivers. The southward flowing rivers such as the Volga and the Dnepr leave their moist forested source areas where they receive large tributary streams, but farther south they cross the more open steppe and wooded-steppe where the higher summer temperatures, the lower rainfall and the hot, dry winds result in the reduction in surface runoff. Intense evaporation occurs, much increased in recent

decades by the additional water surface area created by large storage reservoirs such as the Tsimlyansk reservoir, situated between the rivers Volga and Don.

The northern river systems of European Russia, represented by the Northern Dvina and the Pechora are endowed with extensive water resources. Divided from the Volga system by a long, low morainic ridge, the rivers meander slowly across the undulating glaciated surface of the lowland with its widespread cover of coniferous forest. In the case of the typical regime of the Pechora, snowmelt water is its main feeder; it creates a high spring flood which reaches the deltaic mouths of the river by late May, accompanied by local flooding caused by ice jams. A subsequent low-flow period extends into late summer. Freeze-up begins in November and extends to April, and at this time the run-off is reduced to low levels, maintained by groundwater from water-bearing rock strata. This annual variation is accompanied by a long-term cycle of fluctuation in level of five year intervals, similar to that of the Volga.

Spring flooding is also characteristic of the regimes of the great rivers of western Siberia. The Ob, with its chief feeder, the Irtysh, drains the west Siberian lowland and much of its huge annual discharge of 12 500 m³/s takes place during spring, fed by snow-melt, but high river levels continue into the summer as the break-up of ice begins in the warmer south and gradually advances northwards. Warmer water in spring coming from southern Siberia overflows the ice which is still obstructing navigation in the north and raises the level of the Ob and other rivers in mid-summer. The spring ice flow on the Ob and the Irtysh is often accompanied by numerous ice jams, causing erosion of river banks and damage to riverside installations.

The basins of the great rivers of northern European USSR and of western Siberia contain large quantities of water absorbed and retained in the extensive mire or bog complexes. These attain their greatest development in western Siberia where the swamping tendency of excess precipitation over evaporation is increased by the concavity of relief and the fall in land level produced by tectonic subsidence. The mires, which date from the early post-glacial era, extend by lateral growth, often encroaching upon forested or agricultural land. Water is contained both in the multitude of surface pools, lakes and swamps and also in the peat itself which has a water content of up to 98% of its volume. Peat has a stabilising and protective effect upon the original relief, and these large accumulations of peat produce changes in the discharge of surface water into the rivers; seepage of water into the deeper layers of rock strata is reduced, accompanied by an increase in the horizontal flow of water above and through the soil and into the river network.

In eastern Siberia the hydrological conditions differ from those farther west as a consequence of low levels of precipitation and of the intensely cold winters. The presence of permafrost becomes an important factor in the water regime of the rivers. The two longest rivers, the Lena and the Yenisey, have average discharge rates greater than 15 000 m³/s and they rank among the greatest rivers in the world. But in winter they are frozen for very long periods and the ice becomes exceptionally thick so that some rivers freeze right down to their bottoms. The spring runoff comprises the melting of snow, ice and rainfall, almost all of which is surface water, as the depth of permafrost prevents the infiltration of water to the subsoil. The groundwater contribution to the runoff is thus reduced but surface runoff is intense and 90–95% of the annual discharge occurs in the spring and summer months when river levels some 20 m above the normal cause extensive flooding. Among the southern tributaries of the Yenisey, the powerful Angara is unique, as its source is Lake Baykal, a natural reservoir and one of the world's largest lakes. This endows the Angara not only with abundant water but also with the most uniform discharge of all the eastern Siberian rivers.

The rivers of this area are, however, unable to supply the water requirements of some important industrial and urban areas. Constantly increasing demands for water for power, irrigation, industrial and domestic uses create water deficiencies in the upper mountainous reaches of the tributaries of the main rivers, within basins such as the Kuznetsk and the Minusinsk basins and the Kansk–Achinsk region. In south-east Siberia, adjacent to the zone served by the Baykal–Amur Mainline, water shortages occur due to light, unreliable rainfall and negligible snowfall so that on some rivers winter flow stops completely. Much of the ground water is unavailable, being contained in icing phenomena, caused by the freezing of ground water extruded by hydrostatic pressures.

WATER RESOURCES IN CENTRAL ASIA

The water balance has a critical importance within the arid and semi-arid Turan lowland area east of the Caspian Sea, for its agriculture is almost entirely dependent upon irrigation. Here, the greatest area of land irrigated for arable crops in the Soviet Union is in the Central Asian republics and the southern part of Kazakhstan. The fertile soils there are within access of copious water supplies from streams originating in the abundant precipitation of the lofty mountain systems to the south of the region, the Tyan Shan, the Hindu Kush and the Pamir–Alay system. Most rivers from these highlands become dry as they emerge on to the hot, arid plains, but the two principal rivers, the Amu Darya and the Syr Darya, maintain their flow for the whole year, and after traversing the deserts of Kyzlkum and Karakum, enter the Aral Sea, a large brackish-water lake. These two rivers thus have immeasurable economic importance for the whole area, and the many irrigated oases within their basins have some of the highest rural population densities of the Soviet Union.

Central Asia has a closed hydrological cycle and is an inland drainage system without any stream connection to the ocean. The incoming water supply of the arid lowlands or piedmont areas is provided mainly by surface runoff from the mountains with a groundwater component flowing within alluvial fans in the foothills. Most streams form expanses of swamp in the piedmont areas or are broken up into distributaries for irrigation and end in an intricate fan of irrigation canals. Intense evaporation of surface water occurs during the heat of summers.

The Amu Darya and the Syr Darya, however, have a continuous heavy flow as they rise above the snow line in the mountains at or above a height of 3500 m. They have two high-water periods: in spring (April–May) from the melting snow in the mountains, and in the summer (June–July) when glaciers begin to thaw. These rivers are thus the chief source of irrigation water. Rivers of snow-feeding such as the Angren, rising at lower altitudes, have a more variable discharge.

The Amu Darya is the largest river of Central Asia. It rises on the north slopes of the Hindu Kush and is fed by vigorous tributaries, such as the

Although Siberia is notorious for the severity of its winters the short summers can be very hot. Here bathers are enjoying the cool waters and hot sun in the Tyumen region with the Trans-Siberian Railway in the distance

Vakhsh, coming from high glaciers and snow-fields which provide abundant meltwater in spring and summer. The lowest discharge is in January and February, the highest in July, so that the regime is very favourable for irrigation as water becomes available at the start of the growing season in spring and the supply reaches its maximum in the summer months, just at the time of greatest need for irrigation. Central Asian rivers, however, carry in suspension and solution vast quantities of mineral material. The Amu Darya transports twice as much suspended alluvium as the Nile, most of which is carried during the increased flow in summer. Deposited on bottomlands and terraces, this alluvial deposit is exceptionally productive, being rich in plant nutrients such as lime, potassium and phosphates.

Apart from this surface water, abundant water also exists within the porous rock strata beneath the desert surface, collected from various sources, including precipitation, together with seepage from the rivers, particularly the Amu Darya, the Murgab and the Tedzhen. This ground-water table slopes gradually from east to west, merging westwards near the Caspian Sea coastlands with the saline seawater. The water can be reached by deeply rooted desert vegetation, and also by wells sunk into the flat surfaces of the takyrs, providing drinking water for sheep after the shallow water

of the takyrs has evaporated. Much of the ground water is brackish but desert animals are accustomed to relatively high salinity levels.

WATER RESOURCES OF THE FAR EAST

The principal drainage basin of the Far East is that of the Amur river. It is larger than that of the Volga, and the river also has a greater average flow (Table 5.2). However, unlike the Volga and the great Siberian rivers, the contribution of snowmelt in spring to the Amur's runoff is relatively small, as a consequence of the cold and very dry winter of the catchment area. In the summer the regular monsoon rainfall and frequent cloudbursts produce intense floods and very high water levels, contributed by the tributaries, such as the Zeya and the Bureya of the left bank, and the Ussuri, of the right bank. The lower Amur between Khabarovsk and Komsomolsk is a low-lying marshy, alluvial plain frequently flooded by fast-flowing streams from the Sikhote Alin. In winter, navigation on the Amur is halted by ice which lasts for five months at Khabarovsk and six months at Nikolayevsk, farther north at the mouth of the river.

TABLE 5.2 CHARACTERISTICS OF SOVIET RIVERS

Name of river	Drainage area (thousands of sq km)	length (km)	Average flow (thousands of m³/s)
Ob	2990	3650	12.5
Yenisey	2580	3487	17.5
Lena	2490	4400	15.5
Amur	1855	2824	11.5
Irtysh	1643	4248	3.0
Volga	1360	3531	8.0
Angara	1039	1779	4.2
Aldan	702	2240	5.2
Kolyma	647	2513	3.8
Kama	507	1805	3.8
Dnepr	504	2201	1.7
Amu Darya	309	1415	1.5
Syr Darya	219	2212	1.2
Northern Dvina	357	744	3.5
Pechora	327	1790	4.1

Sources: Atlas SSSR; Lavrishchev (1969).

THE USE OF RIVER RESOURCES

In common with other industrialised countries, the USSR has an immense and rapidly increasing consumption of river water for many essential requirements: for domestic supplies of drinking water and sewage disposal; for industry and the production of electric power; for irrigation, transport of passengers and freight, and for recreation. The increases are particularly heavy in the European part of the country, where, as we have seen, there is least water available. Industrial needs in the more developed areas rise to 300–320 km^3 each year. Many modern products make heavy demands upon water supply: thus cotton textile factories use about 250–300 m^3 of water per tonne of fabric, and in the production of synthetic rubber, each tonne requires about 2000 m^3 of water; and the refining of petroleum and the production of some metals use a great quantity of water.

In the Soviet era these requirements, together with that of flood control have made necessary the modification of the natural regimes of the rivers; this has been the case particularly in the European and Central Asian parts of the USSR. The huge volumes of additional water runoff created by seasonal floods have been stored within vast reservoirs behind concrete dams where it is available for power generation, irrigation and navigation improvement by canal and water diversion systems. Repeated use of much of the same water has allowed the construction of a downstream succession of dams, reservoirs and power stations along much of the course of the Volga and the Dnepr. In the middle and lower reaches of these rivers they traverse the steppe and semi-desert zones where the annual precipitation is light and variable, and it is here that the stored water has been diverted to irrigation systems which are essential for the maintenance and improvement of agricultural output. Farther south, in the true deserts of Central Asia where rainfall is totally insufficient for agriculture, all cropping is completely dependent on irrigation from the Amu Darya and the Syr Darya rivers.

But agricultural operations such as drainage and ploughing have also affected the water balance. North of the wooded steppe zone of European Russia, in the zone of mixed forests and their nonchernozem soils, the relief, the cooler climate and the more abundant precipit-

ation have created extensive wetlands in the western lowland areas such as in Belorussia. They occur in low-lying, badly-drained depressions within the morainic landscapes, and consist of swamps, peat-bogs and lakes of varying sizes and depths. Before and during the Soviet period gradual progress has been made in draining such areas and in cultivating their organic soils, although very large areas of such wetlands remain to be reclaimed. About 1.3 million hectares of such swamplands are located in the upper Dnepr basin in the Polesye of Belorussia, of which half have been drained and used for grain crops. However, the ecological influences of this process have become apparent: the former marshland has lost its natural capacity for water accumulation and discharge, thus affecting the water balance; reduced water levels now occur in the rivers and lakes of the area, including serious diminution of the runoff of the Dnepr river which is fed in part from the Polesye swampland. The hydrochemical properties of the water have also been adversely affected. After the removal of the forest cover the soils after reclamation have been shown to be much less consistently fertile than expected. Large areas are now subject to soil salinisation and dessication and requiring irrigation and protection from erosion by windbreak plantations.

Autumn ploughing in the USSR's European part has replaced the former spring ploughing and has affected the river runoff. Ploughed in the fall, the soil is more able to absorb water from snow-melt or autumn rain on sloping sites than if ploughed at other times so that river runoff decreases, but soil moisture supply and the groundwater component of the water balance are both increased, both being available for plant growth. In the past, spring runoff from the fields took place upon a soil surface compacted by harvesting operations and hence having reduced absorptive capacity. Deeper ploughing by modern techniques, in contrast with the shallower ploughing of former times, together with crop rotation and fertilisation have also been beneficial to ground conditions at the expense of surface runoff. The hydrologic efficiency of autumn ploughing increases in the arable belt from north to south and from west to east and becomes greatest in the forest-steppe and steppe zones.

A similar result is achieved by protective forest plantations in the areas of insufficient precipitation. Strips of woodland or 'shelter belts' are

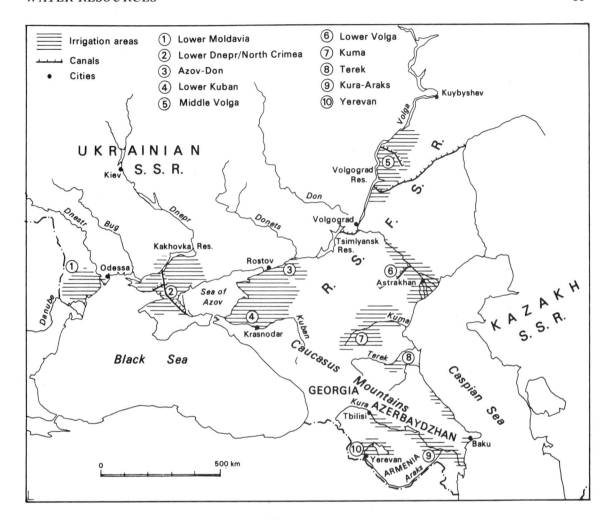

Fig. 5.1 Irrigation schemes, southern European USSR

important in producing deep snow-drifts, thus increasing the groundwater supply from snowmelt. The total result is an increase in crop productivity, and lessened runoff and also reduced erosion and risk of flooding, although low-season flow is enhanced by the addition to the groundwater supply.

THE CASPIAN AND ARAL SEA PROBLEMS

The increasing heavy demands upon the water resources created by the requirements of industry, urban development and irrigated agriculture in the subhumid and desert southern and southwestern areas of the USSR have adversely affec-

ted the water resources of these areas. Particularly heavy losses of water have occurred from evaporation from reservoirs and from seepage, infiltration and evaporation from irrigation systems as well as through transpiration from plant life.

The result has been the widespread disturbance of the ecology of the large inland seas of the south, the Caspian Sea and the Aral Sea. Both seas are the focii of landlocked inland drainage systems, dependent upon their feeding rivers, mainly the Volga of European Russia and the Amu Darya and Syr Darya of Kazakhstan and Central Asia; but the level of water and also its quality in both seas are now threatened by the reduction in the quantity of the inflow.

The level of the Caspian Sea was reduced by 2.5 m during the period from 1960 to 1975, mainly as a result of withdrawal of water for irrigation and losses from evaporation from the large Volga reservoirs. Since reaching its lowest level, in 1977, the sea has begun to rise at an average rate of 12 cm per year, a total of 1.1 m from 1978 to 1986. This is thought to be due to long-term changes in the flow of rivers into the basin, though the Soviet scientists recognise that the factors involved are complex. But the drop in level has reduced the area of the northern part of the sea by 7000 km², causing drying of the shallower areas which are of importance for supporting the ecosystems of the sea. The conditions in the delta channels and the lower reaches of the river have been changed so that navigation and irrigation have become difficult; in addition, the increased salinity resulting from decreased inflow of freshwater has affected fish and fur-bearing animal production.

Environmental changes in the much smaller Aral Sea are more serious. Since the beginning of 1961 the much greater use of the runoff of the Amu Darya and the Syr Darya for irrigation has caused the water level of the sea to fall by 7 m, representing a shrinkage of the sea area by 13 000 km² and a retreat of the shoreline by up to 40 km. There has been a rise in salinity of the water but a decrease in the supply of nutrients provided by the runoff, thus changing the nutrient balance of the sea. This variation, together with a deterioration in the oxygen regime of the water has reduced fish populations. The processes of desertification have become increasingly evident by the increased salinity and dessication of soils in the area and the change in the riverside vegetation from aquatic and freshwater marsh plant life to that associated with xeric and halophytic conditions. Salt deposits, blown from the exposed sea bed have damaged adjacent agricultural land. So extensive are the present levels of withdrawal that runoff from the Amu Darya stops for one to three months for most years and that of the Syr Darya is severely reduced for the whole year.

The Sea of Azov, the shallow, partly enclosed area of the north-eastern part of the Black Sea has also been affected by reduced inflow of river water, giving rise to higher salinity levels and a decrease in oxygen concentration of the water, causing the death of fish populations. The effect of the reduction in the runoff of the rivers of the

south-western area of the USSR, such as the Don, the Terek and the Ural, has been followed by a shallowing and a drying of their delta systems affecting their economic use. Sandbanks and mudflats have been accumulating and the inflow of saline water penetrating the arms of the delta has led to a decline in water quality and to changes in the delta ecology. In the case of the Volga, however, the continuing growth of the delta has provided some compensation by the provision of large numbers of habitats for wetland wildlife, now inhabited by numerous species of geese, duck, swans and other birds. It is an important nature reserve and forms an oasis of life in the desert which adjoins it.

Such situations in the water balance, perceived to be critical in southern and south-western USSR during recent decades, have engendered proposals to improve the supply of freshwater by a southward diversion of a substantial part of the water of the basins of the great northward-flowing rivers of northern European and of Asian USSR. There is a striking geographical contrast between the super-abundance of freshwater resources in the sparsely-peopled north with its poor soils, its arduous climate and relatively low level of development—and the more highly developed, densely-peopled warmer south and south-west; in the latter regions their productivity could be increased by the provision of more water for the irrigation of their fertile soils and for industrial and municipal use. Thus the water diversions appeared to be both an economic necessity and to be technically feasible, particularly in the absence of any marked topographical barrier between the two great drainage basins.

Inter-basin water transfers

Two separate projects were planned, the less extensive to be in European USSR, and the much more ambitious and grandiose to be in western Siberia (Fig. 5.2). In Europe a substantial part of the runoff of each of the rivers flowing into the Barents Sea, the Pechora, the Onega, Northern Dvina and the Sukhona was to be diverted into the Volga basin. Large reservoirs and pumping stations were required to raise the level of the water and to divert it through canal systems over the low intervening watersheds southwards to the Volga. A later stage would transfer part of the water of Lake Onega also to the upper Volga at the expense of the lake's natural drainage to the

Much of the tayga is swampland, with an excess of water useless for cultivation because of the low temperatures that prevail

Baltic via the River Neva. Finally a large freshwater reservoir would be formed in the Gulf of Onega, totally excluding seawater by a barrier and to be linked to a chain of reservoirs along the Onega valley. Hydro-electric stations were an integral part of the scheme, supplying power for the pumps and providing a surplus for regional industry. The water diversions, involving a withdrawal of 19 km³ of water from northern drainage were to be combined with measures to economise in water use in southern regions by improving the efficiency of existing irrigation systems and also to reduce evaporation losses by cutting off the shallow areas of reservoirs and to isolate the shallow Kara–Bogaz gulf of the Caspian Sea.

The Siberian project, far more ambitious in its objectives, was to utilise part of the northward flow of the combined Ob and Irtysh rivers by means of a dam at Belogorye, situated near their confluence. To the south, water from a reservoir on the Ob near Tobolsk was to be pumped over the low Turgay divide into a canal 2500 km in length, 200 m wide and 16 m deep where gravity would carry it along the rest of the route to Central Asia. A volume of 27 km³ was to be taken from the Ob–Irtysh system and the project included reversal of flow of the Irtysh, and storage reservoirs impounded by three low-head dams with pumping facilities. Thirteen billion roubles was the estimated cost of the Siberian diversions.

Extended irrigation in Central Asia and Kazakhstan was the main purpose of the Siberian diversions, increasing the area irrigated by 4.5 million hectares and allowing greater agricultural output and economic growth. This would generate employment opportunities in this area where there is rapid growth of the predominantly Moslem population but minimal emigration to other parts of the USSR.

But in 1985–86 decisions were taken by the Soviet Government to abandon both the

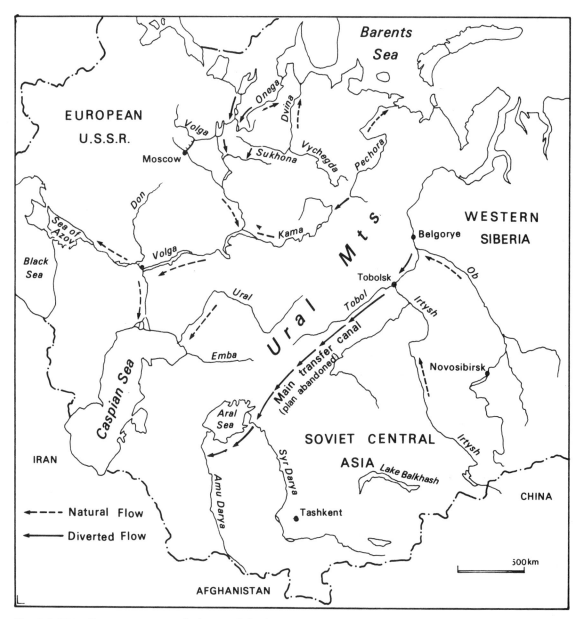

Fig. 5.2 River diversion projects, which were shelved in 1989

European and the Asian water transfer projects briefly described above. The measures of the new Five Year Plan for 1986–90 did not include provision for starting either of the projects which were officially cancelled by a decree approved by the Politburo in August, 1986.

These decisions were made on the basis of results of many protracted surveys and feasibility studies carried out during a period of more than twenty years by authoritative organisations in the USSR, including engineers, economists, geographers and hydrologists of the most eminent expertise. Many different plans and designs for the water transfers were formulated and made public often to be modified in various ways in response to criticisms of their likely effect on the environment. And it was mainly concern for environmental impact that finally produced the shelving of both diversion schemes, together with decisions of the new Soviet leadership with its insistence on the need for efficiency in resource use and on the need for re-appraisal of the costs and the economic feasibility of capital-intensive resource projects.

According to informed opinion in the USSR, the shortage of water in Central Asia could be rectified by improvement in the systems of irrigation in use which do not make the most effective use of the combined flow of 100 km³ provided by the two great rivers of Central Asia, the Amu Darya and the Syr Darya. Thus more than one-third of the water was lost from irrigation canals without linings, causing waterlogging and salinisation of the soils. This, with other defects in the irrigation techniques indicated that about 70% of the water provided was not utilised in the most effective manner possible.

A similar objection to the European project, on the grounds that precipitation in the sub-humid areas of southern and south-western European USSR is adequate for crops, given more efficient farming, irrigation and better use of shelter-belts, made the proposed transfers to the Volga unnecessary, particularly in view of the recent rise in the level of the Caspian Sea. To have provided more water for irrigation would have risked losses in soil fertility through waterlogging and salinisation; and any improvement in crop yields through irrigation would be insufficient to compensate for the costs of providing additional water.

On the other hand, the various design studies of the proposed transfers of water revealed that their possible impacts upon the landscapes and their ecosystems could have been catastrophic, as was indicated by expert opinion when the planning was in a preliminary stage. Experience and knowledge of the consequences of river diversions and impoundments in the USSR had been gained earlier with the various schemes on the Volga and the Dnepr; but the scale of the Siberian and the north European diversions was infinitely greater and affected immense tracts of the landscape with likely repercussions in countries other than those of the Soviet Union.

According to the results of the many feasibility studies of the possible environmental impacts of the diversions, several kinds of disruption were indicated such as of the hydrology and ecology of the areas concerned or of their social or economic geography.

Smaller scale projects previously completed in the USSR have shown that river systems whose volume is reduced by water transfer develop increased levels of pollution by, for instance, organic wastes from pulp and paper mills or from sewage. These types of effluent have a high biological oxygen demand, leading to the deoxygenation of the water, with fatal effect on aquatic organisms, particularly fish. Thus the diversions from, for instance, the Ob system would accentuate the normal winter low water levels, increasing the concentration of such pollutants and posing a serious threat to the valuable fishing of species of white salmon, sturgeon, charr and pike which form 70% of the Siberian fish landings. Fishing would also be affected by the construction of dams and reservoirs, as planned for the Ob, Irtysh, Pechora and Onega, as these obstruct the upstream migration of species to their spawning grounds, thus hindering fish reproduction.

Reduction in the level of rivers by the withdrawals of water would improve the drainage of flood plains, though it would not be likely to reduce the waterlogged condition of many swamps and peat mires such as those of western Siberia, as many of the largest are maintained by precipitation and are beyond the level of groundwater. Some riverine wetland wildlife habitats would suffer, however, from increased drainage. River navigation would require dredging as a result of the deposition of sandbanks formed by the erosion and deposition of material by tributary streams whose power of erosion would be increased by the lowered base-level of the main stream. On the other hand, river navigation would be improved by higher levels on the upper Volga and by the proposed new through waterways such as the Pechora–Kama canal and also by the Siberia–Central Asia irrigation canal.

The withdrawals of up to one-half or more of the freshwater of the northern rivers to the coastal areas of the Kara Sea and the White Sea, and the large coastal gulfs, such as the Gulf of Ob would produce widespread disruption of their hydrology; the levels of salinity would increase, whereas nutrient levels in the water would diminish, being derived from the nutrient input from rivers draining from the land. Both these changes would adversely affect the coastal biology and reduce fish productivity. The reduction of vast volumes of relatively warm river water flowing to the coastal seas would be certain to reduce sea temperatures and thus increase the thickness and duration of ice, with possible wider repercussions on the entire environment of the southern Arctic Ocean. Navigation would be affected, and the Northern Sea Route would be closed by ice for longer periods. The new Soviet leadership has, furthermore, emphasised the need for greater development of industry and communications in this area, and these would be hindered by the changes produced by the water withdrawals.

Some of the socio-economic effects of the diversions were pointed out by objectors: of great concern was the likelihood of damage to ancient towns from the effect of rising water-tables created by reservoir construction, notably in the upper Volga region. Here, in cities such as Vologda, historical and cultural monuments and other artefacts were threatened by the diversion of the Sukhona river. The various engineering projects would also interrupt communications in that area and would require extra labour where there was already a shortage. In the Pechora basin the new reservoirs would flood more than 2000 hectares and require the upheaval and rehousing of many thousands of people; valuable forests would have to be felled and much adjacent land would be lost from the waterlogging of soils.

Although additional water would be provided in Central Asia for irrigation by the diversion of the Ob–Irtysh system, the proposed new north–south main transfer canal would absorb large quantities of water by evaporation and also by percolation and infiltration into the ground, as the canal would not be sealed by a lining. However it would create a new navigable waterway for use in summer only and would provide water for irrigation in adjacent areas. Nevertheless, it would be detrimental otherwise for several reasons, including the inevitable disruption of ecologically sensitive arid environments including soil salinisation and flooding caused by seepages from the canal. In large areas where it traversed land already affected by soil salinisation, the saline water would be drawn into the fresh water of the canal and impair its utility for irrigation. Disruption of wildlife in nature reserves was also likely.

WATER QUALITY

The question of an adequate supply of freshwater

is only one aspect of the problem of adequate management of the water resources of the Soviet Union; of even greater importance than quantity is the maintenance of its quality. But in this respect, despite the much greater control exercised by the State on the development of natural resources than in the West, the spectacular industrial and urban expansion in the USSR during the past fifty years created a deterioration of water quality that became catastrophic in some regions. Thus the waters of the Volga, Kama, Oka, Belaya, Ural, Northern Donets and other rivers lost many of their valuable natural properties. The discharge of toxic effluent from industrial units and domestic sewage from towns and cities into these rivers increased almost twenty times during the period 1930 to 1970 and many lakes have been affected in the same manner.

Poisons in solution occur in the effluents of many kinds of modern industry. They include acids and alkalis, phenols and cyanides from chemical industries and mines. The commonest poisonous inorganic substances are chlorine, ammonia and hydrogen sulphide and the salts of many heavy metals such as copper, lead, zinc, chromium and mercury, very small amounts of which can remove all animal life, including fish, from streams. Deoxygenation of water containing organic waste occurs as a result of the action of bacteria, and this results in the death of many forms of aquatic plant and animal life. Deep, slowly flowing rivers such as those of the Soviet lowlands have naturally low oxygen levels, particularly in winter, but this condition is worsened when they are polluted by organic matter rich in nitrate and phosphate; these nutrients cause abundant algal growths which decay and are then consumed by oxygen-consuming bacteria. Water returned to rivers and lakes in a heated state (as from thermal power stations) is also harmful to aquatic organisms such as fish, as their reproductive processes are often delicately adjusted to the temperature of the water and its natural seasonal changes.

Lake and reservoir water is particularly subject to these influences because it is static or has only a relatively feeble through-flow so that an accumulation of pollutants may develop within it, markedly affecting the water quality of even very extensive bodies of water. The lakes Onega, Baykal, Balkhash and the Ivankovo and Kama reservoirs have been affected by effluent from pulp and paper mills and other industrial plant along their shores. The great size of the Caspian Sea has not saved it from serious contamination by wastes of the oil industry; until fairly recently waste oil was deposited directly into the water from the oil refineries at Baku, and tankers were allowed to discharge their ballast overboard; oil also escaped from offshore drilling operations.

Even in Siberia, where water resources are so abundant, there is concern for the diminishing quality of water, particularly in western Siberia where there has been extensive development of industry concerned with oil and gas and the growth of towns such as Tobolsk, Surgut and Omsk. The dominance of low temperature periods in the climate of the area, the sluggish flow of rivers, and the permafrost all tend to reduce the effectiveness of the natural self-cleaning processes of polluted water by the chemical and microbiological decomposition of the pollutants.

In some Siberian areas of heavy concentration of industry and population where the water resources are much less abundant, there is increasing pressure on the environment. In south Siberian industrial areas such as the Kuzbas, where several million people in many towns occupy a relatively small area with many coal-mining, metal and chemical manufactures, problems are occurring of providing acceptable standards of both air and water quality. Here there is recognition of the urgent need of installation of equipment capable of the treatment of sewage and of industrial toxic wastes by catalytic, biochemical, filtration and other methods. These and other hydrological and ecological conditions are being studied at a wide network of river laboratories dispersed across the USSR, but the number of such water-monitoring stations is considered inadequate to cover satisfactorily the vast areas drained by the various river systems.

LAKE BAYKAL

The conflict between the requirements of resource exploitation and the need for conservation of vulnerable and fragile natural resources is illustrated by the case of Lake Baykal in east Siberia. Public concern for the future of the unique flora and fauna of this lake has not been allayed by repeated official assurances that all possible safeguards for their protection had been taken. During the past 25 years, objectors have emphasised the dangers to the lake ecosystem which have been created by the development of lake-side industry.

This lake has no equal; it is the deepest continental depression on earth, being much more than one mile in depth (1620 m), and is also the world's second most voluminous body of freshwater, holding 23 000 km^3. The water has continuously supported communities of remarkable plants and animals since the formation of the lake within a rift valley in Late Tertiary times, and consequently many of its aquatic organisms are endemic and of outstanding scientific value and interest. Flanked by high mountains, the lake is also of great natural beauty.

Economic activities have created pollutants of both the pure, clear water of the lake and the air in the vicinity: wastes from paper and pulp mills; chemical contaminants leached from floating logs brought to the mills along rivers; eroded soil particles washed or blown into the lake after felling of the lakeside forests; emissions of dust, gas and smoke from industrial units near the lake; unregulated use of motor vessels—all have contributed to progressive pollution and damage to the lake ecosystem. This has threatened the survival of mammals such as the Baykal seal and numerous fish species such as the endemic goby and the omul, with others which include fishes of commercial importance such as the salmon, charr and whitefish. The public outcry has induced the authorities to take some action, ineffective at first but now including reduction of wastes from pulp mills by improved technology. Pollution from floating logs has been largely eliminated by the provision of overland transport for the timber, and there has been a restriction upon further industrial development, including a diversion of the course of the Baykal–Amur–Mainline so that it does not impinge on the lake's environment. It was proposed in 1987 to relocate the pulp mill away from the area of the lake (35 km^2) affected by its effluents to a position farther south, and to discharge the wastes through a 75 km pipeline into the Irkut river; the latter is a tributary of the Lower Angara river which flows from Lake Baykal. This scheme is emphatically opposed by scientific opinion in the city of Irkutsk and also at repeated public demonstrations, as it is seen as a further threat to the ecology of the lake.

THE CONSERVATION OF WATER RESOURCES

The centralised control of resources characteristic of the Soviet planning system has not hitherto included sufficient provision for the conservation of the environment despite laws that exist for that purpose and enforced by ministries such as the Ministry of Recreation and Water Management, or the Ministry of Public Health. This becomes evident from articles published in the Soviet popular press and in Soviet scientific journals. Thus, problems of declining water quality have been described for the Volga basin, the Caspian Sea, the Central Region and for Siberian river systems. These and also the apparently insoluble and persistent problem of Lake Baykal appear to point to the weakness of the laws and the ineffective sanctions that are applied to enforce them.

Western and Soviet observers have commented upon factors that have contributed to the failure in the past of Soviet administrations to provide adequate safeguards for the maintenance of water quality (ZumBrunnen 1984). The Five Year Plans have made little provision for the possible side-effects of various economic projects such as the hydro-electric schemes, exemplified by the unexpected over-enrichments of rivers by organic wastes and the consequent deoxygenation of the shoal areas of rivers such as the Volga and the Dnepr. In addition, the Soviet bureaucratic hierarchy with its multitude of agencies and other authorities lacks adequate provision for mutual coordination and cooperation between and among them, but allows independent scope and sometimes even inter-departmental rivalry. Further, the managerial incentive system places high priority on the quantitative flow of production and the attainment of specific output targets; the introduction of new technology such as would be necessary for water purification treatment would be seen as both a hindrance and an additional cost to production.

However, the published records are indicating that increasing importance is now being given to the subject of water use and water conservation. More than 7 billion rubles were expended during the tenth Five Year Plan (1976–80) for water protection, and 60% of the total water used by Soviet industry had been subjected to systems of water cycling and re-use, with even higher figures recorded for the petro-chemical and gas industries. Polluted waste-water discharges, however, declined by only 20% during the Plan period.

An important turning point was reached in 1982 with the introduction of a water-pricing policy, users of water having to pay in accordance with a scale of charges dependent upon the river basin or on the particular needs of individual industry. Previously, with a few exceptions, water use had been free and unrestricted. This new system placed emphasis on the need for more efficient and economical use of water and encouraged the introduction by enterprises of recycling systems, particularly in areas where water supply is relatively limited, making savings in its use imperative. Penalties were to be imposed for exceeding fixed limits.

Since that time, with the advent of the new administration in the USSR and the Five Year Plan for 1986–90, a much more intensive and efficient use of resources became necessary; this resulted in a tighter control of water. According to press reports the quality of water of the Volga is monitored regularly for its chemistry and for the quantity of water utilised for irrigation and for industrial requirements. Discharge of harmful effluents into the river by state farms or industrial units is followed by the imposition of penalties; initial warnings, if ignored, being followed by fines and finally by the compulsory closure of the enterprise. Reports have stated that in 1985 more than 300 enterprises in the USSR were closed down for the illegal emission of toxic effluents and many thousands of managers and other executives were fined for violation of water regulations or were deprived of their production bonuses, or in extreme cases even imprisoned.

The total expenditure on environmental protection is now stated to be more than 9000 million rubles each year; in the early 1980s nearly 70% of industrial water was stated to be re-circulated, a proportion planned to increase by 1989 to 90%. The eventual objective is to attain a closed–cycle, completely free of waste and sewage, or if that is not technically possible, to achieve the complete neutralisation of effluent pollutants.

Recent examples quoted in the Soviet press include that of the Magnitogorsk steel works which operates closed-circuit water circulation at all stages of production; another is that of the Verkhne–Isetsk works in Sverdlovsk, also in the Urals, which was threatened with closure as the management had ignored the problem of purifying the waste water discharged into the Zhet river. However, a change made in the production process and the installation of water recycling equipment allowed the works to continue operations. East of the Urals, the old city of Tobolsk with its new petro-chemical combine is becoming one of the most important industrial centres in western Siberia. Its effluents, entering the Tobol river, are monitored regularly for pollution levels. The closed cycle purification equipment to be operated in the future is expected to rectify the present problem of effluents and also to yield valuable by-products.

West of the Urals, river water used for cooling purposes in thermal and atomic power stations is becoming a useful resource; when returned to the river in a heated state it is harmful to many aquatic organisms, but it is now being used for the production of carp and other fish adapted for rapid growth in warm water with low oxygen levels. The fish are fed on protein-rich food converted by bacterial action from organic fish wastes. The usual tall water-cooling towers are replaced by a system of plastic piping laid under the soil which, becoming warmed, is utilised in hot-house mushroom and vegetable cultivation. Such fish-rearing pond systems operated in conjunction with power stations are being installed in several parts of European Russia.

IRRIGATION

The southernmost third of the Soviet Union, though endowed with abundant solar radiation and the most productive soils in the USSR, lacks sufficient precipitation for sustained agriculture. It is either marginal and variable, as in the sub-humid black-earth steppe zones, or totally inadequate as in the case of the desert zones of Central Asia and Kazakhstan. These are the zones where irrigation has an essential role in crop production. At present the irrigated land occupies 18 million hectares, of which the greatest area (8.1 million hectares) is in Central Asia and Kazakhstan; next in size is in the steppe zones of the Ukraine and the RSFSR, with 6.9 million hectares; smaller areas are in Siberia, the Soviet Far East, Transcaucasia and the European USSR (Figs 5.2 and 5.3).

IRRIGATION IN THE STEPPE ZONES
During the decades following the Second World

War, large increases in the area of irrigation were established in the chernozem steppes of the Ukraine, the Volga–Don area and its extension south towards the Caucasian foothills. Important impoundments of the major rivers provided multiple facilities including water for irrigation. In the entire steppe area, precipitation decreases from west to east—(600 to 400 mm along the lower Dnepr river to 400–200 mm along the lower Volga river)—and the recurrent droughts make irrigation necessary for more reliable and consistently increased cereal and other crop yields. All the major rivers in this area provide water for the various projects; in the west, a current project is to utilise the water of the rivers Danube, Dnestr and Southern Bug for irrigation and other purposes within part of the north-west coast of the Black Sea in the Moldavian SSR. Farther east, the large Kakhovka reservoir on the lower Dnepr waters large tracts of the floodplain and large areas of the dry Crimean steppe farther south. Beyond the irrigated area associated with the lower Don river, streams rising in the snowfields of the Great Caucasus mountains, the Kuban, Terek and Kuma rivers, bring water to a vast ramified irrigation system on the steppe areas of Stavropol kray and the Nogay steppes. The Kuban is of particular importance, as part of its upper course is diverted northward to increase the flow of the smaller rivers, the Yegorlyk and the Kalaus, which are used to irrigate the plains of the Stavropol foreland. Two important canals, the Terek–Kuma canal and the Kuma–Manych canal distribute water respectively to the Nogay steppe and the 'Black Land' pastures, an important wintering area for sheep. A third major canal, the Volga–Chogray canal has been started in this region to irrigate land in the Kalmyk ASSR from the lower Volga. But this project, together with that of the new Volga–Ural canal may be postponed as both schemes were dependent upon additional water for the Volga from the northern diversions.

The immense reservoirs now occupying the middle and lower Volga, designed for flood control, power, improved navigation and urban water supply, have a vital role to play in the irrigation of the Trans-Volga lowland of the Kuybyshev, Saratov, Volgograd and Uralsk oblasts. Crop failures in that region have been common in the past as a result of droughts which have in some years reduced wheat production to a

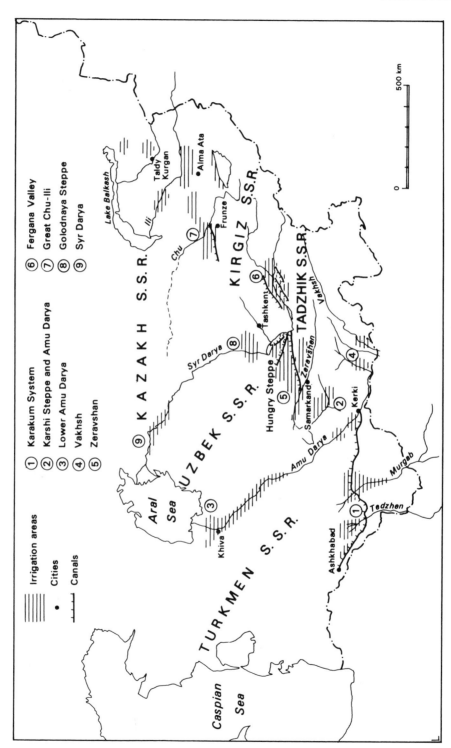

Fig. 5.3 Irrigation schemes in the Central Asian republics and Kazakhstan

small fraction of that of the more favourable years. Hence large irrigation schemes now distribute Volga water to these dry lands to the east of the river, resulting in improved and more reliable crop yields. Water is raised 100–200 m by pumping to the main irrigation canals such as the Saratov canal, from which it flows to the 70 000–80 000 hectares of southern chernozem and chestnut soils of the steppe. Shelter-belts of trees are important in reducing the dessicating effect of summer winds. The Volga–Ural canal is planned to combine the water of the Volga with that of the Ural river for the irrigation of the semi-arid region of western Kazakhstan. Farther south, beyond the Great Caucasus mountains, irrigation is of great importance in the poor steppe and semi-desert regions south of the main ranges extending from eastern Georgia into Azerbaydzhan. These areas form the basins of powerful rivers, notably the Kura and the Araks which feed irrigation systems in their lower and middle courses. The Armenian SSR contains many small irrigated basins and valleys such as that of the Razdan river which flows from Lake Sevan.

An important irrigation project in western Siberia is that of the Kulunda steppe. Begun in 1978, it takes water from the Ob river south of Novosibirsk across the chernozem steppe in a westerly direction. It is designed to irrigate 20 000 hectares and to provide more stable harvests in an area of low and irregular rainfall and dessicating sukhovey winds.

IRRIGATION IN CENTRAL ASIA AND KAZAKHSTAN
The principal area of irrigated land is found in Central Asia and Kazakhstan, where almost 69% of the total of 8 million hectares is found in the two largest republics of Kazakhstan and Uzbekistan. As in the European part of the USSR, considerable increases in the irrigated land followed from the operation of the tenth Five Year Plan (1976–80), although the area of land where there is a potential for irrigation remains very large (Fig. 5.3).

The largest irrigation projects are associated with water withdrawn from the river with the greatest stream-flow, the Amu Darya. Among the most important is the irrigation of the piedmont areas adjacent to the Kopet Dag highlands of arid southern Turkmenistan by the 900 km Karakum canal. This leaves the Amu Darya at Kerki and runs westward through oasis farmlands formerly

dependent upon the intermittent flow of rivers such as the Murgab and the Tedzhen but now having a constant supply from the canal. Lakes situated along the route serve as settling basins. Work is extending the canal from Ashkhabad toward the Caspian coast, increasing its present irrigated area of more than half a million hectares and eventually requiring 70% of the Amu Darya's annual flow at Kerki. However, the river's upper basin is also being used for the irrigation of land along the course of its mountain right-bank tributaries such as the Vakhsh and the Surkhandarya, flowing from the highlands of Tadzhikistan and Uzbekistan. Farther downstream along the Amu Darya is the Karshi steppe system, irrigated by water pumped up from the main river to supplement the more limited supply of the Kashkadarya river; this system was to increase to an area of more than 200 000 hectares allowing greatly increased cotton production.

A further supply of water is provided by another right-bank tributary of the Amu Darya, the Zeravshan, rising in the large Zeravshan glacier; two main canals, the Iski–Angar and the Amu–Bukhara canal distribute the water to oasis centres such as Karshi, Bukhara and Samarkand; they are protected from encroaching wind-blown sand by shelter belts of drought-resistant saxaul trees. Intense utilisation of water in this region of Uzbekistan causes the Zeravshan to fail to reach the main river. Various oases occur along the middle course of the Amu Darya where the river bottom is braided, with many separate channels divided by constantly changing sand or silt islands. Farther downstream in the lower course, a recent irrigation development, completed in 1985, has involved the river's 600 m-wide gorge where a spillway dam built across the Amu Darya provides flood control and is part of the Tyuyamuyun project; the latter includes four separate reservoirs having a combined capacity of 7.8 billion m^3, and providing additional water for irrigated rice and cotton cultivation in the Khiva oasis, part of the Khorezm oblast of Uzbekistan.

The Syr Darya river is similar in regime and general features to the Amu Darya though it is less powerful and has a smaller stream-flow. But like the latter it rises in snowfields and glaciers of the Tyan Shan mountains which supply the vital late-spring and summer floodwaters. Its upper course flows through the Fergana valley, a large intermontane depression where irrigated agricul-

Water being piped to a Crimean vineyard. Irrigation is essential in this southerly area for intensive agriculture

ture has had a very long tradition, controlled before the Soviet period by feudal landowners. One of the earliest and most ambitious of Soviet water-management projects was the building of the Great Fergana canal in 1939. This took water on a major scale from the upper Syr Darya or Naryn river along the southern side of the valley, intersecting the several separate irrigated alluvial fans there and giving a greater and more uniform water supply and thus an extension of cultivation towards the arid land in the centre of the valley. Later two other canals were built, the North Fergana in 1940, and the Central Fergana or Andizhan canal in 1970. Recent developments are the Great Namangan canal built during the tenth Five Year Plan and the Andizhan reservoir situated at the eastern end of the Fergana valley (1983). The latter takes water from the Kara Darya, a tributary of the Syr Darya and is designed to accommodate 1.75 billion m³ of water for irrigating the cotton fields of the valley. The total irrigated area is 800 000 hectares and is shared by Kirgizia, Uzbekistan and Tadzhikistan.

At the western end the Syr Darya is impounded by the Farkhad dam near Bekabad; the reservoir gives storage facilities for the irrigation of the Golodnaya (Hungry) steppe, southwest of

Tashkent, via the South Golodnaya steppe canal, and waters a million hectares of former desert soils. Two other right-bank tributaries of the Syr Darya, the Chirchik and the Keles, provide water for a new irrigation scheme, the Gazalkent project in the Keles valley north-east of Tashkent. Water is impounded in the new Gazalkent reservoir on the Chirchik river and pumped along the 66 mile Uzbek–Kazakhstan irrigation canal into the site of the project in the Keles valley.

Farther east, several rivers flowing northward from the Tyan Shan mountains are used for the irrigation of important oasis areas associated with major cities. Alma Ata, the capital of Kazakhstan is sited where several of these streams leave the mountains and flow onwards to join the River Ili, a main feeder of Lake Balkash. Frunze, the capital of Kirgizia, has an analogous position in the upper basin of the Chu river.

Problems of irrigation

The decision to shelve the plans for diversion of northern rivers has placed great emphasis upon the urgent need to rectify the shortcomings that have beset irrigation development in the USSR. These have been both administrative and technical. Lack of coordination between and among government and administration agencies or departments has caused delay in the implementation of plans, such as those for the preliminary preparation of sites chosen for new irrigation projects and for the provision of equipment, or of infrastructure facilities for incoming farm personnel.

But it is mainly in the actual application of water to the land that the deficiencies of Soviet irrigation technology are apparent, resulting in waste of water, damage to soil and loss of productivity of crops. Traditional practice has been on a simple distribution of water through a network of long unlined canals and a surface application to the crops by gravity through furrows to cotton fields and through flooding in the case of alfalfa rotation. The losses of water through evaporation, plant transpiration and infiltration are heavy in this system, amounting to more than half of all the water conducted by the canals from its sources in the rivers. Faulty installation of points of water distribution from the trunk canals is a further cause of spillage. Much surplus floodwater is allowed to drain into depressions near the main rivers which become large lakes, such as that of the Sarykamysh depression, fed by irrigation drainage and waste water from the Amu Darya and left-bank fields of the Khorezm oasis.

Excessive soil water gives rise to both water-logging and soil salinisation. Desert soils, having evolved within conditions of low effective pre-cipitation and pronounced evapotranspiration for much of each year, tend naturally to develop salt-enriched horizons. The over-watering of soils, long widespread in the Soviet arid zone, has produced secondary salinisation; this takes place when water surplus to plant requirements is al-lowed to seep downwards into the water table, the upper surface of groundwater. The latter is often saline, a condition derived from high salinity levels in the parent rocks, or from an influx of salt-enriched ground-water from distant highland areas, or from nearby salty-water lakes. The ground water, fed by excess irrigation, rises and floods the surface soil, or is lifted to that position by water movement through the soil capillaries produced by evaporation and transpiration. The characteristic soil is a solonchak, having a surface white crust of salts of sodium or calcium or magnesium; such a soil is difficult to use for agriculture until drainage is improved and a pro-cess of flushing is applied, using large quantities of fresh water.

Soil salinisation has been a common feature within the older irrigated areas of Central Asia, such as the Fergana valley where water escaping from unlined irrigation canals or from excess surface application have caused a rise in the water table.

The situation is being improved by more modern irrigation and drainage techniques. In the Golodnaya steppe, enclosed horizontal drainage at a depth of 3 to 3.5 m has been installed, giving greater efficiency and economy in water con-sumption, reducing water withdrawals from $16\,500\ \text{m}^3/\text{ha}$ to $9700\ \text{m}^3/\text{ha}$. Such re-construc-tion is costly and requires the closing of existing systems for up to two years and the consequent loss of production. The Soviet press have re-ported that many innovations in irrigated agricul-ture are being introduced in Central Asia. Ineffi-cient manual labour is being replaced by auto-mated application; soil sensors monitor the soil conditions, such as temperature and humidity, and the information is relayed to a computer so that correctly regulated amounts of water may be applied. Canals are being lined and equipped with water gauges; the drop irrigation method, in which piping with water-drop gear is positioned directly above plant roots, is being experimentally installed; a new laser-based monitoring system, designed in the Frunze research unit may elimin-ate over-watering; maps for the use and pro-tection of water resources are being made from surveys made by remote sensing techniques.

The economic reforms introduced by the Gorbachev regime are also expected to provide conditions promoting greater efficiency in the op-eration of techniques and management of irriga-tion. Thus the reduction of excessive bureaucratic control will allow collective farms to exercise initi-atives and to have greater responsibility in the supervision of irrigation. Common failings in the past have included wastage of water and of cap-ital resources, inefficient land use resulting in low crop yields and shortfalls in the planned output. The imposition of a levy upon water used for irrigation, already applied in some river catch-ment areas and introduced experimentally in Tadzhikistan and Kirgizia, has resulted in sub-stantially reducing water withdrawals and has improved productivity in the collectives of these republics. But despite this favourable result the practice of charging farms for their irrigation withdrawals does not receive full official approval.

Many areas under irrigation in Central Asia make insufficient use of the groundwater reserves. As pointed out earlier in this chapter, the geo-logical structures of the Turan lowland of the region are formations containing porous aquifers in which groundwater accumulates, derived from seepage of surface river run-off. The sources are precipitation and snow or glacier meltwater from the marginal mountains and piedmont foothills. The volume of such usable water has been estim-ated at $15.7\ \text{km}^3$, including $7.65\ \text{km}^3$ in the Amu Darya basin, $7.3\ \text{km}^3$ in the Syr Darya basin, and the remainder in smaller drainage basins. Part of this reserve is important for the supply of drink-ing water for the flocks and herds of the region, but in general it is under-utilised and supplies only about 10% of the withdrawals. Problems in its use are its depth from the surface, its slow rate of recharging and its variable mineralization. Shal-low wells are, however, sunk into the clay surfaces of the shallow pans or takyrs and the water obtained is used for drinking or for irrigating small vegetable gardens.

Water power in Uzbekistan, an old water wheel with a hydro-electric plant in the distance.

Irrigation and reservoir location

The requirements of irrigation are not satisfied by many of the water projects in the USSR established for multi-purpose utilisation. This conflict arises from the opposing demands of irrigation and of hydro-electric production. Irrigation demands greatest water availability during the warmer season of maximal crop growth; the demand for electricity is much more constant and requires the maintenance of an adequate head of water supported by a sufficient volume in the reservoir. This volume cannot therefore be available for irrigation. Furthermore, the power demand is more variable and yet is greater in the non-growing season, giving rise to frequent withdrawals from the reservoir at a time when, during wetter years, there is a maximum need for water to be stored in it for reserve irrigation water. Mountain sites are most suitable for purpose-built reservoirs for irrigation, as these produce reservoirs that are both deep, narrow and sited for maximum catchment; hence losses from evap-

oration are minimised and flooding of adjacent land is reduced.

Moscow's water supply

Many of the towns and cities of western USSR, like Moscow, draw part of their water supply from underground sources, the use of which is expected to be doubled during the next fifty years. A USSR government geological and hydrological survey has recently been completed and will ensure that a balance is maintained between withdrawals and refill of the reserve by percolation, and infiltration from surface runoff. Some cities, such as Baku have temporarily suspended using their groundwater source in order to safeguard this reserve.

The acute and increasing problem of shortage of water resources in European USSR is exemplified by the case of Moscow and its surrounding region. The Soviet capital's water is naturally restricted as a consequence of its geographical position near the principal divide of the Russian plain from which streams radiate away from the city, northwards to the upper Volga and southwards to the basin of the Oka river. The Moscow river and its numerous tributary streams formed the original source of Moscow's water, with the artesian groundwater reserve contained in the basin of Carboniferous and younger sedimentary strata which form the plain on which the city stands. The Moscow river gathers up these small tributaries, many flowing from the north and northwest, draining the southern slopes of the Klinsk–Dimitrov ridge and feeding the many large supply reservoirs, such as the Istrinskoye reservoir. Most of these are now piped or channelled or flow underground below the city streets.

In 1937 the rapid growth of the city and the increasing level of industrialisation in the region resulted in the provision of a supplementary supply of water, adding to that of the Moscow river; this came from the upper Volga, brought by the new 129-km Moscow–Volga canal. The latter is today part of an extensive system including reservoirs, freighting ports, canals and hydro-electric stations which forms a navigable waterway linking the capital to the sea. Despite treatment of Moscow's growing quantity of effluent its volume became large relative to that of the river's runoff. Additional dilution became necessary and was provided in 1978 by the diversion of the Vazuza river, a tributary of the Volga, into the Moscow river. Irrigation in the region makes other demands on the water resources. In the future, however, a further addition to the city's water supply may have to be provided, from Lake Ladoga or from the Andropov (formerly Rybinsk) reservoir on the Volga.

The water balance within the Soviet capital is characteristic of very large cities. Moscow receives a 10% higher precipitation than surrounding non-urban districts. In summer, showers produce a very high runoff from the impermeable surfaces of streets and buildings, but much of this is lost in evaporation (57%) and only about 16% reaches the subsurface. The remainder drains into the Moscow river. In winter, snow is cleared systematically from the impermeable surfaces and brought to temporary dumping areas such as ravines where it is melted artificially. Although Moscow is a conspicuously clean city, runoff from its streets inevitably becomes polluted. However, 90% of the water is subjected to circulation treatments which partially remove contaminants; fully treated wastewater has to be diluted with freshwater for downstream use in order to re-introduce ecological processes and for self-purification.

BIBLIOGRAPHY

Ashkochensky, A. N. (1962), Basic trends and methods of water control in the arid zones of the Soviet Union,' in *The problems of the arid zone*, UNESCO, Paris, pp. 401–410.

Atlas SSSR (1969), Moscow, 2nd ed.

Baydin, S. S. (1980), 'Redistribution of river runoff between sea basins and its role in the environmental complex of seas and river mouths,' *Soviet Hydrology*, **19**, 2, pp. 88–93.

Borovskiy, V. M. (1980) 'The drying out of the Aral Sea and its consequences,' *Soviet Geography*, **21**, pp. 63–77.

Central Intelligence Agency (1974), *USSR agriculture atlas*, Washington DC.

Dreyer, N. N. (1969), 'Water resources of the major economic regions of the RSFSR, and the other Union Republics,' *Soviet Geography*, **10**, pp. 137–145.

Gerasimov, I. P. (1968), 'Basic problems of the transformation of nature in Central Asia,' *Soviet Geography*, **9**, pp. 444–458.

Gerasimov, I. P., Armand, D. L. and Yefron, K. M. (eds.) (1971), *Natural resources of the Soviet Union, their use and renewal,* translation ed. W. A. D. Jackson, Freeman, San Francisco.

Gerasimov, I. P. *et al.* (1976), 'Basic problems in the transformation of nature in Central Asia,' *Soviet Geography*, **17**, pp. 235–245.

Goldman, M. I. (1972), *The spoils of progress; environmental pollution in the Soviet Union*, MIT Press, Cambridge, Mass. and London.

Hynes, H. B. N. (1970), *The biology of polluted waters*, Liverpool.

Kes' A. S. (1980), 'The present stage in the evolution of the Sarykamysh depression,' *Soviet Geography*, **21**, pp. 524–533.

Kes' A. S. *et al.* (1981), 'The present state and future prospects of using local water resources in Central Asia and southern Kazakhstan,' *Soviet Geography*, **21**, pp. 414–425.

Kiselev, V. N. (1984), 'The optimal use of land resources in the Belorussian part of the Poles'ye swamps,' *Soviet Geography*, **25**, pp. 572–583.

Kosarev, A. N. and Makarova, R. E. (1988), 'Changes in the level of the Caspian Sea and the possibiliy of forecasting them,' *Soviet Geography*, **29**, pp. 617–624.

Kovda, V. A. (1961), 'Principles of the theory and practice of reclamation and utilisation of saline soils in the arid zones,' in *Salinity problems in arid zones,* UNESCO, Paris, pp. 201–213.

Kuznetsov, N. T. (1977), 'Geographical aspects of the future of the Aral Sea,' *Soviet Geography*, **18**, pp. 163–171.

Kuznetsov, N. T. and Lvovich, M. I. (1971), 'Multiple use and conservaton of water resources,' in Gerasimov *et al.* (1971), pp. 11–39, San Francisco.

Lappo, G., Chikishev, A. and Bekker, A. (1976), *Moscow, capital of the Soviet Union*, Moscow.

Lavrishchev, A. (1969), *Economic geography of the USSR*, Moscow.

Leontyev, O. K. (1988), 'Problems of the level of the Caspian and the stability of its shoreline,' *Soviet Geography*, **29**, pp. 608–617.

Lupachev, Yu. V. (1980), 'Hydrologic conditions of the mouth area of the Pechora river and their possible changes after withdrawal of part of the runoff from the basin,' *Soviet Hydrology*, **19**, 2, pp. 94–102.

Lvovich, M. I. (1977), 'Geographical aspects of a territorial distribution of water resources in the USSR,' *Soviet Geography*, **18**, pp. 302–312.

Lvovich, M. I. and Tsigel'naya, I. D. (1981), 'The potential for long-term regulation of runoff in the mountains of the Aral Sea drainage basin,' *Soviet Geography*, **22**. pp. 471–483.

Micklin, P. P. (1972),'Dimensions of the Caspian Sea problem,' *Soviet Geography*, **13**, pp. 589–603.

Micklin, P. P. (1978), 'Irrigation development in the USSR during the 10th. Five-Year Plan (1976–1980),' *Soviet Geography*, **19**, pp. 1–24.

Micklin, P. P. (1986), 'The status of the Soviet Union's north–south water transfer projects before their abandonment in 1985–6,' *Soviet Geography*, **27**, pp. 287–325.

Micklin, P. P. (1987), 'Irrigation and its future in Soviet Central Asia: a preliminary analysis,' in *Soviet geography studies in our time*, (1987), eds. Holzner, L. and Knapp, J. M., Milwaukee, pp. 229–261.

Poladzade, P. (1984), 'Water economy of the USSR: man and the desert,' *Soviet Union*, **10**, pp. 18–21.

Pryde, P. R. (1972), *Environmental pollution and environmental quality in the Soviet Union*, London.

Shabad, T. (1981), 'Uzbek-Kazakhstan irrigation canal under construction,' *Soviet Geography*, **22**, News Notes p. 128.

Shabad, T. (1983), 'The Tyuyamuyun irrigation project in Central Asia is working,' *Soviet Geography*, **24**, News Notes, p. 465.

Shabad, T. (1983), 'The Andizhan reservoir in the Fergana valley completed,' *Soviet Geography*, **24**, News Notes. p. 467.

Shabad, T. (1985), 'Work begins on the irrigation canal in Kalmyk ASSR,' *Soviet Geography*, **26**, News Notes, p. 771.

Shabad, T. (1986), 'Geographic aspects of the new Soviet Five Year plan 1986–80,' *Soviet Geography*, **27**, pp. 1–16.

Shabad, T. (1986), 'Soviet decree officially cancels north-south water transfer projects,' *Soviet Geography*, **27**, News Notes, pp. 601–604.

Vendrov, S. L. and Dyakonov, K. N. (1976), *Vodokhranilishcha i okruzhayushchaya prirodnaya sreda*, Moscow.

Voropayev, G. V. *et al.* (1982), 'The problem of redistribution of water resources in the midlands region of the USSR,' *Soviet Geography*, **23**, pp. 713–728

Vozresenskiy, A. N., Gangardt, C. G, and Gerardi, I. A. (1975), 'Principal trends and prospects of the use of water resources in the USSR,' *Soviet Geography*, **16**, pp. 291–307.

ZumBrunnen, C. (1984), 'A review of Soviet water quality management, theory and practice,' in *Geographical studies in the Soviet Union*, Demko, G. J. and Fuchs, R. J., Univ. of Chicago Research paper, No. 211.

6 Population

ETHNIC COMPOSITION

As has been described in Chapter One, the present territory of the USSR was inherited, in 1917, from its political forerunner, the Russian Empire which, by a process of expansion lasting some 400 years, had spread out from its original nucleus around Moscow to cover vast areas of eastern Europe and northern Asia. Like the other European-dominated empires established in the same period, the Russian Empire came to include, in addition to the Russian themselves, a large and varied collection of other peoples, so that the Soviet Union, inheriting the lands of the Empire, also inherited its extremely diverse population. Thus, the peoples of the present-day USSR differ from each other in their racial, cultural, historical and religious backgrounds and in the languages they speak.

The Soviet census of 1979 recognised no fewer than 92 distinct national groups within the population of the USSR. Many of these groups were very small: 16 had fewer than 10 000 members each and 39 were between 10 000 and 250 000 strong. This leaves 37 larger nationalities, ranging in size from 287 000 to 137 million, which together accounted for 98.7% of the Soviet population (Table 6.1).

The 92 nationalities defined in the 1979 census fall into some 10 major ethnic groups, the distribution of which is illustrated in Fig. 6.1. It should be realised that there has been a great deal of migration within the Russian Empire and Soviet Union by members of these groups. They have thus become intermingled to a progressively greater extent and there are now relatively few

areas inhabited by a single group to the exclusion of all others. Nevertheless, each ethnic group has an area or areas in which it forms the majority of the local population or in which the bulk of its members are to be found.

Since language is the main criterion by which the Soviet Union distinguishes its national groups, the distribution of these groups is best described on a linguistic basis. The USSR contains representatives of two of the world's major language families, the Indo–European and the Ural–Altaic, both of which originated in the Eurasian interior in remote prehistoric times. Languages of the Indo–European family, which is much the larger of the two, are indigenous over a zone extending from the Atlantic in the west to northern India in the east. Ural–Altaic languages have spread to Siberia and the Pacific in one direction and to eastern Europe and the Middle East in the other. Within the Soviet Union there are several areas in which languages belonging to each family are spoken by peoples living in adjacent districts.

Indo-European ethnic groups

By far the largest sub-division of the Indo–European family found in the USSR is the Slav group of tongues, spoken by roughly three quarters of the population. Dominant among the Slavs are the 137 million *Russians* proper, or Great Russians, who make up some 52 per cent of the inhabitants of the Soviet Union. The traditional home of the Russians is the mixed forest zone of the USSR, centred on Moscow, and it is here that the bulk of them are still to be found. As

TABLE 6.1 THE LARGER NATIONALITIES OF THE SOVIET UNION
POPULATION (000s) AT THE CENSUS OF 1979 (PERCENTAGE OF TOTAL POPULATION IN
BRACKETS)

Russians	137 397	(52.42)	Bashkirs	1371	(0.52)
Ukrainians	42 347	(16.16)	Mordovs	1192	(0.45)
Uzbeks	12 456	(4.75)	Poles	1151	(0.44)
Belorussians	9 463	(3.61)	Estonians	1020	(0.39)
Kazakhs	6 556	(2.50)	Chechens	756	(0.29)
Tatars	6 317	(2.41)	Udmurts	714	(0.27)
Azerbaydzhanis	5 477	(2.09)	Mari	622	(0.24)
Armenians	4 151	(1.58)	Osetins	542	(0.21)
Georgians	3 571	(1.36)	Avars	483	(0.18)
Moldavians	2 968	(1.13)	Koreans	389	(0.15)
Tadzhiks	2 898	(1.11)	Lesghians	383	(0.15)
Lithuanians	2 851	(1.09)	Bulgars	361	(0.14)
Turkmen	2 028	(0.77)	Buryats	353	(0.13)
Germans	1 936	(0.74)	Greeks	344	(0.13)
Kirgiz	1 906	(0.73)	Yakuts	328	(0.12)
Jews	1 811	(0.69)	Komi	327	(0.12)
Chuvash	1 751	(0.67)	Kabardins	322	(0.12)
Latvians	1 439	(0.55)	Karakalpaks	303	(0.12)
			Dargintsy	287	(0.11)

Source: Vestnik Statistiki 1980 (2), 24–5, Moscow.

the dominant force in the growth of the Empire, however, they played a leading role in migration eastwards into Siberia, which has been going on since the seventeenth century. Owing to the small size of the indigenous population of Siberia, Russians are now in a majority over the greater part of the inhabited zone, where they are now found not only in urban areas but also forming the bulk of the rural population wherever settled agriculture has been established. Russians have also moved in large numbers into Kazakhstan, where they now outnumber the indigenous Kazakhs. In the other four Central Asian republics, however, they are largely confined to urban areas and make up only about 15% of the total. A somewhat higher proportion of Russians is found in the Ukraine, particularly in the eastern part, where they assisted in the industrialisation of the Donbas. Russians have also moved in appreciable numbers into Latvia and Estonia since 1945, and account for 28% of the population of those republics. Areas with relatively few Russian settlers, less than 10% in each case, include Lithuania and the Transcaucasian republics of Georgia and Armenia.

The *Ukrainians,* of whom there are more than 42 million, are the second largest national group,

both among the Slavs and in the Soviet population as a whole. Speaking a language closely allied to Great Russian, the Ukrainians originated as a distinctive group in the wooded steppe and steppe lands of the European south, where they were much affected by the Tatar invasions of the medieval period. Like the Russians, the Ukrainians carried out widespread colonisation in Siberia, where they are to be found in considerable numbers, particularly in the west and in the extreme far east.

The third major Slav group, the *Belorussians,* is much smaller (9 million) and is largely confined to the western part of the country. As a result of this location, they were subject to Lithuanian and Polish influences and many Belorussians became Roman Catholics in contrast to the predominantly Orthodox Russians and Ukrainians.

The *Poles* form a Slav minority group of considerable size in the western parts of Belorussia and the Ukraine, areas which were at various times under Polish control and where the population remains very mixed.

The *Lithuanians* and *Latvians* speak Baltic languages which, though remotely related to Russian, are quite distinctive, employing Roman as distinct from Cyrillic script. These two groups,

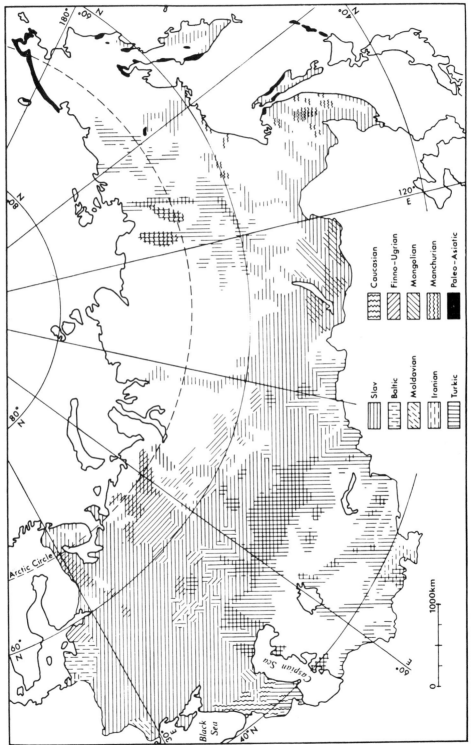

Fig. 6.1 Distribution of major ethnic groups. Unshaded areas are very thinly populated, mainly by small tribes

numbering about 4 million, have been much affected by western contacts; the Lithuanians, as a result of their long connection with Poland, adopted Roman Catholicism, while the Latvians received the Lutheran form of Christianity from German immigrants.

In addition, there are several non-Slav Indo–European groups within the European USSR. These include the 3 million *Moldavians* of Bessarabia, who speak a Romance language allied to Rumanian. The Teutonic sub-division is represented by 1.9 million *German*-speakers, descendants of farmers and artisans who entered Russia from the eighteenth century onwards, often as refugees from religious or political persecution. Before the Second World War, they were mainly concentrated along the west bank of the Volga, where they had their own Volga–German Autonomous Republic, but during the war this was abolished and the population dispersed to Central Asia and Siberia. The number of *Jews* in the Soviet Union (1.8 million) is much below the pre-war level owing to the effects of the German occupation and some recent emigration. Although a Jewish Autonomous Oblast was established in the Far East as long ago as 1934, it has failed to attract many settlers and is the home of only one per cent of Soviet Jews. The majority are found in the European part of the country, but there are also a number of very ancient Jewish communities in Central Asia. Most of the 344 000 *Greeks* live in old trading cities along the Black Sea coast.

Finally, among the Indo–European peoples, we must note the presence of Indo–Iranian groups, speaking languages allied to Persian, in Central Asia (*Tadzhiks*) and the Caucasus (*Osetins* and numerous smaller groups), most of whom are Moslems.

The Caucasus, and particularly Transcaucasia, is a region of very great ethnic diversity. In addition to a number of Indo–European groups, the region contains the Turkic-speaking *Azerbaydzhanis* who belong to the Ural–Altaic family. Two large groups, the *Georgians* (3.6 million) and *Armenians* (4.2 million), have their own separate languages which are related neither to the Indo–European nor to the Ural–Altaic, nor to each other. Each has its own highly distinctive script, quite different from both Roman and Cyrillic. The Georgians are Orthodox Christians and the Armenians follow the Gregorian rite. The

Children in the Soviet north are used to going to school in snow and blizzards, as at this farm kindergarten in the Nenets Autonomous Okrug

region has many other mutually unintelligible languages spoken by such peoples as the *Chechens, Lesghians* and *Kabardins;* many of the smaller Transcaucasian groups adhere to Islam.

Ural–Altaic ethnic groups
There are at least four main sub-divisions of the Ural–Altaic language family represented in the Soviet Union: the Finno–Ugrian, Turkic, Mongolian and Manchurian.

Finno–Ugrian peoples are distributed over a broad zone which embraces much of western Siberia and the northern half of the European USSR, where they were present before the arrival of the Russians. Over large areas they have been assimilated into the Slav majority but there are several areas where Finno–Ugrian languages are still spoken by the majority of the population. One such zone is in the extreme north-west of the Soviet Union, where the one million *Estonians*, much influenced by cultural contact with the Germans, and Protestant in religion, form a compact block of Finno–Ugrian speech. The 138 000 *Karelians* are closely related to the neighbouring Finns, but are now heavily outnumbered in their homeland by Russian settlers. A second zone in which Finno–Ugrian groups can be identified lies immediately west of the Urals, extending from the Arctic Ocean to the middle Volga. In the extreme north are the *Nentsy* reindeer-herders (30 000) and the *Komi* (327 000), many of whom are now employed in mining, lumbering and farming. Between the Volga and the Urals live the *Mordovs, Udmurts* and *Mari*, who together number 2.5 million. These groups are culturally more advanced than their northern neighbours: they have adopted Orthodox Christianity and settled agriculture and are now involved in the industrial development of their territories. Finally, to the east of the Urals, the west Siberian lowland, a wilderness covered by swamps and forests, is thinly occupied by the *Khanty* (21 000), *Mansi* (7700) and other small Finno–Ugrian tribes.

Turkic-speakers, who number about 40 million and thus constitute some 15% of the Soviet population, are the largest non-Slav element and are widely distributed in the regions east of the Volga. 23 million Turkic-speakers belong to the four major groups of Soviet Central Asia: the *Uzbeks, Kazakhs, Turkmen* and *Kirgiz*. The Uzbeks are by tradition settled farmers, engaged mainly in irrigated oasis cultivation, and have been much influenced by Persian culture. The Kazakhs were predominantly nomadic pastoralists but, since the Revolution, have for the most part been settled. Turkmen have included both pastoralists and farmers. The Kirgiz, like their Indo–Iranian Tadzhik neighbours, have a background of mountain stock-rearing but also include valley cultivators. All these groups are Moslem, and Islamic culture is dominant throughout this part of the Soviet Union, especially in rural areas. Most of the towns however, many of which are of very ancient origin, now have large communities of Russian settlers. The *Azerbaydzhanis* (5.5 million), on the west side of the Caspian, form an extension of this Central Asian Turkic zone and are quite distinct from the adjacent Georgians and Armenians.

Further large Turkic groups living between the Volga and the southern Urals include over 6 million *Tatars*, 1.8 million *Chuvash* and 1.4 million *Bashkirs*. Smaller nationalities speaking Turkic languages occur in various parts of Siberia, but do not amount to much more than 650 000 people in all. Most important are the *Yakuts* (328 000) of the Lena basin, the *Tuvinians* (166 000) of the Sayan Mountains and the *Khakass* (71 000) of the Altay.

Peoples of Mongolian origin form a small minority of the population of Siberia, rising to a majority in a few southern areas, the most important group being the *Buryats* (353 000) of the Lake Baykal region. These people are Buddhists, as are their cousins the *Kalmyks* (147 000) who reached the steppe to the west of the lower Volga in the eighteenth century.

A number of small Manchurian groups live in various districts of eastern Siberia and the Far East. The largest of these, the *Evenki*, number only 25 000. The *Koreans*, once an important minority near the border with Manchuria, now live mainly in Soviet Central Asia, though a few are still to be found in the Amur lowlands.

Paleo–Asiatic peoples

Probably pre-dating the arrival of the Ural–Altaic groups are a number of small, rather primitive tribes who live in the extreme north-east of the country. These Paleo–Asiatic peoples, each numbering only a few thousand, include the *Chukchi* of the Anadyr peninsula, who are related to the American Eskimo, the *Koryaks* of Kamchatka and the *Ainu* of Sakhalin and the Kurile Islands. The *Aleuts* of the Komandorskiye Islands, off the east coast of Kamchatka, had the distinction of being the smallest national group (441 people) recorded in the 1970 census.

The larger national groups retain their own languages for official purposes and newspapers and books are published in them. Russian is, however, taught in all schools and is the second language of all educated members of the minority groups. Several of the less important languages, spoken only by small numbers of people, are dying out and being replaced by Russian. In 1979, 82% of the Soviet population claimed the ability to speak Russian as either their first or second language, including about 60% of people of non-Russian nationality. Ability to speak Russian was highest among dispersed groups such as the Jews (97%) and Germans (94%) and among the peoples of the Volga–Ural zone, such as the (Chuvash (83%), Mordovs (93%) and Udmurts 88%). At the other end of the scale, ability to speak Russian was lowest among the Moslems of the south; barely 40% of the Uzbeks, Tadzhiks, Turkmen, Kirgiz and Azerbaydzhanis spoke Russian. It should be noted that in many areas where Russians are in a minority they are often to be found in positions of influence, and Russians play a leading role in the affairs of the Soviet Union to an even greater extent than their numerical superiority suggests.

POPULATION DISTRIBUTION

The total population of the Soviet Union at the beginning of 1989 was 286 717 000. Only two countries, China (over 1000 million) and India (800 million), have larger numbers. The population of the USSR exceeds that of the USA (246 million) by 16% and is over five times that of the United Kingdom (56 million). In relation to the size of the Soviet Union, however, its population is a good deal less impressive. The USSR has an area of 22.4 million square kilometres (8.6 million square miles) and covers one-sixth of the world's land surface but it contains only one-seventeenth of the world's population. Its area is about two-and-a-half times that of the United States (9.4 million km²) and China (9.6 million km²) more than six times that of India (3.3 million km²), more than six times that of India (3.3 million km²) and no less than 92 times that of the United Kingdom (245 000 km²). Thus the average population density of the USSR (12.8 people per km²) is much lower than in any of the other countries mentioned: UK 229, India 242, China 110, USA 26 per km². These comparisons emphasise the fact that the Soviet Union is a thinly settled country; closer examination reveals that vast areas are virtually uninhabited, contrasting vividly with the relatively limited areas of densely settled territory.

This is illustrated in Fig. 6.2 and Table 6.2 which show the distribution of the population among the various republics and economic regions in 1986 (the latest date for which details were available at the time of writing). Of the 286.7 million inhabitants of the USSR, 186.1 million or 66.7% live in the European part of the country, which accounts for only 24% of the national territory. This relatively densely settled area is continued southward into the Transcaucasian republics, whose population of 15.3 million represents 5.5% of the total living in 0.8% of the area. A further 46.5 million, 16.7% of the population, live in Kazakhstan and the four Central Asian republics, which together constitute 17.9% of Soviet territory. All these regions and particularly the European USSR, are in striking contrast to Siberia and the Far East. These vast territories make up 57.3% of the land area but contain only 11.1% of the population; in an area of 12.8 million km² there are only 30.9 million **people. Most striking of all is the Far East region,**

an area as large as Europe with a population (7.6 million) roughly equal to that of Austria.

Fig. 6.3 shows in some detail variations in population density throughout the USSR, and from this map it is possible to identify areas in which density is significantly above or below the national average. Immediately apparent is the relatively restricted area falling into the 'above average' category. This includes what may be termed the main settled zone, a triangular area with its base along the western frontier and its apex in the Urals. There is a southward extension of this zone into the North Caucasus and Transcaucasian regions. Outside these areas, districts with above-average density are small and discontinuous. One such covers parts of south-west Siberia and continues eastward in a narrow belt along the Trans-Siberian Railway to the Pacific at Vladivostok; another occurs ·in parts of Soviet Central Asia. For a fuller picture of the situation and of the factors at work, this pattern must be discussed on a regional basis.

European USSR

The location of the main settled zone of the Soviet Union is indicated by densities above 25 people per km². The most densely populated districts of all (over 100 people per km²) are in and around the major cities: Moscow, Kiev and the Donbas industrial complex, for example, stand out clearly on Fig. 6.3. However, very high densities (by Soviet standards) are also recorded in some predominantly rural areas where conditions are especially favourable for agriculture, as in the western Ukraine. The agricultural productivity of the Ukrainian steppe and the wooded steppe as a whole is reflected in a broad belt of territory where densities are between 50 and 100 people per km², despite the fact that the urban proportion rarely exceeds the national average of 66% and is often well below that level. As an example we may note the Cherkassy oblast, with an area of 20 900 km² and a population of 1.5 million of whom 747 000 are classed as rural, giving an overall density of 73.1 people per km² and a rural density of 35.7 people per km². Rural densities are lower in the eastern Ukraine but overall density is raised by the presence of major industrial cities; thus the Donetsk oblast has an area of 26 500 km² and a population of 5.3 million, 90% urban; its rural density is only 19.5 people per km², but the

TABLE 6.2 POPULATION OF THE USSR, 1986, BY REPUBLICS AND MAJOR ECONOMIC REGIONS*

Region or Republic	Area 000 km²	%	Population 000	%	Density per km²
USSR	22 402.2	100.0	278 784	100.0	12.4
RSFSR	17 075.4	76.2	144 080	51.7	8.4
North	1 465.3	6.5	6 003	2.2	4.1
North-west	196.5	0.9	8 134	2.9	41.4
Centre	485.2	2.2	29 795	10.7	61.4
Volga–Vyatka	263.2	1.2	8 362	3.0	31.8
Black Earth Centre	167.7	0.8	7 652	2.7	45.6
Volga	536.4	2.4	16 081	5.8	30.0
North Caucasus	355.1	1.6	16 340	5.9	46.0
Ural	824.0	3.7	19 980	7.2	24.2
West Siberia	2 472.2	11.0	14 358	5.2	5.8
East Siberia	4 122.8	18.4	8 875	3.2	2.2
Far East	6 215.9	27.7	7 651	2.7	1.2
Ukraine	603.7	2.7	50 994	18.3	84.5
Donets–Dnepr	220.9	1.0	21 568	7.7	97.6
South-west	269.4	1.2	21 938	7.9	81.4
South	113.4	0.5	7 488	2.7	66.0
Baltic	189.1	0.8	8 616	3.1	45.6
Lithuania	65.2	0.3	3 603	1.3	55.3
Latvia	63.7	0.3	2 622	0.9	41.2
Estonia	45.1	0.2	1 542	0.6	34.2
†Kaliningrad oblast	15.1	0.1	849	0.3	56.2
Transcaucasia	186.1	0.8	15 304	5.5	82.2
Georgia	69.7	0.3	5 234	1.9	75.1
Azerbaydzhan	86.6	0.4	6 708	2.4	77.5
Armenia	29.8	0.1	3 362	1.2	112.8
Central Asia	1 279.3	5.7	30 456	10.9	23.8
Uzbekistan	449.6	2.0	18 487	6.6	41.1
Kirgiziya	198.5	0.9	4 051	1.4	20.4
Tadzhikistan	143.1	0.6	4 051	1.4	20.4
Turkmeniya	488.1	2.2	3 270	1.2	6.7
Kazakhstan	2 715.1	12.1	16 028	5.7	5.9
Belorussia	207.7	0.9	10 008	3.6	48.2
Moldavia	33.7	0.2	4 147	1.5	123.1

* For the boundaries of these regions, see Fig. A, page 2
† Figures for the Kaliningrad oblast are also included in the RSFSR totals
Source: Narodnoye khozyaystvo SSSR v 1985 godu, Moscow, 1986

overall figure is 201 people per km², one of the highest in the country.

Away from this high-density zone, densities diminish both northwards into the mixed forest and southwards into the drier parts of the steppe, though in the grainlands of the North Caucasus economic region high densities are common. North of a line from Leningrad to Perm, there is an abrupt change to areas where the average density is below 10 per km² and in many places below 1 per km². In these northern districts, agriculture is of little or no importance and the dominant activities are lumbering and mining. Consequently, the urban proportion is abnormally high. An extreme example is provided by the Murmansk oblast. This has an area of 144 900 km² and a population of 1 100 000, giving an overall density of 7.6 per km². However, 92.5% of the population is urban, 426 000 living in the port city of Murmansk alone, with a rural population of only 83,000 and a rural density of only 0.6 per km².

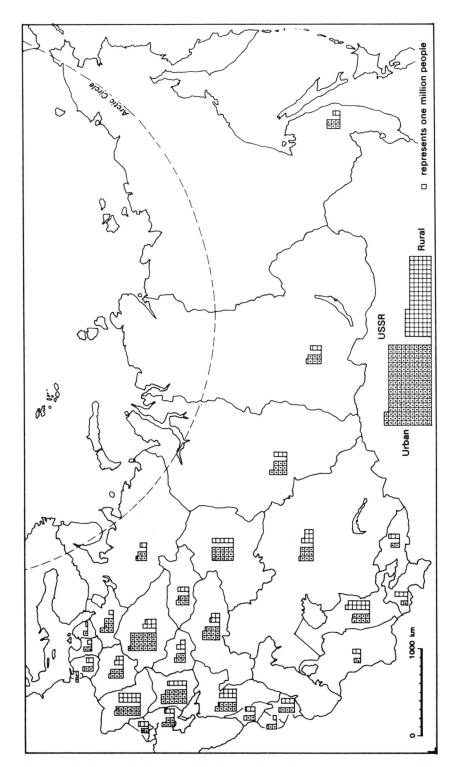

Fig. 6.2 Distribution of the Soviet population in 1986, by republics and major economic regions. Each small square represents one million people, 0.36% of the total population

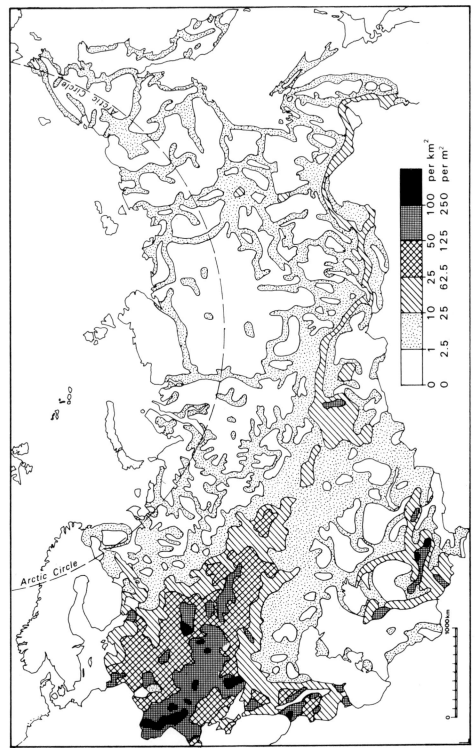

Fig. 6.3 Population density. White and stippled areas have densities below the national average

Eastwards towards the Urals, the densely settled zone becomes narrower as the negative areas of tayga and tundra to the north and dry steppe and desert to the south become wider with increasing continentality. A pocket of densities in the 25–50 people per km² range marks the position of the main industrial area of the central Urals. Here again, agricultural potential is limited and the rural population is small. In the Sverdlovsk and Chelyabinsk oblasts, 7 million out of a population of 8.2 million are classed as urban and the rural density is only 4.5 people per km².

The Caucasus

The main Caucasian range forms a narrow belt of thinly settled territory on either side of which are productive lowlands where population density is several times the national average. This applies particularly to Georgia and Armenia, which have densities of 75.1 and 112.8 people per km² respectively. Highly productive sub-tropical agriculture in western Georgia supports rural densities exceeding 100 people per km² in places.

Kazakhstan and Central Asia

In the steppelands of northern Kazakhstan, despite the expansion of the cultivated area achieved under the virgin lands scheme of the 1950s, overall densities are generally well below 10 people per km² and rural densities less than 5 people per km²; for example the Tselinograd oblast has 6.8 persons per km² and a rural density of 2.6 persons per km². Populations are predominantly rural save in the Karaganda oblast where there has been large scale industrial development but conditions are too dry for productive farming (87% urban, rural density 2.1 people per km², overall density 15.9).

South of the Kazakh steppe lie the vast expanses of the Turanian desert, where densities are below 1 person per km² over large areas. Beyond the desert is the densely-populated mountain-foot zone where, although there are a number of large and ancient cities, some 60% of the population is rural and very high rural densities, based on irrigated farming, are common. In the Fergana basin, for example, an area of 19 200 km², live nearly 5 million people, of whom 3.2 million (65.3%) are classed as rural, giving a rural density of 167 people per km². Narrow belts of high density also extend out across the desert along the valleys of the Amu Darya and Syr Darya rivers.

Siberia and the Far East

As already indicated, these are the most thinly settled regions of the USSR. West Siberia, East Siberia and the Far East have overall densities of 5.8, 2.1 and 1.2 persons per km² respectively. Only in West Siberia is there a zone of appreciable size in which densities exceed 10 people per km²; this embraces the Kuzbas industrial area and the agriculturally more productive parts of the Altay kray and the Novosibirsk and Omsk oblasts. Despite recent developments in oil and gas production, the central and northern parts of the West Siberian lowland are very thinly populated indeed; covering most of this region is the Tyumen oblast which has an area of 1435 000 km² but a population of only 2.7 million, giving an average density of only 1.9 people per km².

In East Siberia and the Far East, densities rise above 10 people per km² only in a narrow belt along the Trans-Siberian railway. Away from this zone, settlement is mainly confined to river valleys and coastal fringes, between which there are vast areas of virtually uninhabited territory. The Taymyr and Evenki autonomous okrugs, for example, have a population of only 75 000 in an area of 1.6 million km² (0.05 people per km²); one person to every 20 km² of land. Over many of these sparsely populated areas of northern Siberia, the urban proportion is very high; in Magadan oblast, for example, 437 000 out of a total population of 542 000 (81%) are classed as urban.

The striking regional variations in population density outlined above are to a large degree the result of physical factors. The low densities prevailing in the Siberian tundra and tayga and in the deserts of Central Asia are obvious cases in which the nature of the physical environment is hostile to the development of a large population. At the same time, historical and economic factors are also of major importance. The long-established role of the European part of the country as its agricultural and industrial heartland has been responsible for the fact that it contains two-thirds of the Soviet population. A great potential for future economic development exists in the Asiatic parts of the Soviet Union, particularly in Siberia, and despite the physical difficulties there are many areas which can, and no doubt eventually will, support much larger numbers than at present. Indeed, the movements of people from the European to the Asiatic parts of the country is a persistent theme in the history of population change in the USSR, even though counter movements occur.

POPULATION CHANGE

This topic will be discussed in two stages: a brief general statement of developments in the period up to the census of 1959, and a more detailed examination of recent trends as exemplified by the changes which occurred between 1959 and 1986.

Population change 1897–1959

To give precise data on the growth of the Soviet population during the present century is by no means a simple task, for a variety of reasons. In the first place, censuses have been carried out on only seven occasions: in 1897, 1926, 1939, 1959, 1970, 1979 and 1989. For years other than these, it is necessary to make use of estimates from a variety of sources, not all of which are wholly reliable. Secondly, there have been major changes in the boundaries of the Soviet Union. Following the First World War, large areas, including Finland, the Baltic states, Poland and Bessarabia, were detached from the USSR; parts but not all of these territories were recovered during and after the Second World War when the Soviet Union also gained additional territory from Germany and Czechoslovakia. In addition, the two World Wars, the Revolution and the Civil War which followed it, together with other internal upheavals, had a profound effect on population growth. These events not only caused a large number of extra deaths but also resulted in reduced fertility so that the number of births was much smaller than that of more normal times. It has been estimated that, had it not been for these setbacks, the population of the USSR would now be approaching 400 million as against an actual total of 286.7 million in 1989. Thus the figures given in Tables 6.3 and 6.4 are, for years prior to 1959, only approximate; nevertheless they give a reasonably good impression of the changes which have occurred since the late nineteenth century.

Prior to the First World War, the population of the Russian Empire was steadily increasing at a rate of about 1.6% per annum. Fertility was still very high, but this was largely of offset by high mortality. Despite this somewhat primitive demographic régime, there was an increase of nearly 35 million (27.8%) in the 16 years between 1897 and 1913. In marked contrast, the next 13 years (1913–26) saw a net gain of only 5.8 million (3.6%) and an annual growth rate of barely 0.3%. These figures indicate the disastrous results of war, revolution and civil war. Had the pre-war rate of natural increase continued down to 1926, the population at that date would have been about 193 million. In fact it was only 165 million, indicating a 'population deficit' of 28 million for the period. This deficit comprised 16 million extra deaths, a shortfall of 10 million births and a net migration loss of 2 million.

Officially-published data for the inter-war period, exemplified by the figures for 1926 and 1940 in Tables 6.3 and 6.4, provide conflicting evidence of population trends. The total increase of 29.1 million (17.6%) in fourteen years indicates an average annual growth of about 12 per 1000, yet the natural increase rate was 23.7 per 1000 in 1926 and 13.2 per 1000 in 1940, this decline being due mainly to a fall in the birth rate. The fact that the average rate of population growth between 1926 and 1940 was lower than the natural increase rate at the beginning and end of the period can only mean that, at some time during this period, the rate of natural increase was well below the 1940 level. This situation may be attributed to the effects of forced collectivisation and low agricultural production which in some areas led to famine in the early 1930s. By the eve of the Second World War, these difficulties had been overcome and the Soviet population was expanding at a rate much more rapid than that of any west European country.

The Second World War proved a setback to Soviet population growth even more serious than that of 1913–26. Between 1940 and 1950 there was an actual decline of 15.6 million (8%) and, since this decade included five post-war years of rapid growth, the decline during the war must have been very much greater, possibly in the region of 30 million. Once again it is revealing to compare the actual situation, this time in 1950, with what might have been expected had there been no war. Even allowing for a continuation of the decline in birth rates visible before the war, the population of the USSR should by 1950, have reached 224

TABLE 6.3 POPULATION CHANGE, 1897–1985
(ALL FIGURES REFER TO THE PRESENT TERRITORY OF THE USSR)

Year	Population (millions)			Change (%)		
	Total	Urban	Rural	Total	Urban	Rural
1897	124.6	18.4	106.2	—	—	—
1913	159.2	28.5	130.7	+ 27.8	+ 54.9	+ 23.1
1926	165.0	33.0	132.0	+ 3.6	+ 15.8	+ 1.0
1940	194.1	63.1	131.0	+ 17.6	+ 91.2	− 0.8
1950	178.5	69.4	109.1	− 8.0	+ 10.0	− 16.7
1955	194.4	86.3	108.1	+ 8.9	+ 24.4	− 0.9
1960	212.3	103.8	108.5	+ 9.2	+ 20.3	+ 0.4
1965	229.2	121.7	107.5	+ 8.0	+ 17.2	− 0.9
1970	241.7	136.0	105.7	+ 5.5	+ 11.8	− 1.7
1975	253.3	153.1	100.2	+ 4.8	+ 12.6	− 5.2
1980	264.5	166.2	98.3	+ 4.4	+ 8.6	− 1.9
1985	276.3	180.1	96.2	+ 4.5	+ 8.4	− 2.1

Sources: Narodnoye khozyaystvo SSSR (Moscow, various years); Bond, A. R. and Lydolph, P. E.; 'Soviet Population Change and City Growth, 1970–79,' *Soviet Geography*, **20**, (October 1979).

TABLE 6.4 VITAL RATES 1913–1985
(ALL FIGURES PER 1000 OF THE POPU-LATION)

Year	Birth Rate	Death Rate	Natural increase
1913	47.0	30.2	16.8
1926	44.0	20.3	23.7
1940	31.2	18.0	13.2
1950	26.7	9.7	17.0
1955	25.7	8.2	17.5
1960	24.9	7.1	17.8
1965	18.4	7.3	11.1
1970	17.4	8.2	9.2
1975	18.1	9.3	8.8
1980	18.3	10.3	8.0
1985	19.4	10.6	8.8

Source: Narodnoye khozyaystvo SSSR za 60 let, (Moscow, various years).

million. Instead, it was only 179 million, indicating an enormous wartime deficit of 45 million. No detailed figures of their war losses have been published by the Soviet authorities, but close examination of the age and sex data in the 1959 census suggests that the deficit of 45 million involved 10 million deaths in the armed forces, 15 million extra civilian deaths (11 million men and 4 million women) and a deficit of births, resulting from low fertility and high infant mortality, of 20 million.

These events have had fundamental effects on the composition of the Soviet population in the post-war years, some of which are shown by the age-sex pyramids in Fig. 6.4. These effects were most marked in 1959 but were still visible in the 1980s. The Soviet population has a large excess of females over males. In 1959 there were 114.8 million women, but only 94 million men, a ratio of 122 women to every 100 men. By 1986 the position had improved considerably, but there were still 113 women to every 100 men and it will be some years before a more normal sex-ratio is achieved. This lack of balance between the sexes is visible in all age-groups born before 1940 and reaches its maximum among those born between 1909 and 1923, the majority of whom were of military age during the Second World War. Among those aged between 55 and 69 in 1979, there were only 52 men to every 100 women, a difference much too great to be accounted for solely by the fact that, even in peacetime, women live longer than men. Low fertility during the war years is reflected in the small numbers of both sexes in the 30–39 age group in 1979, while the large numbers aged 20–29 indicate accelerated natural increase in the early post-war years. The inset at ages 10–14 reflects fertility decline in the 1960s (see below). As a result of these factors, the Soviet population is a relatively youthful one, though less so than 20 years ago. In 1979 there

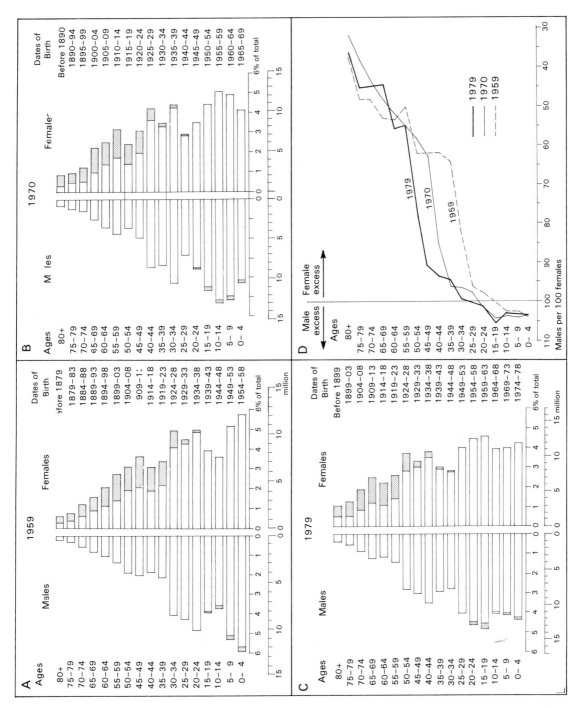

Fig. 6.4 A, B and C show the age and sex structure of the Soviet population in 1959, 1970 and 1979 respectively; D shows the sex ratio (males per 100 females) for each age group at the three census dates

were 65 million people (25%) below the age of 15 and only 25 million (9.5%) aged 65 or over.

A further general point is that, in common with most other countries, the USSR is becoming increasingly urbanised. The proportion of the population classed as urban rose steadily from the very low level of 15% in 1897 to 48% at the 1959 census. The landmark of 50% urban was passed in 1961 and 65.6% was reached in 1986. The rural population is now declining in numbers but remains very much larger (95.9 million in 1986) than is the case in other developed countries.

With the publication of the 1959 census results, an assessment of change over the previous 20 years became possible and the regular publication of demographic data in Soviet statistical handbooks permits a more detailed examination of more recent trends. Between 1939 and 1959, the total population of the USSR grew by only 18.1 million or 9.5%, an annual growth rate of little more than 0.4%. This overall change involved a large decline during the war and a rapid growth in the post-war years. Rural population declined in many areas and, over much of the European part of the country, more than offset the urban growth of the period, giving a net decline in places of 10% or more. Total growth well above the national average, however, occurred in the Asiatic regions. The population of Kazakhstan, for example, grew by 50.8% and that of the Far East by 69.7%.

Natural increase in the early post-war years rose well above that of the inter-war period; while the birth rate declined from 31.2 per 1000 in 1940 to 24.9 per 1000 in 1960, the death rate fell dramatically from 18.0 to the remarkably low level of 7.1 per 1000 at the latter date. Thus the natural increase rate throughout the 1950s was around 1.75% per annum, half as much again as in the inter-war years.

The net result of these trends, combined with large-scale migration from European to Asiatic areas, was a further increase in the proportion of the Soviet population living in Siberia, the Far East, Kazakhstan, Central Asia and the Transcaucasus. Between 1939 and 1959, the population of the European USSR rose by only 2.9%, while that of Transcaucasia grew by 18.4%, Central Asia by 29.8%, Siberia and the Far East by 36.9% and Kazakhstan by 50.8%. The European share of the total fell from 78.9 to 74.1%, that of Asiatic areas rose from 21.1 to 25.9%. In terms of absolute numbers, there was an increase of only 4.3 million in the population of the European sector, whereas the population of the Asiatic USSR grew by 13.8 million.

Population change 1959–1986

Detailed data for the period 1959–1986 are displayed in Tables 6.5 and 6.6. Over this 27-year period, the Soviet population grew by 70 million

TABLE 6.5 BIRTH, DEATH AND NATURAL INCREASE RATES (PER 1000) BY REPUBLICS, 1950–1985

	Birth Rate				Death Rate				Natural Increase Rate			
	1950	1965	1975	1985	1950	1965	1975	1985	1950	1965	1975	1985
USSR	26.7	18.4	18.1	19.4	9.7	7.3	9.3	10.6	17.0	11.1	8.8	8.8
RSFSR	26.9	15.7	15.7	16.5	10.1	7.6	9.8	11.3	16.8	8.1	5.9	5.2
Ukraine	22.8	15.3	15.1	15.0	8.5	7.6	10.0	12.1	14.3	7.7	5.1	2.9
Lithuania	23.6	18.1	15.7	16.3	12.0	7.9	9.5	10.9	11.6	10.2	6.2	5.4
Latvia	17.0	13.8	14.0	15.2	12.4	10.0	12.1	13.1	4.6	3.8	1.9	2.1
Estonia	18.4	14.6	14.9	15.4	14.4	10.5	11.6	12.6	4.0	4.1	8.3	2.8
Belorussia	25.5	17.9	15.7	16.5	8.0	6.8	8.5	10.6	17.5	11.1	7.2	5.9
Moldavia	38.9	20.4	20.7	21.9	11.2	6.2	9.3	11.2	27.7	14.2	11.4	10.7
Georgia	23.5	21.2	18.2	18.7	7.2	7.0	8.0	8.8	15.9	14.2	10.2	9.9
Azerbaydzhan	31.2	36.6	25.1	26.7	9.6	6.4	7.0	6.8	21.6	30.2	18.1	19.9
Armenia	32.1	28.6	22.4	24.1	8.5	5.7	5.5	5.9	23.6	22.9	16.9	18.2
Kazakhstan	37.6	26.9	24.1	24.9	11.7	5.9	7.1	8.0	25.9	21.0	17.0	16.9
Uzbekistan	30.9	34.7	34.5	37.2	8.8	5.9	7.2	7.2	22.1	22.8	27.3	30.0
Kirgiziya	32.4	31.4	30.4	32.0	8.5	6.5	8.1	8.1	23.9	24.9	22.3	23.9
Tadzhikistan	30.4	36.8	37.1	39.9	8.2	6.6	8.1	7.0	22.2	30.2	29.0	32.9
Turkmeniya	38.2	37.2	34.4	36.0	10.2	7.0	7.8	8.1	28.0	30.2	26.6	27.9

Source: Narodnoye khozyaystvo SSSR, Moscow, various years

TABLE 6.6 POPULATION CHANGE 1959–1986 BY REPUBLICS AND MAJOR ECONOMIC REGIONS

Region or Republic	Numerical change (000) Urban	Rural	Total	Percentage change Urban	Rural	Total
USSR	+ 82 952	− 12 995	+ 69 957	+ 83.0	− 11.9	+ 33.5
RSFSR	+ 43 657	− 17 111	+ 26 546	+ 70.9	− 30.6	+ 22.6
North	+ 2 061	− 666	+ 1 395	+ 81.3	− 32.1	+ 30.3
North-west	+ 2 531	− 652	+ 1 879	+ 56.4	− 36.9	+ 30.0
Centre	+ 9 137	− 5 060	+ 4 077	+ 59.8	− 48.3	+ 15.9
Volga–Vyatka	+ 2 483	− 2 374	+ 109	+ 77.3	− 47.1	+ 1.3
Black Earth Centre	+ 2 379	− 2 496	− 117	+ 112.4	− 44.2	− 1.5
Volga	+ 5 557	− 2 115	+ 3 442	+ 91.6	− 30.8	+ 27.2
North Caucasus	+ 4 393	+ 346	+ 4 739	+ 88.6	+ 5.2	+ 40.8
Ural	+ 4 674	− 2 215	+ 2 459	+ 46.1	− 30.0	+ 14.0
West Siberia	+ 4 596	− 1 489	+ 3 106	+ 80.3	− 26.9	+ 27.6
East Siberia	+ 2 975	− 573	+ 2 402	+ 87.1	+ 18.7	+ 37.1
Far East	+ 2 598	+ 219	+ 2 817	+ 79.6	+ 14.0	+ 58.3
Ukraine	+ 14 543	− 5 418	+ 9 125	+ 76.0	− 23.8	+ 21.8
Donets–Dnepr	+ 5 710	− 1 908	+ 3 802	+ 50.8	− 29.3	+ 21.4
South-west	+ 4 716	− 1 815	+ 2 901	+ 86.8	− 13.3	+ 15.2
South	+ 2 481	− 59	+ 2 422	+ 100.6	− 2.3	+ 47.8
Baltic	+ 2 725	− 721	+ 2 004	+ 82.8	− 21.7	+ 30.3
Lithuania	+ 1 343	− 451	+ 892	+ 128.4	− 27.1	+ 32.9
Latvia	+ 680	− 151	+ 529	+ 57.9	− 16.4	+ 25.3
Estonia	+ 428	− 83	+ 345	+ 63.3	− 15.9	+ 28.8
†Kaliningrad oblast	+ 274	− 36	+ 238	+ 69.4	− 16.7	+ 39.0
Transcaucasia	+ 4 369	+ 1 431	+ 5 800	+ 100.2	+ 27.8	+ 61.0
Georgia	+ 1 120	+ 70	+ 1 190	+ 65.4	+ 3.0	+ 29.4
Azerbaydzhan	+ 1 850	+ 1 161	+ 3 011	+ 104.7	+ 60.2	+ 81.4
Armenia	+ 1 399	+ 200	+ 1 599	+ 158.6	+ 22.7	+ 90.7
Central Asia	+ 7 685	+ 9 103	+ 16 788	+ 161.0	+ 102.4	+ 122.8
Uzbekistan	+ 5 016	+ 5 365	+ 10 381	+ 183.8	+ 99.8	+ 128.1
Kirgiziya	+ 911	+ 1 074	+ 1 985	+ 130.1	+ 78.4	+ 96.1
Tadzhikistan	+ 907	+ 1 761	+ 2 668	+ 140.4	+ 132.0	+ 134.7
Turkmenia	+ 851	+ 903	+ 1 754	+ 121.4	+ 110.8	+ 115.7
Kazakhstan	+ 5 156	+ 1 562	+ 6 718	+ 126.8	+ 29.8	+ 72.2
Belorussia	+ 3 838	− 1 885	+ 6 953	+ 154.7	+ 33.8	+ 24.2
Moldavia	+ 1 253	+ 10	+ 1 263	+ 195.2	+ 0.4	+ 43.8

† Figures for Kaliningrad oblast are also included in RSFSR totals
Source: Narodnoye khozyaystvo SSSR , Moscow, various years

or 33.5%, an average annual increase of about 1.1%. These figures denote a slow rate of growth since 1959 compared with that of the 1950s, mainly as a result of diminished fertility. Until about 1960, the birth rate was declining relatively slowly and this trend, combined with a continuing reduction in the death rate, gave an accelerating natural increase, which reached a peak of 17.8 per thousand in 1960. After 1960, however, the death rate edged slowly upwards and the fall in the birth rate accelerated. Thus, by 1969, the rate of natural

increase had fallen to 8.9 per thousand, a level significantly below that of the inter-war years and precisely half that of 1960. During the 1970s and early 1980s there has been a slight recovery in the birth rate but the death rate has also continued to rise; as a result, the rate of natural increase has changed very little: it was 8.8 per thousand in both 1975 and 1985.

One result of these trends has been an actual reduction in the number of people added to the Soviet population each year. Between 1950 and

1960, this annual increment averaged 3 380 000, with a peak of 3 900 000 in 1959. During the 1960s the average was 2 940 000 and it fell again to an average of 2 280 000 during the 1970s. There was a slight increase to an average of 2 383 000 over the period 1980–86. Any forecast of future trends presents considerable difficulties, but a further reduction in fertility seems likely, at least in those Asiatic areas now characterised by exceptionally high birth rates. At the same time, mortality will probably show a further slow increase as the population continues to age. Thus it seems likely that, within the next decade, the rate of natural increase in the Soviet Union will decline still further, possibly even to west European levels.

Within these general trends, there are marked regional contrasts as well as differences between urban and rural populations. These contrasts are illustrated by Fig. 6.5 as well as by Tables 6.5 and 6.6. Over the country as a whole rapid urban growth continued, the urban population increasing by 83% in 27 years. The rural population, on the other hand, showed a net decline of 11.9%, compounded from decreases of more than 40% in some areas (Centre, Volga–Vyatka and Black Earth Centre regions) and increases of 100% or more in others (Uzbek, Tadzhik and Turkmen republics). Table 6.7 summarises numerical and percentage changes and shows that the net addition to the Soviet population between 1959 and 1986 was divided about 46/54 between the European and Asiatic sections of the country.

By 1986, the proportion of the Soviet population living in Europe had fallen still further to 66.7%, compared with 69.2% in 1979, 73.2% in 1959, 78.6% in 1939 and 82.1% in 1913. Migration played an important role in bringing about this change, but regional variations in the rate of natural increase were also of major significance as a closer examination on a regional basis clearly illustrates.

EUROPEAN USSR

As already indicated, this part of the country has experienced a rate of growth appreciably below the national average; between 1959 and 1986 its population increased by only 21%. This was the result of a relatively slow natural increase, due mainly to low fertility, combined with emigration to other parts of the country. European USSR includes three regions, the Centre, Urals and Donets–Dnepr, which have throughout the Soviet period contained the bulk of the country's industrial capacity and about a quarter of its population. All three had rates of growth below the national average; between 1959 and 1986. This applied even to their urban populations, indicating that industry in these areas, while still expanding, was doing so a good deal less rapidly than in areas of more recent industrial development such as the Volga region. The highest rates of urban growth were in fact recorded in such regions as the Black Earth Centre, Belorussia, Lithuania and Moldavia, formerly industrial backwaters which, over the past 25 years, have been the scene of numerous industrial developments.

European USSR is also characterised by a very marked decline in rural populations. This reached its extreme in the Centre and Volga–Vyatka regions, long the most densely settled parts of the mixed forest zone, and was considerable not only in regions of low agricultural potential, like the North-west but also in highly productive agricultural regions such as the Black Earth Centre. Indeed in the Volga–Vyatka and Black Earth Centre regions, though urban growth was up to the national level, total population has shown practically no growth since 1959. Rural depopulation also occurred, though to a less marked degree, in the Ukraine, the Baltic republics, Belorussia and the Urals. Three European regions provide exceptions to this general picture of rural decline. The South (i.e. the southern Ukraine) experienced only a small reduction in its rural population, despite a low natural increase. In the Crimea, the rural element grew by nearly 70%, partly as a result of immigration into that area. The North Caucasus had a small rural increase which may have involved some migration gain but was also due to a high natural increase among non-Slav minorities. Moldavia, a region of intensive and prosperous agriculture, was also favoured by a high birth rate.

SIBERIA AND THE FAR EAST

These regions show trends very different from those of earlier periods. Much of their population growth over the past 100 years has been the result of immigration from the European USSR and the high fertility associated with a youthful immigrant population. Thus their rate of increase in the interwar years was twice, and that between 1939 and 1959, five times, the national average.

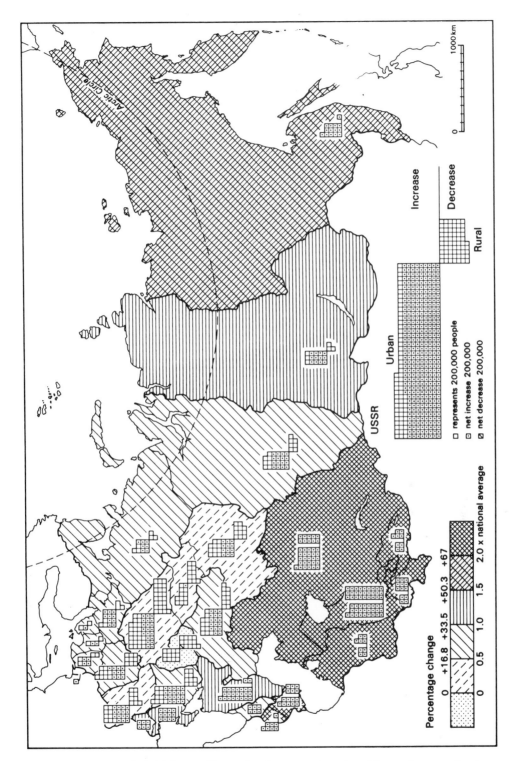

Fig. 6.5 Population change, 1959–86, by republics and major economic regions. The small squares (each representing 200 000 people or 0.7% of the total population) show numerical increases (above the line) and decreases (below the line) in the urban (left) and the rural (right) populations for each area. Squares with dots indicate the net increase. The shadings show **percentage** increase in relation to the national average of 33.5%.

TABLE 6.7 POPULATION CHANGE, 1959–86 BY MAJOR GEOGRAPHICAL DIVISIONS

	Urban 000	%	Rural 000	%	Total 000	%
European USSR	+ 55 574	+ 74.7	− 23 248	− 29.2	+ 32 326	+ 21.0
Asiatic USSR	+ 27 378	+ 106.9	+ 10 253	+ 34.8	+ 37 631	+ 68.4
Siberia and Far East	+ 10 168	+ 82.0	− 1 843	− 18.1	+ 8 325	+ 36.9
Kazakhstan	+ 5 156	+ 126.8	+ 1 562	+ 29.8	+ 6 718	+ 72.2
Central Asia	+ 7 685	+ 161.0	+ 9 103	+ 102.3	+ 16 788	+ 122.8
Transcaucasia	+ 4 369	+ 100.2	+ 1 431	+ 27.8	+ 5 800	+ 61.0
USSR	+ 82 952	+ 83.0	− 12 995	− 11.9	+ 69 957	+ 33.5

Soviet citizens and tourists watch the guard leaving Red Square after the changing-over ceremony at Lenin's tomb. In the background is GUM, the State Universal Store which occupes a pre-revolution building

Since 1959, however, their rate of increase has been only a little above that of the USSR as a whole, involving below-average growth in West Siberia and rapid increase in the Far East. The present growth rate in Siberia and the Far East as a whole represents an annual addition to the population of these vast territories (57% of the land area of the USSR) of only about 400 000.

Such a slow build-up of population presents serious problems in the development of the regions' vast natural resources.

Despite important agricultural developments, notably the virgin lands scheme in south-west Siberia, there was a decline of 1.8 million in the rural population of Siberia and the Far East between 1959 and 1986. Furthermore, despite major industrial developments, urban growth has been slightly below the national average and can be accounted for largely by natural increase and rural-urban movement. From published data and from numerous comments in the Soviet press, it appears that, since the labour controls characteristic of the Stalin period were relaxed, the rate of migration to the eastern territories has declined drastically and the Soviet authorities are finding it increasingly difficult to maintain in these regions a permanent labour force large enough to carry out plans for their economic development. Furthermore, the natural increase, which until 1960 was well above the national average, mainly because of high fertility, has also undergone a marked decline. In 1950, birth rates in Siberia and the Far East were well above the national average; today they are little above those of European areas.

KAZAKHSTAN AND CENTRAL ASIA

These five republics provide a strong contrast with the areas discussed so far, for their populations are expanding more rapidly than those of any other part of the Soviet Union. Between 1959 and 1986, the population of Kazakhstan and Central Asia rose by 102%, from 23 to 46.5 million. In these regions there is no question at present of rural population decline; in fact, since 1959, the increase in rural numbers (10.7 million) has not been very far behind urban growth (12.8 million) overall, and has actually exceeded urban growth in the four Central Asian republics. This unique situation may be attributed mainly to the high rate of natural increase which, throughout practically the whole of the region, is double and in places three times the national average. The underlying cause is simple: while death rates have been brought down to levels similar to those prevailing in other parts of the Soviet Union, birth rates remain extremely high and, except in Kazakhstan, have shown little or no decline over the last 30 years. The fact that these regions are inhabited mainly by non-Slav peoples is the prime

The Soviet people are avid readers. Bookshops and kiosks are constantly busy as at this state farm in the Murmansk region. The abacus is still widely used though electronic tills are in use in city stores.

reason for this situation. High fertility is part of the unique, essentially non-European culture which prevails, at least in rural areas, throughout Central Asia. Even so, natural increase alone cannot account for the very high rates of population growth, averaging 3% per annum. There has, in addition, been considerable immigration from other parts of the country, mainly to urban areas but, in Kazakhstan at least, to rural areas as well.

Kazakhstan shows somewhat different trends from the other four republics yet, at the same time, clearly illustrates the factors at work in the region as a whole. Between 1959 and 1986, the population of Kazakhstan rose by 72.2%, the urban element increasing by no less than 126.8% and the rural by 29.8%. Its birth rate was much lower than those of the other four republics, reflecting the fact that Russians are the largest ethnic group, and the rate of natural increase declined from 3% in 1960 to 1.7% in 1985. Considerable immigration has been an important element in the growth of the republic's population. The other four republics still have birth rates well above 30 per thousand and natural increase rates of 2.4 to 3.3% per annum.

Thus Kazakhstan and Central Asia have experienced not only rapid natural increase but also, in some years at least, a sizeable net migration gain. As a result, their share of the total Soviet population rose from 8.6% in 1939 to 11% in 1959 and 16.7% in 1986. Since rapid natural increase is found mainly among non-Slav groups, it follows that their relative size is also increasing.

TRANSCAUCASIA

Transcaucasia shows a rate of population growth second only to that of Central Asia with a 61% rise between 1959 and 1986, including 27.9% rural and 100% urban increase. In this case, however, rapid natural increase among predominantly non-Slav populations, particularly those of Azerbaydzhan and Armenia, accounts for practically all the increase, and immigration has played only a minor role, taking place mainly to oil-producing districts. The Transcaucasus has fewer Slavs among its population than any other major region of the USSR. Total numbers are, of course, quite small, but the economic importance of Transcaucasia is indicated by its high density and rapid growth and the fact that it contains twice as many people as the whole Far Eastern region.

Types of population change

Regional contrasts in the nature and scale of recent population trends are further illustrated by the typology of population change shown in Fig. 6.6, where 28 areal units—the republics and major economic regions which constitute the USSR—have been allocated among seven types according to their rates of total, urban and rural population change in relation to the average values of the Soviet Union as a whole. The seven types, and the areas belonging to each, are listed in Table 6.8, and selected data are given for each type in Table 6.9.

Type 1. Between 1959 and 1986, the nine areas in this category recorded a total increase of 34.1 million (79.5%) which, in numerical terms, constituted slightly less than half the entire growth of the Soviet population. Both rural and urban elements grew rapidly, the former by 47.9% and the latter by no less than 127.2%. Practically all this growth occurred in the essentially non-Slav areas of Central Asia and the Transcaucasus, supported by high rates of natural increase and, in Central Asia at least, by a significant net migration gain. The only part of the European USSR included in

TABLE 6.8 TYPES OF POPULATION CHANGE IN THE USSR, 1959–1986

1. *Total increase above average; urban increase above average; rural increase*: Kazakh SSR, Kirgiz SSR, Tadzhik SSR, Turkmen SSR, Uzbek SSR, Armenian SSR, Azerbaydzhan SSR, Moldavian SSR, North Caucasus region (RSFSR)
2. *Total increase above average; urban increase below average; rural increase*: Far East region (RSFSR)
3. *Total increase above average; urban increase above average; rural decrease*: East Siberian region (RSFSR), Southern region (Ukraine)
4. *Total increase above average; urban increase below average; rural decrease*: Kaliningrad oblast (RSFSR)
5. *Total increase below average; urban increase above average; rural decrease*: Lithuanian SSR, Belorussian SSR, Black Earth Centre and Volga regions (RSFSR), South-western region (Ukraine)
6. *Total increase below average; urban increase below average; rural increase*: Georgian SSR
7. *Total increase below average; urban increase below average; rural decrease*: Northern, North-western, Central, Volga–Vyatka, Ural and West Siberian regions (RSFSR), Donets–Dnepr region (Ukraine), Latvian SSR, Estonian SSR.

this category—the North Caucasus region—experienced a relatively small rural growth of about 5%. In all Type 1 areas, rapid urban growth in regions still characterised by rather low levels of urbanisation was combined with continuing agricultural development. The trends in these regions represent a distinct southward shift of the centre of gravity of the Soviet population. Type 1 regions' share of the total rose from 20.6% in 1959 to 27.6% in 1986; particularly striking was the rise from 23.7% to 39.9% in their share of the rural element.

Type 2. The only region to fall into this category was the Far East. Urban growth below the national average reflects the rather slow pace of industrial development; the 'rural' population rose by 14%, but much of this increase occurred in non-agricultural settlements too small to be placed in the 'urban' category.

Type 3 applies to two very different areas, the large East Siberian region and the small Southern section of the Ukraine. In East Siberia, pronounced rural decline combined with urban growth marginally above the national average to produce total population growth very slightly in excess of

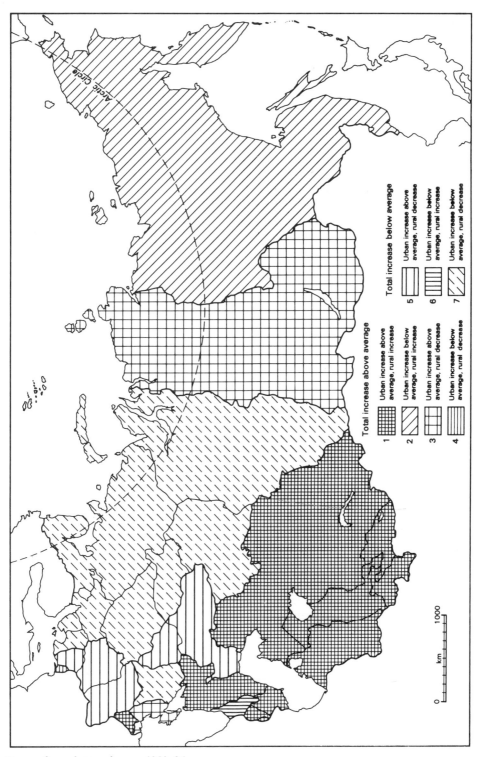

Fig. 6.6 Types of population change, 1959–86.

the rate for the USSR as a whole. In the South, the total growth rate was a good deal higher, involving a doubling of urban numbers and a very small rural decline.

Type 4 contains only one area, the Kaliningrad oblast, which is recorded separately as a result of its unique status as a detached section of the RSFSR. Such a combination of growth rates can only occur in highly urbanised districts—the oblast was 65% urban in 1959 and 79% urban in 1986. Other such cases would no doubt be visible from an analysis of data at the oblast level.

Type 5. The areas in this group experienced a rural decline of 10.4 million (31.4%) between 1959 and 1986 and thus, along with those in Type 7, constituted a zone of marked rural depopulation. This process affected not only regions of low agricultural potential, like Lithuania and Belorussia, but also some of the most highly productive agricultural areas such as the Southwest, Volga and Black Earth Centre regions. Urban growth, however, was rapid, amounting to 113.5% for the group as a whole and over 150% in Belorussia. This rapid urban growth was associated with low levels of urbanisation at the beginning of the period. Type 5 regions had only 34% of their population living in towns in 1959; by 1986 the figure was 62%, not far below the national average.

Type 6. The Georgian SSR was the only area of this type. Along with the rest of Transcaucasia, the republic continued to experience rural growth, though at a much slower rate than in Azerbaydzhan and Armenia, while urban growth, in contrast with that of the other two republics, was well below average. As Table 6.5 shows, Georgia has a low rate of natural increase and there is little immigration.

Type 7. In 1986, some 40% (112.4 million) of the Soviet population lived in Type 7 regions, a significant reduction from the 45% living in the same regions in 1959. This group contains many of the most densely settled and heavily industrialised parts of the country; 77.2% of the population lived in towns by 1986. Particularly striking is the fact that this zone of slow growth, with its low rate of natural increase and overall net migration loss, included not only the main areas of nineteenth-century economic development, such as the Centre and Donets–Dnepr regions, but also the Urals and West Siberia, which had been areas of particularly vigorous growth prior to 1959

Type 7 regions had a higher rate of rural depopulation and a slower rate of urban growth than any other category. As a result, total growth, at 18.7%, was less than a quarter as rapid as that of Type 1 regions, which had a numerical increase (34.1 million) roughly double that of Type 7 (17.7 million). Despite their 45% share of the Soviet population in 1959, Type 7 regions have contributed only a quarter of total growth since that date.

Summary of trends

To summarise, we may note the following main points:

(i) The population of the Soviet Union is still expanding at a significant rate but, since the early 1960s, there has been a marked falling off in the rate of growth, which is now very close to that of the United States and may well, in the not-too-distant future, approach that of the majority of countries in Europe. The slowing down of population growth is now presenting serious problems to Soviet economic planners, and these are likely to become greater with the passage of time. In particular, the size of the annual addition to the labour force has now begun to decline and economic expansion will depend to an increasing extent on improvements in labour productivity more than on increases in the numbers at work.

(ii) There are pronounced regional contrasts in the rate of population growth which are due, to a large degree, to differences in the birth rate between the Slavs and the other Soviet peoples. As a result, the latter, though still very much in a minority, are becoming a progressively larger proportion of the total.

(iii) Migration movements also play a very important role. Throughout the country, there is large-scale movement from the countryside towards the towns, so that urban populations are growing more rapidly than rural almost everywhere. At the same time, there are major interregional movements, mainly from the European to the Asiatic section of the country, though the direction of these movements has changed. Until the 1950s, migration on a massive scale took place to Siberia, but movement in this direction in the 1960s and 1970s has been on a much smaller scale and has taken second place to movements into Kazakhstan and Central Asia.

(iv) The combined effects of natural increase and migration have produced a clear regional

TABLE 6.9 DATA FOR THE SEVEN TYPE-REGIONS IDENTIFIED IN TABLE 6.8

	1	2	3	4	5	6	7	Total
Total Population								
1959 (000)	42 923	4 834	11 539	611	50 211	4 044	94 665	208 827
1959 (% share)	20.6	2.3	5.5	0.3	24.0	1.9	45.3	100.0
1986 (000)	77 041	7 651	16 363	849	59 282	5 234	112 364	278 784
1986 (% share)	27.6	2.7	5.9	0.3	21.3	1.9	40.3	100.0
1959–86 change (000)	+ 34 118	+ 2 817	+ 4 824	+ 238	+ 9 071	+ 1 190	+ 17 699	+ 69 957
1959–86 change (%)	+ 79.5	+ 58.3	+ 41.8	+ 38.9	+ 18.1	+ 29.4	+ 18.7	+ 33.5
Urban Population								
1959 (000)	17 091	3 265	5 879	395	17 146	1 713	54 489	99 978
1959 (%)	39.8	67.5	50.9	64.6	34.1	42.4	57.6	47.9
1959 (% share)	17.1	3.3	5.9	0.4	17.1	1.7	54.5	100.0
1986 (000)	38 827	5 863	11 335	669	36 615	2 833	86 789	182 931
1986 (%)	50.4	76.6	69.3	78.8	61.8	54.1	77.2	65.6
1986 (% share)	21.2	3.2	6.2	0.4	20.0	1.5	47.4	100.0
1959–86 change (000)	+ 21 736	+ 2 598	+ 5 456	+ 274	+ 19 469	+ 1 120	+ 32 300	+ 82 953
1959–86 change (%)	+ 127.2	+ 79.6	+ 92.8	+ 69.4	+ 113.5	+ 65.4	+ 59.3	+ 83.0
Rural Population								
1959 (000)	25 832	1 569	5 660	216	33 065	2 331	40 176	108 849
1959 (%)	60.2	32.5	49.1	35.4	65.9	57.6	42.4	52.1
1959 (% share)	23.7	1.4	5.2	0.2	30.4	2.1	36.9	100.0
1986 (000)	38 214	1 788	5 028	180	22 667	2 401	25 575	95 853
1986 (%)	49.6	23.4	30.7	21.2	38.2	45.9	22.8	34.4
1986 (% share)	39.9	1.9	5.2	0.2	23.6	2.5	26.7	100.0
1959–86 change (000)	+ 12 382	+ 219	− 632	− 36	− 10 398	+ 70	− 14 601	− 12 996
1959–86 change (%)	+ 47.9	+ 14.0	− 11.2	− 16.7	− 31.4	+ 3.0	− 36.3	− 11.9

dichotomy within the USSR. Central Asia, Kazakhstan and most of Transcaucasia have high natural increase rates, a net migration gain and rapid growth of both rural and urban populations. The greater part of the European USSR, together with much of Siberia, have low fertility, net migration loss, rural depopulation and relatively slow urban growth.

(v) Urbanisation continues at a rapid pace, but the rural element, though now declining in absolute as well as in relative terms, remains much larger than in most other industrially-developed countries.

(vi) For a long time now, the population of the Asiatic regions has grown more rapidly than that of the European zone. Nevertheless, the European USSR remains overwhelmingly predominant in population as it is in industrial and agricultural production, and this balance is changing only slowly. One major problem now facing Soviet economic planners is that of establishing, and maintaining, in Siberia and the Far East, a population large enough to permit the development of the vast resources of those eastern regions.

BIBLIOGRAPHY

Akiner, S. (1983), *Islamic peoples of the Soviet Union*, Kegan Paul International, London.

Azrael, J. R (ed.) (1978), *Soviet nationality policies and practices*, Praeger, New York.

Anderson, B. A. (1977), 'Data source in Russian and Soviet demography,' in Kosinski, *op. cit.*

Baldwin, G. (1979), *Population projections by age and sex for the republics and major economic regions of the USSR, 1970–2000*, Bureau of the Census Washington D. C.

Ball, B. and Demko, G. J. (1978), 'Internal migration in the Soviet Union,' *Economic Geography*, **54**, pp. 95–113.

Bond, A. R. and Lydolph, P.E. (1979), 'Soviet population change and city growth, 1970–79: a preliminary report,' *Soviet Geography*, **20**, pp. 461–488.

Bond, A. R. (1981), 'Some comments on Soviet population change, 1970–1979,' *Soviet Geography*, **22**, pp. 532–537.

Brook, S. I. (1980), 'Demographical and ethnogeographical changes in the USSR according to post-war data up to 1979,' *Geojournal* Supplementary Issue **1**, pp. 7–21.

C.S.P.P. (Centre for the Study of Population Problems, University of Moscow) (1973), *Problemy narodonaseleniya*, Statistika, Moscow.

Czap, P. (1977), 'Russian history from a demographic perspective,' in Kosinski, *op. cit.*

Demko, G. J. (1977), 'Demographic research on Russia and the Soviet Union,' in Kosinski, *op. cit.*

Demko, G. J. and Casetti, E. (1970), 'A diffusion model for selected demographic variables: an application to Soviet data,' *Ann. Assoc. Amer. Geogr.*, **60**, pp. 533–539.

Desfosses, H. (1977), 'The USSR and the world population crisis,' in Kosinski, *op. cit.*

Dewdney, J. C. (1971), Population changes in the Soviet Union, 1959–1970, *Geography*, **56**, pp. 325–330.

Dewdney, J. C. (1982), 'Recent population changes in the Soviet Union,' in A. Findlay (ed) *Recent National Population Change*, pp. 69–87. Population Geography Study Group, Institute of British Geographers.

Eason, W. (1979), 'Demographic divergences at republican level' in NATO Economic Directorate, *Regional Development in the USSR*, pp. 119–138, Brussels.

Feshbach, M. (1979), 'Propects for outmigration from Central Asia and Kazakhstan in the next decade,' *Soviet Economy in a Time of Change*, US Govt. Printing Office, Washington DC.

Feshbach, M. (1984) and Rapaway, S. (1976) 'Soviet population and manpower trends and policies,' *Soviet Economy in a New Perspective*, US Govt. Printing Office, Washington DC.

Feshbach, M. (1984), 'The age structure of Soviet population: preliminary analysis of unpublished data,' *Soviet Economy*, **1**, pp. 177–193.

Field, N. C. (1963), 'Land hunger and rural depopulation in the USSR,' *Ann. Ass. Amer. Geogr.* **53**, pp. 465–478.

Garry, R. (1962), 'Réflexions sur l'évolution de la population sovietique de 1939 a 1959,' *Rev. Canadienne Geog.*, **16**, pp. 3–12.

Grandstaff, P. J. (1977), Estimates of USSR population by republics, 1951–73, *Soviet Geography*, **18**, pp. 258–261.

Harris, C. D. (1970), *Cities of the Soviet Union*, Rand McNally, Chicago.

Harris, C. D. (1970), 'Population of cities of the Soviet Union in 1897, 1926, 1939, 1959 and 1967,' *Soviet Geography*, **11**, whole issue.

Harris, C. D. (1971), 'Urbanization and population growth in the Soviet Union, 1959–1970,' *Geographical Review*, **41**, pp. 102–124.

Konstantinov, O. A. (1981), 'Some shifts in the population of the USSR' *Soviet Geography*, **22**, pp. 407–418

Kosinski, L. A. (ed.) (1977), *Demographic developments in Eastern Europe*, Praeger, New York.

Kovalev, S. A. (1976), Farewell to the rural scene, *Geographical Magazine*, **48**, pp. 427–432.

Kovalev, S. A. and Kovalskaya, N. Ya (1971), *Geografiya naseleniya*, University of Moscow.

Kozlov, V. I. (1980), 'Ethnic processes and trends in the ethnic compostion of the population in the USSR,' *Geojournal* Supplementary Issue **1**, pp. 23–30.

Krotki, K. J. (1977), 'Fertility and KAP surveys in eastern Europe and the Soviet Union,' in Kosinski, *op. cit.*

Kvasha, A. and Kiseleva, G. P. (1973), 'The impact of age structure on the growth of population in the USSR,' in *CSPP op. cit.*

Larmin, O. V. *et al.* (1972), 'Social-demographic aspects of urbanization in the USSR,' *Soviet Geography*, **13**, pp. 99–108.

Lewis, R. A. (1969), 'The post war study of internal migration in the USSR,' *Soviet Geography*, **10**, pp. 157–166.

Lewis, R. A. and Rowland, R. H. (1969) 'Urbanization in Russia and the USSR, 1897–1966,' *Ann. Assoc. Amer. Geogr.*, **59**, pp. 776–796.

Lewis, R. A., Rowland, R. H. and Clem, R. S. (1976), *Nationality and population change in Russia and the USSR: an evaluation of census data, 1897–1970*, Praeger, New York.

Listengurt, F. (1976), 'Soviets seek the city lights,' *Geographical Magazine*, **48**, pp. 492–496.

Lorimer, F. (1946), *The population of the Soviet Union—history and prospects*, League of Nations, Geneva.

Lydolph, P. E. (1972), 'Manpower problems in the USSR,' *Tijdschrift voor Econ. en Soc. Geografie*, **63**, pp. 331–344.

Lydolph, P. E., Johnson, R. and Mintz, J. (1978), 'Recent population trends in the USSR,' *Soviet Geography*, **19**, pp. 505–539.

Matthews, M. (1972), *Class and society in Soviet Russia*, Lane/Penguin, London.

Mickiewicz, E. (1973), *Handbook of Soviet social science data*, Free Press (Collier-Macmillan), New York.

Moiseyenko, V. M. (1973), 'The role of migration in the formation of urban population in the USSR,' in CSPP, *op. cit.*

Parker, W. H. (1973), *The Russians*, David anc Charles, Exeter.

Perevedentsev, V. I. (1969), 'Contemporary migration in the USSR,' *Soviet Geography*, **10**, pp. 192–208.

Pokshishevskiy, V. V. (1971), *Geografiya naseleniya SSSR*, Prosveshcheniye, Moscow.

Pokshishevskiy, V. V. (1972), Evaluation of the Soviet population census, 1970, *Geoforum*, **9**, pp. 3–60.

Pokshishevskiy, V. V. (1980), 'Soviet cities: progress in urbanization in the seventies,' *Geojournal* supplementary Issue **1**, pp. 35–44.

Pressat, R. (1985), 'Historical perspective on the population of the Soviet Union,' *Population and Development Review*, **11**, 315–334.

Rapaway, S. (1979), 'Regional employment trends in the USSR, 1950 to 1975,' *Soviet economy in a time of change*, US Govt. Printing Office, Washington DC.

Rapaway, S. and Baldwin, G. (1985), 'Demographic trends in the Soviet Union, 1950–2000' in *Soviet Economy in the 1980s*, US Govt. Printing Office, Washington DC.

Roof, M. K. and Leedy, F. A. (1959), 'Population redistribution in the Soviet Union,' *Geographical Review*, **49**, pp. 208–221.

Rowland, R. H. (1982), 'Regional migration and ethnic Russian population change in the USSR,' *Soviet Geography*, **23**, pp. 557–583.

Rowland, R. H. (1986), 'Changes in metropolitan and large-city population of the USSR, 1979–85,' *Soviet Geography* **27**, pp. 638–658.

Rybakovskiy, L. L. (1975), 'Inter-regional migration analysis,' *Soviet Geography*, **16**, pp. 435–452.

Sagers, M. J. (1984), 'Regional distribution of industrial employment in the USSR', *Soviet Geography*, **25**, pp. 166–176.

Shabad, T. (1975), 'Intercensal migration,' *Soviet Geography*, **16**, pp. 466–472.

Shabad, T. (1979), 'Preliminary results of the 1979 Soviet census,' *Soviet Geography*, **20**, pp. 440–456.

Shabad, T. (1980), 'Ethnic results of the 1979 Soviet census,' *Soviet Geography*, **21**, pp. 440–487.

Shabad, T. (1985), 'Population trends in Soviet cities', *Soviet Geography* **26**, pp. 109–153.

Slater, P. B. (1975), 'A hierarchical regionalisation of RSFSR administrative units using 1966–69 migration data,' *Soviet Geography*, **16**, pp. 453–465.

Smith, H. (1976), *The Russians*, Sphere Books, London.

Sokoloff, A. N. (1972), 'Rural and urban society in the USSR,' *FAO Monthly Bulletin of Agricultural and Economic Statistics*, **21**.

Stanley, E. (1968), *Regional distribution of Soviet industrial manpower, 1940–60*, Praeger, New York.

Thomas, C. (1972), Urbanisation and population change in Russia, *Scottish Geographical Magazine*, **88**, pp. 196–207.

Urlanis, B. Ts. (1974), *Problemy dinamiki naseleniya SSSR*, Nauka, Moscow.

Valenti, D. I. (ed.) (1973), *Narodnonaseleniye: prikladnaya demografiya* Statistika, Moscow.

Wädekin, K. E. (1966), 'Internal migration and the flight from the land in the USSR, 1939–59,' *Soviet Studies*, **18**, pp. 131–152.

Wixman, R. (1981), 'Territorial russification and linguistic russianization in some Soviet republics,' *Soviet Geography*, **22**, pp. 667–675.

Wixman, R. (1984), 'Demographic trends among Soviet Moslems,' *Soviet Geography*, **25**, pp. 46–60.

Wixman, R. and Caro, P. (1981), 'Territorial differences in population growth in the USSR, 1970–79,' *Soviet Geography*, **22**, pp. 155–161.

7 Agriculture

Throughout the history of the Soviet Union and the preceding centuries, farming has been a troubled sector of the economy and modernisation attempts have failed repeatedly to make agriculture efficient. To achieve success where earlier government efforts have failed would be a major success for Mikhail Gorbachev's perestroika campaign.

In the development of the reconstruction of the economy in the late eighties the emphasis has been on giving greater opportunity to farm managers and workers to rely on their own initiative in developing farm enterprises to meet the needs of the population.

The difficulties facing reconstruction are enormous, the long history of peasant subjection and the mismanagement of previous attempts at reform being grafted on to the severity of the physical conditions. Thus before looking at present patterns on the land it is necessary to consider the legacy of the past.

Agriculture has been practised in lands now included in the Soviet Union since very early times, the oases of central Asia having supported sedentary farming while stock rearing was carried out in the mountain lands, both long before farming spread about 5000 BC from the Mediterranean and Near East lands to southern Europe. Early cultivation in the afforested lands was based on forest-fallow methods in which natural or re-grown vegetation was burnt to clear it and fertilise the ground with the ashes, and seeds sown with the aid of digging sticks to turn the soil. After a few crops the land would be allowed to revert to rough pasture or scrub while fresh clearings were prepared. Similar methods were used in the more open country where the burning of the grass would precede cultivation, but the exposure of the steppes of Russia to the invasions of successive waves of nomads, described in Chapter 1, delayed the development of agriculture in those areas.

Thus it was in the forest-steppe area, somewhat more protected from raids than the steppes, and possessed of easily worked loess soils, that agriculture prospered more steadily, based on primitive wheats, peas, lentils and flax, oxen, sheep and pigs. Throughout the Neolithic Age, as already noted (Chapter 1) and the Bronze Age (roughly 3000–2000 BC) there was a slow diffusion of improvements, especially in the southern European areas. The light plough or *ard* of the Bronze Age was followed in the Iron Age of the last millenium BC by a wide range of iron tools, the use of which spread also from southern Europe into the interior. By the second century AD such improvements had been carried as far as western Siberia, so that diffusion of farming had spread in a great semi-circle from the eastern lands in southerly latitudes and back round to the east in more northerly and harsher climates, but with new technology added on the way. The continuing evolution of agricultural methods in Kievan Rus and further north in the forest lands has already been mentioned (Chapter 1), leading to the establishment of the typically Russian form of rural organisation centred on the commune or *mir*, in which land use was managed on a co-operative basis. The elders of the commune regulated the use of meadows, forests, rivers and other resources as well as supervising the strip cropping in the arable fields, including periodical rotation of the strips among the villagers, and the primitive rotation of crops. The latter involved mainly an autumn-sown cereal, which was wheat in the more southerly areas and rye elsewhere, and a spring-sown crop, barley.

It was only as serfdom was being gradually eliminated in western Europe that it became fully entrenched in Russia, following the granting by the tsars of increased powers and lands to favoured gentry in return for military support and colonisation, which required greater control and subjection of the peasantry. Serfdom was strengthened by successive rulers of Russia during the seventeenth and eighteenth centuries by, for example, giving owners the right to pursue and recapture runaway serfs, and by extending the practice into areas like the Ukraine where it had previously not prevailed. The landowners were legally required, from 1734, to assist their serfs through periods of famine, but the advantages of the system lay clearly with the owners, whose land was worked by the serfs. Two main forms of relationship between serf and owner evolved. Under the *obrok* system the serf made payments to the landowner in return for an allotment of land for his own use, whereas under *barshchina* the serf earned his allotment solely by labour. In either case the owner's land was cultivated by the serfs.

There were many rebellions, some of which, like that led by Pugachev (1773–75), threatened the whole fabric of the Russian state, but most were minor affairs, easily suppressed. Many serfs managed to escape to the freer lands of Siberia, or the southern steppes, where they joined Cossack bands, but there was little fundamental change until the latter half of the nineteenth century. The three-field system continued to prevail, the commune continued to regulate local affairs and the landowner continued to hold almost total power over his serfs or 'souls' as they were called. As the nineteenth century progressed, however, it became increasingly apparent to the more perspicacious landowners that serfs did not, and could not be expected to provide efficient labour, and that with increasing commercialisation, including opportunities for the export of crops abroad, as well as sales in the expanding industrial towns, their interest would be served by a change in the system. The entrepreneurs in the industrial towns had even more reason for wanting an end to serfdom, because it restricted the supply of labour to the towns.

Proposals for the emancipation of the serfs were eventually approved by Tsar Alexander II. The Emancipation Act of 1861 conferred legal freedom on the majority of the 47 million (out of a total population of 74 million) serfs and compelled the landlords to give up certain portions of their estates to the serfs. The areas of land allotted to the peasants were, however, too small to give them economic security and independence and many soon fell into debt to the landlords, or to neighbouring peasants who were more successful. Some of the latter were able to buy up holdings that fell vacant and become more prosperous, but even in 1917 about one third of the land remained in the hands of the gentry. Furthermore, the small holdings allotted to the freed serfs, who had to make payments for them against loans advanced for 49 years by the state, did not become the property of the peasants themselves—land titles were vested in the commune The smallness of the holdings, the high valuation put upon them, and the vesting of the land in the commune made the peasants feel that they had been cheated out of the land which they felt should be theirs. Many left the land to seek employment in the growing industrial centres but many more remained to fall into debt to the landlords and the more successful peasants who developed into the *kulak* class which was later to become the chief target of the revolutionaries.

The reforms of 1906
Continued unrest led eventually to further reforms. The vigorous Prime Minister, Stolypin, brought in legislation to assist modernisation of farms in 1906. Stolypin's aim was to reduce the power of the commune and to divide up the land into compact holdings, for in most cases the land was still held by the commune in the three-field system, in which each peasant worked a number of strips of land scattered over a large area. Other reforms included making finance for development easier to obtain and migration easier, so that more holdings became available for sale to the more prosperous peasants. It is possible that Stolypin's reforms would eventually have resulted in a satisfactory pattern of agriculture but the entry of Russia into the First World War in 1914 soon placed an intolerable burden on the peasants, as on the rest of the country.

As the men were drafted to the fighting fronts, more and more of the struggle to keep up production fell on the women, children and old folk. Arrangements for transport and marketing grew steadily worse, and hunger in the big cities contributed materially to the rise in opposition to the Tsar and his conduct of the War.

The Revolution

When, in 1917, the February Revolution ejected the Tsar, the peasants widely assumed that the land would at last become theirs; but the provisional Government postponed action on this vital question until a Constituent Assembly could be held, meanwhile continuing the war. The soviets, or councils of workers, soldiers and others constituted a rival authority. Those holding the Bolshevik point of view held that the war must be ended and the land distributed to the people at once. When the Bolsheviks carried out the October Revolution, which created the first communist state, the peasants again assumed that the land would finally become theirs and proceeded to appropriate the estates. Without delay, the new revolutionary government nationalised all land and instituted strict controls to endeavour to restore food supplies to the starving cities. When the Red Army finally emerged victorious from the civil war, the countryside was largely devastated and its occupants utterly demoralised and bewildered. The peasants had assumed the land would finally be theirs after the revolution but there was great confusion and with the currency almost worthless those who were able to raise crops kept them for their own survival, so the position in the towns remained serious. Lenin's New Economic Policy gave greater freedom to trade and a measure of stability to the countryside, and for a time, it looked as if the Bolsheviks might be willing to accept a countryside of small, peasant farmers, but local soviets began to press for the establishment of communes. These communes bore limited resemblance to those of the historic Russian kind, which had, indeed, taken control again in many areas after the Revolution, but the emphasis was on the confiscation of the land and equipment of the former landowner and the wealthier peasant (kulak) for the benefit of the poorer classes. Hence, the Bolsheviks drew their main support in the countryside from the impoverished peasants. Voluntary collectivisation made progress is some areas, but in many cases the earlier communes were disbanded before full-scale collectivisation was ordered. Continuing difficulty with food supplies for the industrial towns, allied to the belief that the peasantry constituted a survival of conservative and capitalist forces, led to Stalin's decision in the late 1920s to enforce collectivisation.

COLLECTIVISATION

In 1928 there were about 25 million peasant holdings which averaged about 15 hectares in size. Initially, the collectivisation movement involved the grouping of typically about 75 such holdings into a collective farm (kolkhoz) administered by a chairman and his assistants. The members were required to devote half or more of their time to the collective enterprise and the rest they could spend on their personal plots of land or gardens, which were allocated on the basis of up to half a hectare per family. Collectivisation was bitterly resisted by the wealthier peasants, by the Cossacks with their traditions of independent frontiersmen, by the semi-nomadic Kazakhs and by others who were opposed to agrarian communism. Many of the opponents were arrested and deported to labour camps. The peasants themselves destroyed crops and slaughtered livestock to prevent them falling into the hands of the collectives.

Nevertheless, there were, as in the earlier attempts at collectivisation, many peasants who genuinely believed in the movement. The power of the government and the Communist Party was behind these, so the transformation of the countryside went on in spite of opposition. By 1940, only about 3% of the peasants remained outside the collectives, mostly in remote areas. Many of the original collectives had been amalgamated to form still larger units, and the average size had risen to about 500 ha of sown land averaging 81 households. In total there were 235 000 collective farms spread across the Soviet Union, with about 75 million people living on them. By this time, some of the advantages of collectivisation had become apparent, especially through the mechanisation programme. There had been hardly any machinery on the farms in 1928 and few of the peasants could have handled tractors. To assist in modernising the farm processes (and at the same time to provide cells of politically reliable socialist workers in the generally conservative or apathetic countryside), Machine–Tractor Stations were set up to handle all the larger and more complex mechanical operations and maintain the machines. In spite of many failings, the new system could carry out ploughing and sowing much more quickly, taking advantage of suitable weather conditions, and it was easier for trained specialists to influence the farming work in the direction of more up-to-date methods.

Just as the Stolypin reform programme was interrupted by the First World War, so the increases in production and productive resources under collectivisation were halted by the Second World War. During the war years, 1941–45, some 25 million Soviet people died and the peasants were called on to adopt a 'scorched earth' policy in front of the German armies to deny them food and shelter for as long as possible. After the war, it took about six years even to catch up with the production level of 1941, when the sown area had been 150 million hectares, compared with 113 million in 1928. The livestock numbers in 1941 had not recovered to pre-collectivisation levels and in 1945 the numbers of cattle and sheep were some 13% less and the number of pigs was almost halved from the pre-war figure.

As in the early days of collectivisation, much improvisation was necessary during the years of post-war reconstruction. The supply of machinery was slow because factory production was overwhelmingly devoted to yet more urgent needs. Black market trade of farm produce for spare parts and materials was common. Slowly recovery was achieved. There were more tractors and combines in use in 1950 than in 1941, though the level of mechanisation remained far short of that in many western countries. Since that time there have been immense improvements in mechanisation, electrification and other aspects of modernisation though the overall pattern is still one of less sophisticated equipment available for working the land than in the more advanced countries of the capitalist world, and correspondingly more reliance on manual labour.

Collective and state farms (*kolkhozi* and *sovkhozi*)*

The land of a collective farm is, like all land in the Soviet Union, owned by the state, but is leased

* Abbreviations for kollektivnoye khozyaystvo and sovetskoye khozyaystvo.

permanently by title deeds to the workers of the farm, who form an association governed by a farm committee. The chairman, a nominee of the Communist Party, occupies a powerful position and is responsible for ensuring that the farm operates in accordance with the annual targets of the national Five Year Plan, broken down into regional and local levels. Remuneration varies with the profitability of the farm but the state now guarantees regular cash payments. Wage rates vary considerably with the more responsible and technical jobs counting higher than ordinary manual labour. The farm is required to deliver to the state a stipulated amount of produce of various kinds at fixed prices. The remainder of the output can be used on the farm, sold to the state (usually at higher prices than for the compulsory deliveries), or sold in the kolkhoz markets which exist in most towns. Here too, the individual collective farm member can sell surplus produce from his personal plot to augment his earnings on the collective. The farm worker may own his own cow and calf, one or two pigs, up to 10 sheep, plus poultry, rabbits and beehives, and his house as well as all personal possessions. Other resources are owned collectively by the kolkhoz and these now include all items of machinery, the Machine–Tractor Stations having been disbanded in 1958, when it was assumed that the farms could manage their own mechanical equipment. The farms have set up workshops for running repairs, but district maintenance stations do major jobs. These are run by a rural technical organisation (*Selkhoztekhnika*).

So far, description of the Soviet farm has been confined to the kolkhoz or collective farm, but the state farm or *sovkhoz* has become increasingly important. The workers on these farms are state employees and they usually earn more than do collective farm workers. State farms were set up to pioneer new methods and break in new land besides setting higher standards generally in the countryside, and they have been greatly extended in recent years. In 1940 there were only 4200 state farms (compared with 235 000 collectives) but by 1986 there were 22 929 state farms compared with 26 300 collective farms. Not only have some collectives been transformed into state farms but the collectives themselves have been amalgamated into larger and larger units so that they have fallen in numbers.

In 1986 the average size of the collective farms was 6456 hectares in area, with over 1950 head of

Maize is an important crop, grown in the south for ripe corn but elsewhere for making silage.

cattle, over 1100 pigs and 1670 sheep. Each farm included approximately 500 households. The average size of state farms was still larger: 15 900 hectares (total area), of which 4706 were sown land; with about 1870 cattle, 1200 pigs, some 3000 sheep and over 460 workers. These are averages for all types of farms, and farms specialising in grain production are often much larger, while sheep rearing farms with a large amount of pasture may have well over 100 000 hectares and ten of thousands of sheep. Even so the sovkhoz average size are rather smaller than they were a few years ago, partly because of the conversion of collectives into state farms.

Because of the great size of these farms they are divided into several units, and the labour force is organised into 'brigades' responsible for particular jobs, such as cultivation, livestock husbandry, and operation and maintenance of machinery Each farm has several specialists, including economists and accountants, field and livestock specialists and veterinary surgeons. Most are large enough to have the amenities of a large village, such as a cinema, library and perhaps a ballet school, as well as ordinary educational facilities and a small hospital. In 1988 moves to give families and other small groups opportunities to farm with some degree of independence were announced.

CLIMATE AND SOIL

Soviet farmers have to contend with some of the worst climates faced by farmers anywhere, and even the best areas suffer badly from long winters or frequent droughts. Such generalisations must, of course, be modified for different regions. The Baltic republics and adjacent parts of the RSFSR and Belorussia are relatively well endowed with moisture, reflected in the importance of dairying in their agriculture, but inadequate drainage poses problems. Southerly areas from the Ukraine to the Far East enjoy high summer temperatures but in few areas does summer rainfall offset high evaporation and transpiration, so that irrigation is necessary for intensive farming. Only in the so-called sub-tropical areas of Transcaucasia and in the Maritime kray in the Far East do high summer temperatures coincide with ample rainfall. Most areas have summer maxima of precipitation but only rarely is the total amount satisfactory for agriculture (Figs 3.1 and 3.2).

Warmth, moisture and light are the principal components of the climatic resources available for agriculture, other climatic features, such as wind and cloudiness, being of subsidiary importance, weakening or reinforcing the effect of the fundamental components. These latter factors may assume dominant importance locally, when they become extreme, as do winds which cause or exacerbate drought and the removal of snow from the soil when its cover is needed, but warmth and moisture provide the basic indices used in agroclimatology and the delimitation of agricultural regions (Shashko, 1962).

The importance of accumulated temperatures, i.e. the sum of daily temperature increments above 10°C in the growing season, is shown in Table 7.1, which gives the requirements of various crops.

Table 7.1 shows that the sum of temperatures required varies considerably according to the variety of crop and the objective for which the crop is being raised. The lower requirements for eastern Siberia and the Far East are explained by smaller fluctuations permitting finer limits and by the clearer continental conditions. In some cases, as in the sub-tropical crops, distribution is conditioned less by accumulated temperatures than by wind conditions. In general, the temperatures shown should ensure satisfactory results in nine out of ten years.

TABLE 7.1 CROP REQUIREMENTS IN ACCUMULATED TEMPERATURES (ABOVE 10°C) FOR RIPENING IN COMMERCIAL CONDITIONS

Crops	European USSR and western Siberia day-degrees C	Eastern Siberia and Far East day-degrees C
Vegetable crops on sheltered ground	400	400
Turnips, cabbage (e)	800	700
Barley (e), winter rye (e) in warmer locations	1000	800
Oats (e), barley (m)	1400	1200
Spring wheat (e), winter wheat, maize (m) for green feed, sugar beet for fodder	1600	1400
Spring wheat (l), sunflower (e) for seed, sugar beet for sugar	2000	1800
Maize for grain (e), beans (e), millet (l)	2200	2000
Maize (m-l) for grain, rice (m), grapes	3200	—
Soya beans (l), ground nuts (e) sorghum (l), figs	3600	—
Cotton (e), lemons, tangerines	4000	—
Cotton (m), rice (l), grapes (l)	4400	—
Olives, oranges, jute	4800	—

e = early,   m = medium,   l = late repening varieties.

Sources: Shashko, D.I.; (1962), 'Climate resources of Soviet agriculture', *Akademia Nauk SSSR* pp. 378–445 and Davitaya, F. F. and Sapozhnikova, S. A.; (1969) 'Agroclimatic studies in the USSR', *Bulletin of the American Meteorological Society,* 50(2) pp. 67–74.

In addition to accumulated temperatures, the availability of warmth at critical periods of plant growth, the risk of damage from heavy frosts and the number of days available for growth are important aspects of temperature. Much depends on the combination of climatic elements in any given year, or of climate and the soil conditions described in Chapter 4.

The main area of agricultural land stretches from the western frontiers, (where the Soviet Union borders the Baltic Sea in the north, and then marches with Poland, Czechoslovakia, Hungary and Rumania) eastwards in a belt of gradually diminishing breadth across the southern parts of Siberia, overlapping the north of Kazakhstan and tapering out in the mountains of east Siberia (Fig. 7.1). Many of the basins in the mountains and the maritime areas beside the Pacific Ocean provide detached areas of temperate climates, while the oases and irrigated areas of Central Asia are particularly valuable for subtropical crops such as cotton. South of the main range of the Caucasus mountains is the one other area in the Soviet Union which has a nearly subtropical climate, and Transcaucasian specialities are grapes, citrus fruits and tea.

The temperate agricultural lands comprise, in Europe, the areas roughly south of Leningrad and extending southward to the Black Sea, and in Siberia, the narrower belt of south Siberia and north Kazakhstan. These are devoted to various types of mixed farming, as will be described later in more detail. The climate of the more northerly areas makes it difficult to get high yields of grain but wheat, rye, barley and oats are grown either for human or animal consumption. The cool, relatively moist climate in the westerly parts favours root crops such as potatoes and green crops such as cabbage, as well as grasses and clovers. The podzolised soils require considerable dressings of fertilisers and, in some areas, extensive drainage networks. This is particularly true of the Belorussian lands which include the Pripyat marshes.

In the southerly parts of European Russia and the Ukraine, the physical conditions are reversed. Evaporation exceeds precipitation and, as noted in Chapter 4, the soils are of the chernozem type (Fig. 4.2). During the eighteenth and nineteenth centuries much of this black-earth land was ploughed up and developed as the main grain producing area of Russia.

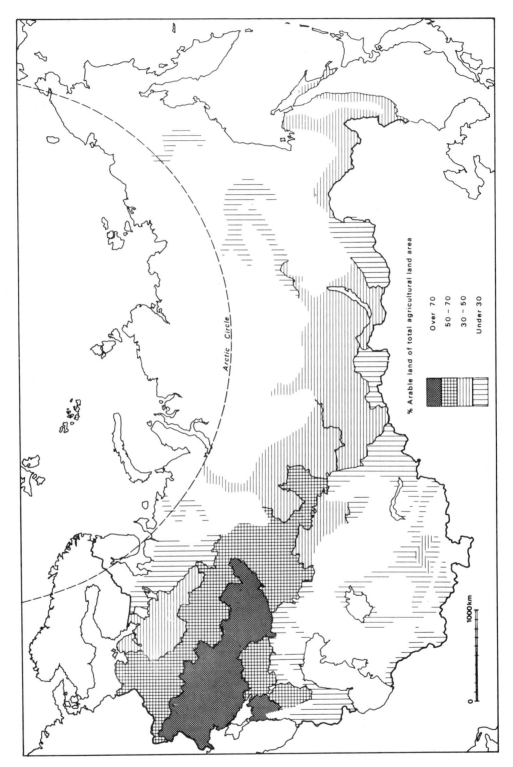

Fig. 7.1 Agricultural land. The distribution of arable land is shown as a percentage of the total agricultural land, which includes pasture and land for hay. Arable land is that which is suitable for ploughing but the total ploughed in any one year may be substantially less in some areas

Between the area of podzolised soils in the northerly areas and the chernozems of the south is the intermediate zone of the wooded steppe, in which the vegetation is also transitional from forest to grassland, which stretches eastward from about Kiev to Kuybyshev, and beyond the Ural mountains from Chelyabinsk to Novosibirsk. Conditions for agriculture here are relatively favourable. Neither the frequency of droughts, nor the excess of cold and damp conditions encountered respectively to south and north hinder cultivation. Grains, including maize, sugar beet, sunflower and other oil crops, dairy cattle, pigs and sheep are among the principal products, but the range is very wide. Irrigation can boost production, but it is less essential than in the more southerly regions. The winter is less severe than further north, so that although snow lies on the ground for about three months, winter wheat can be sown in preference to spring wheat which yields less highly. This does not apply, however, east of the Ural mountains where conditions are drier and there is insufficient snow cover to protect the seed in the ground, or the snow may be blown away to expose the ground to the exceedingly severe frosts so that spring sowing is necessary.

In the Far East, along the Pacific Ocean coast and in the valleys of the Amur and Ussuri rivers, a modified kind of monsoonal climate prevails and, in this relatively mild and humid area, rice is one of the crops grown, along with a wide selection of temperate crops and fruits. The agricultural zone here is, however, narrow, because of the mountains which, as previously mentioned, dominate east Siberia. Summers are short, but cultivation can be carried out in the valley lands, even on permafrost, where the surface of the soil thaws. Improved plant breeding has enabled agriculture to spread in the more favoured valleys right to the Arctic coasts, but production in such areas is necessarily limited to the hardier grains, grasses and livestock. Apart from this pioneer fringe of cultivation, most of the tundra lands are grazed by vast herds of reindeer, herded nowadays, at least in part, from the air. Yet, alongside the aeroplane and helicopter, the sledge hauled by dogs or reindeer is still invaluable as a means of transport in winter.

Land Amelioration

The difficult physical conditions that characterise the Soviet Union, together with Soviet belief in the ability of man to overcome problems posed by the environment, have led to much land improvement work, especially irrigation in water-deficient areas and drainage in humid areas. Before the Revolution, about 3.5 million ha were irrigated, mainly in Central Asia. By 1970, nearly 11 million ha were irrigated and additions have continued at the rate of 600–700 000 ha per year, to make the total 20 194 000 ha in 1987. Almost one-half of this total is in Central Asia and Kazakhstan (Fig. 5.2), with the Ukraine and southern parts of the RSFSR (Fig. 5.1) notable among recent increases, there being increasing emphasis on irrigation of grain and fodder crops.

By contrast, lands requiring drainage are naturally to be found in the northern, moist areas where evaporation is low. Schemes to improve drainage are proceeding at a rate similar to those for irrigation, the total being over 19.4 million hectares in 1987 compared with 10.2 million in 1970 and 8.4 million in 1956. Much of the work continues to be in Belorussia and adjacent areas of the Ukraine, and the Baltic Republics, but a recent emphasis on drainage in the RSFSR has pushed it into first place in the statistics. Drained lands actually in use agriculturally amounted in 1987 to 14.9 million hectares, rather more than one half (8.4 million) being under crops, and most of the remainder (5.9 million) under pasture and for hay.

For many years, addition to the cultivated area was the main concern of agricultural planners. Severe soil erosion, especially in the virgin lands, however, eventually became recognised and conservation measures were put into effect and excessive cropping has given way to more balanced land use policies, especially involving fodder crops. Marked erosion can, however, still be seen in mountainous, hilly and semi-arid areas and further measures are needed.

THE REGIONAL PATTERN OF AGRICULTURE

Climatic and associated soil conditions limit the types of agriculture that are practicable, but in any region large urban centres stimulate production of agricultural products for consumption in the towns. While climatic and soil conditions continue to limit the local specialisation to suitable products, the nearness to markets and consequent low costs of transport, as well as the reduced risk of fresh produce perishing on the way to market, make it practicable for such areas to incur higher costs of production. Thus, more may be spent on fertilisers, housing for dairy cows, glass and heating for greenhouses, and other requirements for intensive agriculture. In the Soviet Union, this is particularly true of the Central Industrial region, which includes Moscow and its surrounding towns, the Dnepr–Don industrial areas of the Ukraine, the Leningrad area, and the industrial regions of western and central Siberia. The Moscow area, especially, has a number of large state farms specialising in vegetable, fruit and dairy products. These interrupt the pattern of regions described below and illustrated in Fig. 7.2.

1. Reindeer rearing and hunting region

This most northerly region extends from Kola to Kamchatka and includes tayga, wooded tundra and tundra. Reindeer rearing has few links with true agriculture but is closely integrated with hunting of fur animals, fishing and catching of seals. The lands of two main natural zones are grazed—the tundra for summer pasture and the northern parts of the tayga as winter pasture, while in mountainous regions, the high unforested lands serve as summer pasture. To the south, reindeer become of subsidiary importance to other branches of agriculture. Scattered hearths (*ochagi*) of cultivation—mainly under glass— occur near mining and transport settlements.

2. Reindeer rearing, hunting and agricultural region

This region includes the northern part of the tayga zone in the European parts of the USSR and almost all the tayga zone in Siberia and the Far East. Accumulated temperatures range from 1000 to 1800 day-degrees. Some cropping and rearing of animals occurs in islands of cultivation in forest and marsh. Heavy dressings of fertilisers are needed. Agricultural work is commonly combined with forestry, fishing and hunting.

3. Northern dairying region

Within the tayga zone, a more developed form of agriculture is found in some European areas, notably near Leningrad and Murmansk and in

west Siberian areas, with specialisation, generally, in cattle raising and dairying based on the natural fodder of water meadows and low-lying pastures. Cultivation is less important, but fast-growing and frost-resistant varieties of grain (mainly barley and rye) are grown, and a larger share of the arable area is occupied by potatoes and fodder crops. Extensive areas of marsh still exist.

4. Livestock rearing and cultivation region of Yakutia

In eastern Siberia more continental conditions prevail and soils are less leached. In the central Yakutia lowlands livestock are reared on natural pasture and hay found in forest clearings, but also crops can be grown at unusually high latitudes. Because of the winter drought only spring crops are practicable, barley, spring rye and spring wheat being predominant. The fodder available and the extensive nature of the husbandry result in a bias towards beef cattle, with dairying subsidiary.

5. North-central dairying and arable region

In the southern part of the tayga and the greater part of the zone of mixed forests from the Baltic republics to the Ural mountains, the cultivation of grains is not seriously restricted by the amount of warmth, accumulated temperatures being 1600–3200 day-degrees. The number of plants that can be raised is, therefore, much greater than in the previously described regions.

The podzolised soils require heavy dressings of lime and fertilisers, and there is still a high proportion of forest and marsh land as yet underdeveloped. Over the greater part of the region arable land varies between 20 and 40% of the agricultural area.

Potatoes and grains (including maize and other crops for silage) are important crops. Roots are used for fodder and these restore fertility to the soils, but liming and fertilising are no less important in this region of surplus moisture and leaching. With their use and with drainage, sown pastures can be highly productive.

A good supply of pasture and fodder facilitates dairying with low costs, and also makes for a wider variety of livestock than in the more northerly regions. This region has been stimulated by large urban and industrial centres, including Moscow, and relatively good communications, which have promoted intensification of

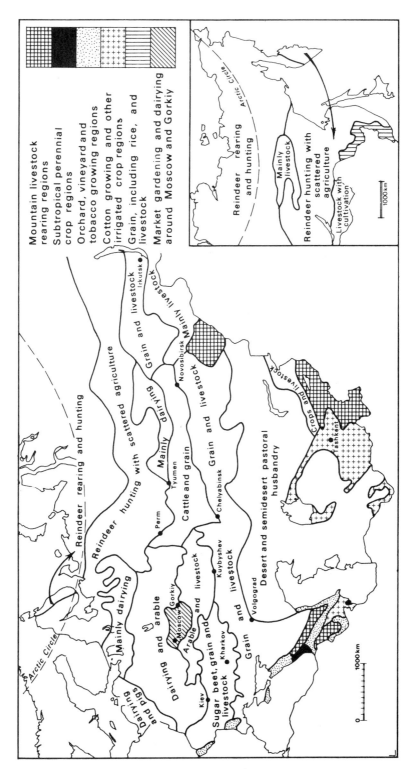

Fig. 7.2 Agricultural regions of the USSR. The regions shown are simplified, boundaries are very much generalised and regions normally merge gently into each other

agriculture, particularly in dairy, pig and poultry and vegetable production. Pig rearing is based on potatoes, grain and the waste products from milk processing. Many large sovkhozes have been created in this region, especially around Moscow, in order to meet the demand for meat, milk, vegetables and fruit.

Further from industrial centres, farming combinations include grain and potatoes with livestock rearing, and this is the most important region for flax.

6. The Moscow–Gorkiy market gardening and dairying region

Immediately surrounding the capital and in a belt extending westwards to include Gorkiy is a region within which the influence of the urban markets is especially discernible. Specialised dairy farms are now numerous, with corresponding emphasis on fodder crops. Vegetable growing has become more specialised and output increased though less land is now devoted to field vegetables and potatoes, a traditional crop on the light soils of the Meshcherskiy area south-east of Moscow.

Flax production has also been cut back sharply to make more room for the products required for food supplies in the urban areas.

7. Baltic dairying and pig rearing region

The greater part of the Latvian, Estonian and Lithuanian republics, the western part of Belorussia and the Kaliningrad oblast of the RSFSR, enjoy relatively moderate temperatures. Lithuania and the coastal belt around the Gulf of Riga have a frost-free period of more than 150 days. Accumulated temperatures range from 1600 to 2200 day-degrees C.

The chief crops are potatoes, fodder crops and bread grains. Potatoes are produced for industrial purposes and as a food crop, but above all for fodder. Natural pastures are less important, arable and permanent hay meadows more important, than in Region 5. Marshlands have largely disappeared under cultivation. Dairy cattle and pigs dominate livestock rearing, especially in Lithuania where numbers of both have more than doubled since 1941 and were 60% more in 1986 than in 1966, but showed no further increase up to 1988.

Hay being stacked on an Estonian collective farm. Harvesting calls for intensive labour from regular and supplementary workers.

8. West-central sugar beet, grain and animal husbandry region

This region occupies the western parts of the wooded-steppe zone, and extends into the southern parts of the mixed forest and northern parts of the steppe zones, including the Central Chernozem and parts of the S. W. Ukraine and Donets–Dnepr economic regions. Pasture and hay meadows are found in moist lowlands, in river valleys and among the mixed forests, but the further south, the less the land that is occupied by forest, waste-lands and commons, and the more by arable use, which totals 75 to 80% of the whole area.

The balance between precipitation and evaporation is better in this region than in the areas to the north (too humid) or to the south (too dry), but severe droughts occur once or twice in four years and so irrigation is valuable. Accumulated temperatures range from 2400 to 3000 day-degrees C. The soils, though rich, are easily washed or blown away and shelter belts are planted to provide protection from erosion.

The majority of farms have a mixed pattern of husbandry, with grain growing and livestock rearing dominant and sugar beet important in the west. In the north-western part, the rotation includes more than 60% grains, especially winter wheat, esparto grass and clover, and sugar beet is the leading industrial crop, commonly occupying between 10 and 25% of the sown acreage. Though sugar beet is restricted in frequency in any one field, it leads to intensive cultivation because it requires deep ploughing, frequent loosening of the soil between rows and heavy fertilising. The remaining crops of the rotation gain also from the weed-free fields. The main winter crop is wheat, but maize, for which good conditions prevail, has long been grown in the western part of the region.

Sugar beet regions have relatively high intensity livestock rearing. The mass of waste products of the beet processing, maize production and concentrates, together provide much of the fodder supply, and the region has become important for both dairying and pre-slaughter fattening. Whereas cattle from the pastoral fattening regions reach the slaughter house predominantly in autumn, here the main fattening period is that of the beet processing, ie winter.

In the southern, drier parts, good crops of winter wheat require clean fallow in the rotation.

The drought-resistant sunflower is an important crop, yielding both oil and animal fodder.

9. East-central region of arable and animal husbandry

This region is also within the wooded steppe and mixed forest zones, but further north and east than the preceding region, extending from the mid-Russian heights to the Ural mountains. The climate here is much drier with accumulated temperatures of 2100 to 2500 day-degrees C.

Crop rotation is based on bread grains, maize and potatoes. Grain crops occupy 50–60% of the sown area. Formerly the main crops were winter rye, oats and 'groats crops'—millet and buckwheat. In recent years the emphasis has turned to wheat and fodder crops, particularly maize, to support cattle and pigs. Other characteristic crops of this area are potatoes and hemp, these being more important than sugar beet, in contrast to the position in the last region. In some parts of the region, sunflowers are grown. Fruits, especially apples, are also grown here.

Fodder crops yield coarse succulent and green feedstuffs as well as concentrated fodder and silage. On these, intensive livestock rearing is carried out, the most important branch being cattle, but with increasing attention to pig breeding and poultry keeping, which need less pasture.

10. South-central grain and animal husbandry region

In this region, in the steppe zone to the west of the Volga, extending from the Bashkir ASSR to the Black Sea and north Caucasus, the advantages of greater warmth are for many crops offset by the dryness so that yields per hectare tend to be less than in the wooded steppes.

There have, however, been great advances in overcoming drought: protective forest plantations, snow control, construction of ponds and reservoirs, irrigation from local streams and also from the great rivers—Don, Kuban and Dnepr. On the irrigated lands there has been development of cattle and sheep rearing, orchards, vineyards and, in places, rice growing.

This is one of the areas involved in the virgin lands developments in the late 1950s and early 'sixties resulting in increases of 70–80% in the sown areas of some oblasts. Some reductions have taken place with the restoration of fallowing to offset soil erosion and depletion.

The main crops are wheat and other grains, which in all total about 50–70% of the sown area, rising to 76% in Orenburg oblast, with wheat occupying similar proportions of the total grain. In the less dry, western parts of the region, maize has been sown increasingly in connection with a swing towards intensive livestock husbandry, and wheat is mainly sown in winter. Eastward, the climate becomes harsher and spring-sown varieties progressively replace winter wheat. Sunflowers are the most important industrial crop except in the higher rainfall areas of the north Caucasus, where sugar beet predominates. Melons, grapes and other fruits are produced. In the eastern parts, however, unimproved steppe pastures remain over extensive areas and are used for stock rearing, with Stavropol kray notable for specialised sheep breeding.

11. Lower Volga grain and livestock rearing region

This is a drier region with dark chestnut soils, and there is more emphasis on grain crops and live-stock rearing in generally less intensive agriculture, but vegetable crops, melons and mustard are important. The lower Volga valley provides a zone of fertile and intensively cropped soils separating semi-desert areas to east and west. Cattle, mainly beef, and fine-woolled sheep utilise the poorer lands. Large numbers of sheep from the neighbouring parts of the steppe zone are moved for winter grazing to the arid steppe pastures of the lowlands near the Caspian Sea.

12. Ural and west Siberia cattle and grain region

This region includes the plains and foothills rising to the Ural mountains on the European side, but is mostly in western Siberia. The European part is the more developed with a cattle and grain type of agriculture intermixed with forest areas. There is less forest land east of the Ural mountains, and the wooded steppe is used for dairying and grain husbandry. In some parts of the west Siberian lowlands large areas of marshes and saline soils hinder development, but they are partially used for dairy and beef cattle. Accumulated temperatures are 1800–2100 day-degrees C.

The better drained areas are in arable cultivation and the abundance of land led to major developments under the virgin land schemes, so

that this became an important grain producing region with spring wheat as the chief crop.

13. North Kazakhstan and west Siberia grain and livestock rearing region

This region comprises, in the north, chernozem steppe, and in the south, dry steppe with mainly dark chestnut soils, accumulated temperatures ranging from 2100 to 2600 day-degrees C. Widespread development began only at the end of the nineteenth century, but between 1954 and 1960 it was the scene of the main development of virgin lands. Large areas were developed by new state farms being organised on land previously in pastoral use.

The region has developed the most specialised grain farming in the country with advanced mechanisation, although the climatic conditions result in sharp fluctuations in yields. Thus, in 1976–80, Kazakhstan supplied over 16 million tonnes of grain annually to the state purchasers. In 1977 only $8\frac{1}{4}$ million tonnes could be supplied but in 1986 it was 16.7 million tonnes. Livestock rearing has been intensified with significant dependence on root crops, but because in this vast area there are widespread occurrences of saline and other soils unsuitable for ploughing, large areas of natural pastures remain. This is especially true of the driest, most southerly areas of sheep and cattle rearing. Soil erosion is a serious problem (Fig. 4.3).

14. Pastoral husbandry region of the desert and semi-desert

With the transition to semi-desert conditions in, Kazakhstan, Astrakhan oblast and the Kalmyk ASSR, natural pasture becomes the main form of land use in non-irrigated areas. In the southern parts of the region snow cover is light and of short duration so that pasture provides maintenance for livestock throughout the greater part of the winter, whereas in the northern part of the desert and semi-desert, between the Aral Sea and Lake Balkhash, stall feeding is necessary all the winter.

Large areas with sandy soils offer comparatively good pastures in spring, when the ephemeral plants are growing and retention of water in the sands makes possible continued grazing in the early summer, but the reservoirs tend to dry up and fodder supplies fall. Conditions improve again in autumn when absinthe and other plants yield large quantities of fodder.

Sheep rearing is generally dominant, but cattle are important in the semi-desert areas. Many are driven to the heights of the Mugodzhar, Kazakh foothills, Tarbagatay and Tyan Shan ranges in summer to graze on mountain meadow-steppe and meadow pastures.

In contrast to the former nomadic husbandry, the state and collective farms prepare hay in grazing areas and fodder concentrates are transported to the wintering places from arable regions. Water supply to pasture and irrigation of hayfields is frequently ensured by modern techniques, including utilisation of artesian water. The production centres of the farms and permanent settlements are usually located near the wintering places.

In the north and north-western regions of the semi-desert, specialisation in sheep breeding and fattening cattle is based on subsidiary cultivation of irrigated and non-irrigated lands. Hay-making is on a comparatively large scale and there is considerable stall feeding of cattle in winter.

In the southern desert zone, where the possibility of stall feeding is less, pastures less well supplied with water and fodder poor in summer, breeding of Karakul sheep for their valuable fleeces is important, but sheep rearing is here subsidiary to keeping camels.

15. Mountain livestock rearing region of the Caucasus

In the mountain regions, where cultivation is hindered by relief and shortness of the growing season associated with altitude, livestock rearing is dominant above 1500 m and lower in exposed places. On the lower slopes forest and meadows are interspersed, with arable land in the valleys and on gentle slopes. Pasture land is less prevalent in the western, more forested and moister parts of the Caucasus than in the eastern drier parts.

Cattle are brought for summer grazing from the foothills and plains (cotton, grain, orchard-vineyard and tobacco regions) and, in addition, there are livestock-rearing kolkhozes based on the mountain lands themselves. Both forms of utilisation are necessary since the grazing capacity of the mountain lands is much higher in the three to four months of summer than in the rest of the year. In fact there is two-way interchange, livestock from the mountain kolkhozes being taken to the plains in winter. This seasonal exchange extends over a distance of 300–500 km. Sheep are involved in the furthest movement, being

pastured in summer on the high, stony and less well grassed areas, while cattle are kept on the less steep and better lands.

16. Mountain livestock rearing region of Central Asia

This region includes the high mountain areas of the Kirgiz and Tadzhik republics, on both ridges and intermontane basins with altitudes above 1500 m. In contrast with the Caucasus, the livestock rearing is much more concerned with wool and meat than dairying. The orientation is explained by the character of the pastures, of steppe rather than meadow type, and the absence of winter precipitation in the intermontane basins, which limits fodder.

Wool output has been greatly increased as a result of the crossing of the formerly widespread Kurdish sheep with the fine-woolled Tyan Shan breed. On the Pamirs and in the highest intermontane valleys of the Tyan Shan, yaks are found as well as sheep and horses.

17. Mountain livestock rearing region of the Altay

In the inner parts of the Altay, shut off from the west and north by mountain ranges, the climate is dry, and the dominant vegetation is grassland of meadow-steppe, steppe and, in parts, even semi-desert types.

In the central, eastern and south-eastern regions of the Altay, pasture is overwhelmingly the main form of agricultural land use. Beef cattle and sheep, together with yaks, are reared. In the north and north-west, cattle take precedence and, while beef is the main product, cheese and butter factories have been established.

18. Orchard, vineyard and tobacco growing region

The conditions which encourage these crops include high accumulated temperatures, 3000–4000 day-degrees C., and a long frost-free period (190–250 days) with a mild winter. Whereas the relief would hinder cultivation for field crops, it helps towards freedom from frost, good soil drainage and a choice of differently exposed slopes for fruit cultivation. Fruit is also dominant on the flats among the foothills where the many mountain streams are used for artificial irrigation.

Apples, pears, cherries, peaches, apricots, figs, walnuts and grapes of both wine and table vari-

Krasnodarsky Sovkhoz in the Altay region illustrates the spaciousness typical of the steppelands. This farm was created as part of the virgin land scheme in the 1950s.

eties are grown. Vineyards are a specialisation of the areas with drier summers. Tobacco is important on the wetter, leached soils of Transcarpathia, the Moldavian republic, the Crimea and western parts of the Caucasus.

In these areas fodder for cattle is limited. However, on hill slopes and in the dry areas of the eastern Transcaucasus and Dagestan, pastures are available and cattle and sheep rearing has developed with seasonal movement as necessary. Pigs are also kept, in some cases on forest pastures. Silkworms are important in some areas, the cultivation of the mulberry and feeding of silkworm caterpillars integrating well with the local intensive agriculture.

19. Region of subtropical perennial crops
This region includes the hilly Caucasian mountain foreland, which in places reaches to the sea, and the cultivated part of the Kolkhid lowlands.

It is the principal region described in the USSR as 'subtropical', with 240–250 frost-free days, and the mildest winters, in which January temperatures average between 3 and 8°C. Accumulated temperatures in the vegetative period exceed 4000 day-degrees C. Annual precipitation is high, totalling between 1200 and 3000 mm.

The cultivated lands form only a low proportion of the total surface area, which includes both high mountains and marshy lowlands. Tea occupies first place among the crops on the hillsides up to about 800 m. and also over a considerable area on the plain. Citrus fruits are second in terms of area planted. They are tolerant of soil conditions but restricted climatically, particularly by winter minimum temperatures. Plantations of lemons, oranges and mandarins are located mainly near the sea and on steep slopes where the risk of frost is least, frequently on terraces. Mulberry plantations occupy third place. They are less frost-hardy than tea but more resistant than citrus fruits, and require less labour input.

Among annual plants, tobacco is important, its distribution depending on soil characteristics. Essential oils (chiefly geranium) are grown on the plains. Vegetables are grown to utilise the colder parts of the year and to be marketed earlier than those from other regions.

The field kitchen of a cotton-growing farm near Tashkent, where the climate favours eating out of doors.

The chief field crop, and, on many farms, the only one, is maize. The climatic conditions permit two harvests of fodder crops. There is little land in pasture, but some collectives drive cattle to the sub-alpine and alpine pastures in the summer. Far more important in terms of value than ordinary livestock rearing, however, is the breeding of silkworms.

20. Regions of cotton and other irrigated crops

In these regions of Central Asia and Transcaucasia the frost-free period is 180–230 days and accumulated temperatures exceed 3000 day-degrees C. The summer is rainless, with a very high number of sunny days (up to 26 in July and 28 in August). The mountains provide water for artificial irrigation.

Within the zone climatically suitable there are regions of strong specialisation in cotton (including Fergana, the Hungry Steppe and the Zeravshan valley), and others with a considerable development of other branches of agriculture. Where irrigation provides sufficient water in the summer, cotton is given precedence, followed by lucerne in rotation with maize, rice and jute.

Heavy applications of fertilisers are needed and both mineral fertilisers and manure are used, while the irrigation waters themselves deposit silts which contain nutrients.

The use of non-irrigated lands is in many cases integrated with the irrigated areas. In the foothills, many cotton collectives grow crops capable of using the abundant rain in early spring and finishing their growth by the onset of the dry season. Such crops comprise wheat (mainly winter varieties) and barley (spring and winter varieties) but also include flax, mustard and sesame.

Cropping of this type has great value to the farms for the supply of coarse and concentrated fodder. Livestock rearing in the cotton region is, in general, divided into two isolated types, that of the oases, based largely on the surrounding pastures, and that of the desert plains of Central Asia where herds are comprised predominantly of Karakul sheep.

Within the cotton area, a number of supplementary agricultural enterprises are found, including sericulture, fruit growing and vine cultivation. Vineyards are often relegated to lands unsuitable for field cultivation, but their produce and that of the orchards places them in the first rank for exports of fruit to the other parts of the Soviet Union.

The cotton growing region has great potential for development because of its reserves of land and sources of water from the great rivers, the Kura, Syr Darya and Amu Darya.

21. Central Asian regions of intensive crops and livestock rearing

North of the cotton region in Kazakhstan and Kirgizia, an area important for crops slightly less warmth-demanding than cotton extends from the foothills of the Tyan Shan down to about 400 m on to the plain.

Except where irrigated, land is largely desert or semi-desert pasture suitable for winter, spring and autumn grazing. On the foothills, above 500–800 m., hayfields are found, while higher yet is the zone of mountain pastures. These are used chiefly in summer but some are available for grazing also in winter, especially those in valleys and ravines sheltered from the north and west by mountain ranges and therefore little affected by snow. Settlements are commonly located in irrigated areas.

In all parts of these regions livestock have a large role, based on the seasonal availability of different pastures, alpine and sub-alpine in summer, desert and semi-desert in winter, though these involve movement of stock over considerable distances.

Cultivation of sugar beet is combined with grain and fodder growing, and livestock rearing on irrigated land and areas of horticulture occur where favourable climatic and soil conditions are found on the foothills, the most important being near Alma Ata.

22. Grain growing and livestock rearing region of eastern Siberia

A grain and cattle rearing region extends discontinuously along the southern parts of Siberia, including parts of the Buryat ASSR and Chita oblast, bordering on Mongolia and China. Agriculture is found in the warmer intermontane basins, separated by forest-covered heights, and these areas show fairly high degrees of continentality with intense cold and only light snow in winter and most precipitation in the relatively warm summer. Accumulated temperatures reach 2500 day-degrees C.

The character of the agricultural economy strongly reflects the great distance of these regions from the economically advanced and densely populated regions of the USSR. Their economy is founded on the supply of the most easterly regions with products for which there is a continuing demand. The emphasis is on grain, meat and dairy products, with little regional differentiation.

23. Livestock regions of east Siberia and the Far East

In the Sayan and neighbouring mountainous areas of Tuva ASSR and the Buryat ASSR cattle and sheep rearing is based mainly on pasture, of which there are considerable areas in the sheltered valleys. As snowfall is light, grazing can be prolonged late into autumn and even into winter, and semi-nomadic herding is still common.

24. Grain, rice and livestock region of the Far East

This region occupies the lowlands of the Amur and Ussuri rivers. Accumulated temperatures reach 2400 day-degrees C. There is ample rainfall which falls mainly in summer, with a comparatively dry spring. These conditions suit warmth-loving and late-flowering crops, such as soya beans, maize and sorghum. Easy conditions for irrigation facilitate widespread rice growing.

The rapid industrialisation of the Far East and the distance from other regions make it important for it to become largely self-supporting in agricultural products. Today, a large part of the sown area is devoted to bread grains, especially spring wheat and oats. There are large areas of hay meadows and pasture and great potential for livestock development, especially cattle rearing.

PRODUCTION PROBLEMS

To meet the needs of the growing population of the Soviet Union—about 283 million in 1987—it is necessary to keep on increasing agricultural production. By world standards, output per hectare and per man employed in Soviet agriculture is rather low. This is partly explained by the severities of the climate and the large areas which are poorly endowed by nature for cultivation. It is also partly to be explained by shortages of machinery, fertilisers and other requirements for agriculture, which have persisted in spite of considerable improvements. The pace of investment in agriculture has, however, been stepped up within the last few years and better prices have been paid to farmers to encourage them to produce more.

As a result of these efforts, production has risen more rapidly of late. There have also been very substantial additions to the areas under cultivation. Most notable was the virgin lands scheme which was at its height between 1954 and 1960. During this period some 35 million ha, mostly in the eastern parts of European Russia, beyond the Volga, and above all in north Kazakhstan and south Siberia, were converted from steppe grazing lands of very low productivity to arable lands producing spring wheat and other crops. Although harvests from these lands, where the annual rainfall and snowfall is very uncertain, have fluctuated greatly, on average about half the Soviet grain supply comes now from these eastern regions. This, in turn, has freed some of the Ukrainian and other western areas, which formerly were required to produce almost all the country's grain, for more varied production, including sugar beet and cattle rearing.

Analysis of Production

The areas of land under the principal crops and the total number of the main classes of livestock throughout the Soviet Union are given in Tables 7.2 and 7.4 with comparable figures for post-war years and for pre-revolutionary Russia.

Output of all grains averaged 205 million tonnes for 1976–80, but only 180.3 million tonnes for 1981–85, although this recovered to 211.4 million tonnes in 1987 with about 100, 78 and 92 million tonnes of wheat respectively included in these figures.

The 1987 harvest of 211.4 million tonnes is probably close to the maximum likely under present conditions. The average annual amounts of grain and other major products are given in Table 7.3 for representative periods. These are 1976–80 and 1981–85 for recent years, 1946–50 when the recovery from the Second World War was beginning and 1909–13, representing the

TABLE 7.2 AREAS OF PRINCIPAL CROPS

Crop	Million ha						
	1913	1950	1960	1970	1980	1985	1987
Wheat, winter sown	8.3	12.5	12.1	18.5	22.6	18.0	15.3
Wheat, spring sown	24.7	26.0	48.3	46.7	38.9	32.3	31.4
Rye	28.2	23.6	16.2	10.0	8.6	9.4	9.7
Maize (for ripe grain)	2.2	4.8	5.1	3.4	3.0	4.5	4.6
Barley	13.3	8.6	12.1	21.3	31.6	29.0	29.3
Oats	19.1	16.2	12.8	9.2	11.8	12.6	11.8
Other grains	8.8	11.2	8.6	10.1	10.0	11.7	11.4
Total grains	104.6	102.9	115.6	119.3	126.6	117.9	115.2
Cotton	0.7	2.3	2.2	2.7	3.1	3.3	3.5
Sugar beet	0.7	1.3	3.0	3.4	3.7	3.4	3.4
Sunflower	1.0	3.6	4.2	4.8	4.4	4.1	4.2
Flax	1.2	1.9	1.6	1.3	1.1	1.0	1.0
Potatoes and vegetables	5.1	10.5	10.6	9.6	8.6	8.1	7.9
Fodder crops, including sown grasses	3.3	20.7	63.1	62.8	66.9	69.8	73.3
Other crops	1.6	3.1					
Total sown area	118.2	146.3	203.0	206.7	217.3	210.3	211.5

To obtain the total agricultural area, fallow land, hay meadows and pasture lands must be added, these being in 1980 about 14 million, 35 million and 287 million ha respectively, making nearly 554 million ha, excluding reindeer pastures.
Source: Narodnoye khozyaystvo SSSR, various years

TABLE 7.3 OUTPUT OF MAIN AGRICULTURAL PRODUCTS

	Annual Averages of output (Million tonnes)				Output per head (kg)			
	1909–13	1946–50	1976–80	1981–85	1909–13	1946–50	1976–80	1981–85
Grains	72.5	64.8	205.0	180.3	450	360	788	663
Potatoes	30.6	80.7	82.6	78.4	190	450	318	288
Sugar beet	10.1	13.5	88.4	76.4	60	70	340	280
Meat	4.8	3.5	14.8	16.2	30	17	57	59
Milk	28.8	32.3	92.6	94.6	180	180	356	340
Cotton	0.7	2.3	8.9	8.3	4	13	34	33

Source: Narodnoye khozyaystov SSSR, various years.

period before the First World War and the Revolution. These figures, divided by the population of the country for the periods concerned, give the amounts available (neglecting imports or exports) per head of the population.

It will be seen that for some years after the last war outputs of grains and meat *per capita* were below the pre-First World War level, and milk was no higher. All these commodities have since been increased in output *per capita* beyond the 1909–13 level by two-thirds or more, while the industrial crops, sugar beet and cotton, have been increased much more. Nevertheless, the levels were giving cause for concern before the poor performance of the late seventies and early eighties, but the averages *per capita* for the 1981–85 period indicate the severity of the problem of agricultural reconstruction needed in the Soviet Union, which only slightly improved in 1986–87.

From Table 7.2 it will be seen that areas of crops as well as outputs have continued to fluctuate. In 1987 the winter wheat did well and there was little necessity for re-sowing with spring grains, which occurs after a winter with inadequate snow cover for the protection of the land. The barley crop also yielded highly. Adverse climatic conditions in 1985, however, had led to failure to complete the sowing on many farms and the season continued with droughts and bad weather which led to a low harvest.

Considerable emphasis is currently being placed on livestock production to provide for a higher standard of living. In the past, the livestock branches have been particularly weak, with low milk yields per cow, and poor meat yields, and they still have a long way to go to equal performances in advanced livestock-producing countries, but milk yields per cow have improved

from 1853 kg in 1965 to 2682 kg in 1987 on both collective and state farms, with the greatest increases in production in the humid western areas and the developing Siberian regions. The overall growth in livestock numbers since pre-Revolution times is shown in Table 7.4 but vast fluctuations have occurred even in recent times.

The personal plots of the collective farmers and workers yield produce far more than proportionate to their areas. The cultivated land in personal use is only about 3% of the total sown area, but from these plots are produced (1985 figures) nearly two-thirds of all the potatoes, and one-third of the eggs, milk, meat and vegetables produced in the Soviet Union. Formerly the proportions were much larger, but increases in output from the collective lands and state farms have reduced the importance of the contribution from the personal plots. Also, as earnings have improved on the collectives, the members have had more incentive to work longer hours on the collective lands and to rely less on their plots. It should, in addition, be remembered that the proportion of produce from the personal plots which is offered for sale is much smaller than from the collective and state farms, as a large part is consumed by the families or personal livestock of the owners. The large farms are more concerned with providing a stable supply for the urban markets.

Development strategies

In order to increase production while releasing labour for work in other branches of the economy there has been great emphasis on the mechanisation and electrification of Soviet farms. In 1928 about one-third of all livestock fodder was required for draught animals and in spite of the creation of the Machine-Tractor Stations to serve

TABLE 7.4 LIVESTOCK

| | Million head at 1st Jan. | | | | | | |
	1916	1951	1966	1979	1980	1981	1988
Cattle Total	58.4	57.1	93.4	114.1	115.1	115.1	120.6
of which, cows	28.8	24.3	39.3	43.0	43.3	43.4	42.0
Pigs	23.0	24.4	59.6	73.5	73.9	73.4	77.4
Sheep	89.7	82.6	129.8	142.6	143.6	141.6	140.8
Goats	6.6	16.4	5.5	5.5	5.8	5.9	6.5
Horses	38.2	13.8	8.0	5.7	5.6	5.6	5.9

Source: Narodnoye khozyaystvo SSSR, various years.

the needs of the new collectives and a considerable emphasis on tractor production in the first and second Five Year Plans there was still heavy reliance on draught animals until after the Second World War. By 1965, however, 1.6 million tractors were in use, compared with 531 000 in 1940. The MTS had been disbanded and collectives and state farms made responsible for their own maintenance of machinery, with a new organisation, *Selkhoztekhnika* (Agricultural-technical service), providing major overhaul services and special contract work. Complaints about design of machinery and lack of spare parts were, however, rife, and these have by no means disappeared. However, numbers of tractors on the farm inventories had increased to nearly 2.8 million by 1987 and there were 774 000 grain combines compared with 520 000 in 1965 and 182 000 in 1940. Electrification now extends to virtually all farms, whereas in 1950 only 15% used this form of power.

Another bottleneck to increasing production which has been, if not removed, at least greatly diminished in its effect, is the shortage of fertilisers. In 1940 only some three million tonnes of fertiliser were available for all Soviet agricultural land. The continuing shortages after the war was one of the factors in the decision to exploit the accrued fertility of the virgin lands during the period that was required for the construction of new fertiliser plants. In 1965, 27 million tonnes of fertilisers were delivered to farms, and by 1980 this figure was raised to 82.0 million tonnes. There

Crop protection and fertilising from the air make use of both fixed-wing aircraft and helicopters. Booms and nozzles distribute sprays, dusts and spread for hoppers.

has also been considerable improvement in the variety and grades of fertilisers available though there are still many complaints about quality and about erratic deliveries to farms.

The application of fertilisers as well as the spraying of crops against pests and diseases, defoliation of cotton to facilitate mechanical harvesting and other tasks are carried out to a considerable degree by aircraft. About 100 million hectares of land are treated from the air annually. The aerial application of fertilisers is especially valuable when fields are too wet from snow melting in spring for tractors to work on them so there has been a big increase in the use of aircraft in the north-west areas of the RSFSR and the Ukraine. In summer, several thousand aircraft are assembled in Kazakhstan and Central Asia for cotton defoliation.

Encouragement of the labour force has accompanied technological improvement with the introduction in the 1960s of guaranteed regular cash payments for collective farm workers to bring them close to the wages paid to state farm workers, and the payments made by the state for agricultural produce have been much improved. Nevertheless, in spite of all incentives, productivity on Soviet farms remains low compared with the more advanced agricultural systems of the western world. The labour force in agriculture amounted in 1986 to 25.7 million persons, a reduction from 1940 of little more than 17%. Distribution of labour provides problems in that some traditional agricultural areas are still overpopulated whereas regions such as the virgin lands and other parts of Siberia and the Far East suffer from severe labour shortages, especially at harvest time. To meet these needs large numbers of workers migrate from the European areas, but this sometimes upsets the balance in the areas from which they move, as it is the younger and more energetic who are most mobile. Hence, there is still considerable pressure on non-farming people to help with the harvests.

When the crops are harvested they still have to be transported to stores or markets and here again there have been severe strains on the system owing to the lack of vehicles, both road and rail. Many complaints have been registered concerning the deterioration of produce in transit, or while awaiting shipment, and much still needs to be done to eliminate wastage both in transit and from vermin and deterioration in store.

Prompt processing of produce intended for factories helps to eliminate waste and much attention has been given to building and modernising plants to handle fruit and vegetable crops, milk and other products. Over one-half of all agricultural produce is now destined for processing compared with under 30% in 1960. This development has played a large part in the development of new forms of enterprise in the Soviet countryside which accelerated during the 1970s. These enterprises include Sovkhoz Factories, Agro-Industrial Associations and Intercollective Farm Co-operative Enterprises. Sovkhoz Factories have the longest history and are mainly concerned with the processing of the produce of one state farm, usually fruit and vegetable, with wine production one of the traditional occupations. An Agro-Industrial Production Association consists of a group of farms and factories together with support organisations such as laboratories and supply chains. Viticulture is an example of a specialisation lending itself to such development, especially in the Crimea and Transcaucasus regions.

Though associated with processing and stimulated partly by the need to improve utilisation of produce, the Intercollective Farm Co-operatives appear now to be concerned mainly with the provision of workshops, electricity generation, fodder production and land improvement. Individual collective farms are shareholders in such enterprises but remain economically and organisationally independent. Similar organisations exist to link both collective and state farms with varying combinations of enterprises.

The total number of interfarm co-operative enterprises was 7200 in 1987 compared with 4564 in 1970. They involved 132983 farms and enterprises (68721 in 1970). The 1987 figures represent a reduction from the peak numbers following a reorganisation in 1985 when many were reclassi-fied as agro-industrial complexes. These co-operatives are strongest in the Ukraine and European areas of the RSFSR and neighbouring Belorussia. They are also important, relative to the sizes of the republics, in Moldavia and Transcaucasia and in the Baltic republics, and less important in Central Asia. Their development indicates a recognition of the need to regard farming as an industry and to promote its integration with the manufacturing industries on which it is increasingly dependent and those which it supplies, as well as in the provision of necessary services.

Perestroika proposals

The efforts at reorganisation in an industrial manner may have contributed substantially to the higher level of harvests obtained in the late eighties but did not raise production to the level required by the expanding population and higher expectations. In 1987 as part of his *perestroika* programme, Mikhail Gorbachev pressed for liberalisation of agriculture to give more freedom to individuals to develop farming on areas of land leased from the collectives. A decree was published in April 1989 permitting families and small groups to rent land, equipment, machinery and buildings. The produce would be owned by the leasing contractor and could be used or sold as he chose.

The leases were to be for five to fifty years with rights of renewal for the farmer and his family who would thus have security of tenure for two or more generations. The state or collective farm is required to agree unless it has good reason for not doing so, but the state will retain ownership of the land. The lessor will, however, have the right to claim for improvements to land or buildings. The decree was stated to be in the nature of an experiment with a more comprehensive law to follow by mid-1990.

BIBLIOGRAPHY

Akademiya Nauk SSSR (1962), *Pochvenno-geografi-cheskoye rayonirovaniye SSSR trans.* by A. Gourevitch, *Soil-geographical zoning of the USSR (in relation to the agricultural usage of lands)* IPST, Jerusalem, 1963.

Bater, J. H. (1989), *The Soviet scene, a geographical perspective*, Edward Arnold.

Blum, J. (1961), *Lord and peasant in Russia from the ninth to the nineteenth century*, Princeton University Press, Princeton.

Bridger, S. (1987), *Women in the Soviet countryside*, Cambridge.

Cohen, S. F. *et al.* (eds.) (1980) *The Soviet Union since Stalin*, Macmillan, London.

Davies, R. W. (1980), *The socialist offensive: the collectivisation of agriculture, 1929–1930*, Harvard University Press, Cambridge, Mass.

Davitaya, F. F. and Sapozhnikova, S. A. (1969), 'Agroclimatic studies in the USSR', *Bulletin of the American Meteorological Society,* **50**, (2) pp. 67–74.

Hedlund, S. (1984),*Crisis in Soviet agriculture*, Croom Helm

Jensen, R. G. (1973), 'Regional pricing and the economic evaluation of land in Soviet agriculture,' in V. N. Bandera and Z. L. Melnyk (eds.), *The Soviet economy in regional perspective*, Praeger, New York, pp. 305–27.

Kaplan, C. S. (1987), *The party and agricultural crisis management in the USSR*, Cornell, Ithaca and London.

Khan, A. R. and Ghai, D. (1979), *Collective agriculture and rural development in Soviet Central Asia*, Macmillan, London.

Khinchuk, K. (1987) 'The agricultural labor force in the Soviet Union,' *Soviet Geography*, **28**, pp. 90–115.

Lewin, M. (1968), *Russian peasants and Soviet power*,Allen & Unwin, London.

Lydolph, P. E. (1979), *Geography of the USSR; topical analysis,* Misty Valley, Elkhart Lake, Wisconsin.

McCauley, M. (1976), *Khrushchev and the development of Soviet agriculture: the virgin land programme 1953–64*, Macmillan, London.

Mathieson, R. S. (1975), *The Soviet Union; an economic geography,* Heinemann, London.

Millar, J. R. (ed.) (1971), *The Soviet rural community*, University of Illinois Press, Urbana.

Narodnoye khozyaystvo SSSR v. . . (various years), *Statisticheskiy ezhegodnik,* Statistika, Moscow.

Pallot, J. (1987), 'Continuity and change in village planning from the 18th century', in *Soviet geography; studies in our time*, eds Holzner, L. and Knapp, J. Milwaukee, pp. 319–349.

Rostankowski, P. (1980), 'The nonchernozem development program and perspective spatial shifts in grain production in the agricultural triangle of the Soviet Union,' *Soviet Geography*, **21**, pp. 409–19.

Shashko, D. I. (1962), 'Climate resources of Soviet agriculture,' in *Akademiya Nauk SSSR*, pp. 378–445.

Smith, R. E. F. (1959), *The origins of farming in Russia,* Mouton, Paris.

Stebelsky, I. (1988), 'Milk production and consumption in the Soviet Union,' *Soviet Geography*, **29**, pp. 459–475.

Strauss, E. (1970), 'The Soviet dairy economy,' *Soviet Studies*, **21**, pp. 269–296.

Stroyev, K. F. (1975), 'Agriculture in the nonchernozem zone of the RSFSR,' *Soviet Geography*, **16**, pp. 186–96.

Stuart, R. D. (1972), *The collective farm in Soviet agriculture*, D. C. Heath, Lexington, Massachusetts.

Symons, L. (1972), *Russian agriculture, a geographical survey*, Bell, London.

Volin, L. (1970), *A century of Russian agriculture*, Harvard University Press, Cambridge, Massachusetts.

Wädekin, K. E. (1973), *The private sector in Soviet agriculture*; trans. K. Bush, ed. G. Karcz, 2nd ed. University of California Press, Berkeley.

8 Minerals, Fuel and Power Resources

The wealth in minerals of the Soviet Union is related to the immense area of the country and the wide variety of geological formations that occur within its boundaries. The widespread distribution of the minerals, however, creates difficulties of exploitation; many are located in the more inaccessible parts of Siberia where inhospitable conditions produce problems of extraction, transport and labour supply and also hinder the creation of a balanced economy in the various economic regions.

The close relationship between geological structures and the occurrence of economic minerals may be illustrated by the case of combustible minerals, such as coal, compared with metal ores. Coal and the other fossil fuel deposits are, of course, organic in origin and are all associated with sedimentary rocks formed since the early Paleozoic era. These formations occur mainly in the extensive lowlands or plateaus of the Soviet Union, in situations where they have been largely unaffected by disturbances such as geological folding or intrusion by igneous rocks. By contrast, many of the metallic mineral ores are found in areas of high seismic activity, such as intense folding and penetration of molten material, exemplified by non-ferrous metal ores such as lead, tin, silver and gold. Such deposits are characteristic of the Pre-Cambrian buried platform or shield, which underlies much of the sedimentary rock cover of the USSR. This ancient eroded mountain mass exposes its crystalline rocks from beneath the cover of younger sedimentary strata in certain places so that its ore deposits become accessible to mining operations as, for instance in the Angara region of Siberia, or in Karelia, in north-western USSR.

THE SOVIET ENERGY SUPPLY

Coal dominated the energy supply during the drive for industrialisation and electrification before the Second World War and for more than a decade after its end. It was the basis of all heavy industry and railway transport. Relatively minor contributions to the energy mix came from hydro-electricity, peat, oil and wood. In the early 1950s coal accounted for as much as 66% of all fuel produced.

A marked change in emphasis appeared in the later 1950s in favour of the hydrocarbon fuels, oil and natural gas, following a belated appreciation by Soviet planners of the greater advantages of these fuels; these included their higher calorific energy, their ease of transport (by pipeline), their cleanliness and the wide range of refinery by-products that became available for the petro-chemicals industry. Within 25 years oil and gas fuels became the dominant fuel in thermal power stations, accounting in the early 1970s for two-thirds of the fuel production. Oil and gas also became important Soviet exports, providing hard-currency earnings and forming an important contribution to the economic development of satellite countries in eastern Europe.

However, it became apparent in the mid-seventies that some of the older oil and gas fields in the European territory of the USSR were failing to reach output targets and were nearing exhaustion.

In view of the vast reserves of solid fuels, especially coal, known to be available, they were again to be given an important role in thermal power generation. Thus the annual production of coal which had increased only slowly during the 1960s has now been stepped up again, despite such problems as those posed by conversion of some thermal power stations from the use of fuel oil to coal. Together with this new emphasis on coal went the decision to proceed with the rapid development of nuclear power, particularly in the European region which had become a fuel-deficient area and where the demand for electricity was constantly increasing.

OIL

Before the Revolution, most Russian oil came from the Baku field, near the western coast of the Caspian Sea, together with smaller outputs from a chain of separate oilfields extending along the north Caucasus piedmont region. This area continued to provide the bulk of the oil supply of the Soviet Union until the Second World War. A much larger oilfield had, however, become known, occupying a huge area of Paleozoic sedimentary rocks extending mainly east of the Volga as far as the Ural mountains. This oilfield, the 'Second Baku' or Volga–Urals field, was rapidly developed during and immediately after the war, and by the early 1950s had surpassed the production of the Baku–north Caucasus area, most of the output coming from the Bashkir and Tatar ASSRs, together with the Kuybyshev oblast. By the middle sixties the field accounted for almost three-quarters of the total crude-oil production and reached a record peak in 1975 of 179 million tonnes, providing much fuel for the post-war economic recovery of the Soviet Union and of her satellite countries in eastern Europe. Today, reserves are much diminished in this field and the principal output of oil now comes from the vast newer oilfields of western Siberia (Table 8.1 and Fig. 8.1).

The latter occur within a great sedimentary basin drained by the Ob river and its tributaries, an area characterised by difficult waterlogged terrain, extensive bogs, swamps, tayga forests, harsh sub-arctic climate and ground affected by permafrost. Rapid development from the 1960s onwards was achieved despite these conditions, with drilling teams being flown in by helicopter from permanent settlements, working a short shift system, then returning for a rest period.

Natural gas in vast quantities is also found within the west Siberian basin, mainly the northern part of the Tyumen oblast, whereas the crude oil deposits extend across a large area of the middle Ob basin covering the Tyumen, Omsk and Novosibirsk oblasts.

The oilfield consists of numerous separate fields, many of which are in the Tomsk and Tyumen oblasts. The supergiant field of Samotlor yielded 140 million tonnes of crude oil in 1977–8, but there are many others of giant capacity; production is controlled by administrative centres which have become large towns based upon the separate fields, such as the Glavtyumenneftegaz association of the six oilfields of Nizhnevartovsk, Surgut, Yugansk, Noyabrsk, Krasnoleninsk and Varyegansk. Within each field several thousand wells are drilled and many new drillings are planned in the current Five Year Plan ending in 1990.

In 1985 western Siberia produced 62% of the total USSR output of 595 million tonnes and this fell considerably short of the target of 628 million tonnes and was accounted for by declining production of the giant Samotlor field. However, production recovered in 1986 and the target of 616.7 million tonnes was reached, an improvement attributed to an increase in drilling, the completion of wells and the expansion of production in the Varyegan district centred on the new oil town of Raduzhnyy, north of Nizhnevartovsk. If the difficulties presented by the environment and by shortages of manpower and lack of infrastructure facilities are overcome, the projected plan target of 635 million tonnes is likely to be reached by 1990. This figure would represent 70% of the total of the USSR oil output being provided by western Siberia but it is dependent upon many small fields in the region being successfully exploited now that the Samotlor field is declining.

In other regions of the Soviet Union, the smaller and mostly older oilfields have slowly diminishing outputs, creating the problem of greater dependence upon west Siberian oil. In European Russia, the Volga–Urals field produced less than 22% of the USSR total output in 1986. However in this field the exploitation of natural bitumen in the Tatar ASSR is regarded as a possible substitute for conventional oil. Farther north, increased output is planned for the Komi ASSR; extraction is in progress from wells on the island of

TABLE 8.1 GEOGRAPHICAL DISTRIBUTION OF SOVIET OIL PRODUCTION (*million tonnes a year*)

	1980	1984	1985	1986	1987	1990 Plan
USSR	603	613	595	617	624	625–640
RSFSR	547	561	542	564		560–575
European Russia	149	(107)	(98)			
Komi ASSR	19	19	18			
North Caucasus	19	12	10			
Volga	111	76	70			
Urals	82	73	69			
Siberia	316	381	375			(450–465)
West Siberia	313	378	372	398		(445–460)
Tyumen	303	(366)	(360)			442–437
Tomsk	10	(12)	(12)			(24)
Sakhalin	3	3	3			(3)
Outside RSFSR	56	52	53	53		65
Ukraine	7.6	(7)	(6)			
Belorussia	3	(3)	(3)			
Georgia	3.2	(2)	(2)			
Kazakh SSR	18.7	19.5	22.8			(30)
Turkmen SSR	8	(7)	(6)	(5)		(5)
Other Central Asia	1.5	(1.5)	(1.5)			
Azerbaydzhan	14	(12)	(12)			14–15

The Ural figures include those of the Bashkir ASSR with about 33 million tonnes in 1985.
Soviet oil statistics include both crude oil and gas condensate.
The figures in brackets are estimates.
The 1987 total production was 624 mmt, no details available.
Sources: Narodnoye khozyaystvo SSSR, RSFSR, various years and from Shabad (1986) estimated from Pravda and Sovetskaya Rossiya.

Kolguyev in the Barents Sea. Komi produced 19 million tonnes of oil and gas condensate in 1986. The remaining oilfield in the main republic, the RSFSR, is situated on Sakhalin island in the Soviet Far East, but its very small output of 2 million tonnes yearly is insufficient to meet the needs of the region which are supplied from western Siberia.

Outside the RSFSR, oil output is contributed from Kazakhstan, Central Asia, Transcaucasia, the Ukraine and Belorussia, the total remaining at about 55 million tonnes, during the 1980s. The leading republic is Kazakhstan, where in the north Caspian basin, the Karachaganak project produces gas condensate (classed as oil in the USSR); farther south, along the eastern shores of the Caspian Sea are the Tengiz deposits, lying at great depth below salt domes; geologically similar is the Zhanazhol oilfield in the Oktyabrsk oblast, having a high content of sulphur and gas condensate which requires processing at the centre of Karaton. Technical difficulties in these fields are retarding development, although production is planned to start in 1988. Farther south, the old Mangyshlak or Uzen field continues a modest output of 18 million tonnes, and in the same area a newer field on the Buzachi peninsula is increasing production and reached 4.7 million tonnes in 1985.

Azerbaydzhan, the third largest oil-producing republic, continues to exploit the Baku oilfield, including its extensions eastwards beneath the Caspian Sea. Most of the output is obtained from the offshore rigs east of Baku and is planned to reach 18 million tonnes in 1988 in the current Five Year plan.

Ukrainian oil comes from two separate producing areas; in the western Ukraine there are three oilfields near the Carpathian mountains, the Borislav, Dolina and Bytkov fields; in the east the oil is in the Dnepr–Donets depression, near Chernigov and Poltava.

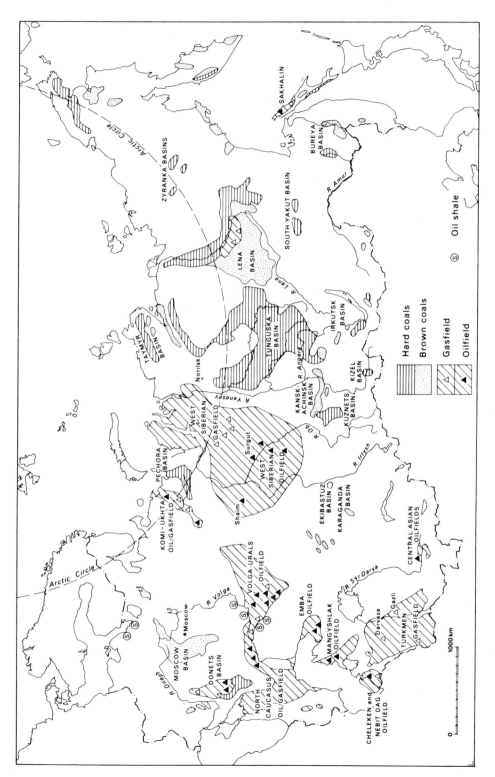

Fig. 8.1 Coal, oil and natural gas; the distribution of the major energy resources of the USSR

Oilfield plant in the Siberian tayga, Tyumen in the Ob river basin

Other oilfields of relatively low output and of only local importance are in the Central Asian republics of Uzbekistan, Turkmenistan, Kirgizia and Tadzhikistan, and in the Transcaucasian republic of Georgia.

There has been continuing speculation in the West, based upon information derived from a wide range of Soviet sources, that the fuel and power supply industries of the USSR are faced with serious problems, particularly those of sustaining the recent recovery of petroleum oil output. No information is given by the USSR on the extent of oil reserves but it appears that the expected recovery of large quantities of oil associated with the giant gas deposits discovered in the Arctic far north have not materialised, despite intensive exploration in that region.

Considerable potential output still exists within the west Siberian oil province, but the big fields being exploited at present are ageing rapidly. They are demanding increasingly large outlays of investment for the replacement of worn or obsolete equipment and for the development of smaller fields in areas of difficult access and arduous geographical conditions of climate and terrain. Therefore the ever-rising costs of the effort to maintain or increase oil output are resulting in diminishing returns and are thus becoming a drain on the national resources. As an example of comparative capital outlays, those for the entire west Siberian gas and oil region which has only 1.3% of the Soviet population are equal to those accorded to regions such as Central Asia and southern Kazakhstan with ten times the population.

On the other hand, Soviet efforts to economise in the use of oil for fuel consumption in, for example, electricity power generation, have not met with desired results and in fact the consumption for this purpose rose sharply in 1984–5 by nearly 4% per annum. The substitution of other fuels for oil is hindered by reduction in the quality of coal supplies and by the inadequate facilities for the storing and distribution of natural gas. An additional and unexpected extra need for oil fuel supplies was created by factors resulting from the aftermath of the Chernobyl disaster such as the temporary slowing down of the nuclear power programme in the industrial west of the USSR. However, the current Five Year Plan restores the former emphasis upon future nuclear installations.

The transport of oil (Fig. 8.2)

On the oilfields, systems of oil-gathering pipelines collect the liquid from the well-heads. After cleaning by oil-treatment plants, pumping stations deliver the oil to storage-tank farms. A nation-wide network of long-distance pipelines then conveys the oil to refineries and to consuming centres.

Prior to the development of the pipeline system, crude oil had been shifted from the oilfields mainly by rail and by water. The main direction of movement had been to the north or northwest from the Baku–north Caucasus oil-bearing areas in southern USSR to the consuming areas such as the Central Industrial region. The Caspian Sea

Oil rigs in the Caspian Sea

and the Volga river played an important role in oil transportation, allowing the fuel to be shipped from Baku to Astrakhan and then transferred to barges for transport upstream to refineries situated at rail crossings at Saratov and Gorkiy. From there it was taken by rail to consuming centres. In 1955 nearly one-third of the Volga freight tonnage was of oil or oil products. Pipelines were used to link the various oilfields of the north Caucasus area to the Black Sea ports of Tuapse and Batumi where refineries were situated and the products shipped to the Ukraine.

With the development in the Volga–Urals field and the increasing demand for oil and oil products in the industrial and urban areas in the west of the USSR, the pipeline system grew rapidly, linking the new field with its markets in two dominant directions. Towards the west, oil was piped to Gorkiy, Ryazan, Moscow and Cherepovets, eventually reaching Leningrad. Another branch ran to Bryansk and thence to Polotsk, Riga and to the Latvian export terminal at Ventspils, with an extension to Klaypeda in Lithuania. Crude oil was also pumped from Bryansk westwards along the 'Friendship' line to refineries at Plock (Poland), Schwedt (East Germany), Bratislava (Czechoslovakia) and to Szazhalombatta in Hungary. Another great artery of oil ran eastwards to serve the Ural manufacturing areas and continued across Siberia to Omsk, Novosibirsk, Krasnoyarsk and Irkutsk (Fig. 8.2).

Further extension and expansion of the pipeline network has followed the increasing production of oil from the new west Siberian fields since 1964. The objective has been to provide adequate pipeline capacity from the Siberian fields to refineries in the European USSR and to western oil-export terminals. The westward flow of oil from these fields in 1970 was only about 4% of the total national production but by 1975, when the total Soviet output was 491 million tonnes, the flow had increased to 23% of the national production. It was expected that by 1980 the Siberian fields would supply the western refineries with 240 million tonnes of crude oil, or 37% of the total output of 640 million tonnes. Pipelines of larger diameter have been necessary to carry these increasing loads.

Pipelines now run from the oilfield centres such as Nizhnevartovsk, Surgut, Yugansk, Samotlor and Noyabrsk, with the bulk of the flow directed south and west through Omsk and Kurgan, cross-

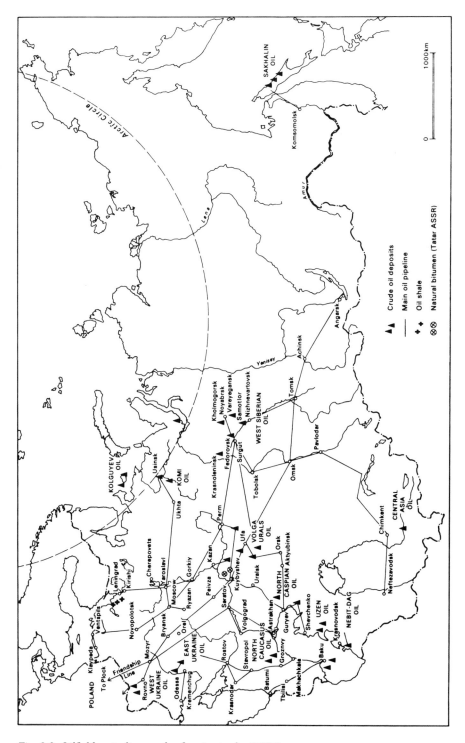

Fig. 8.2 Oilfields, pipelines and refineries in the USSR

ing the Urals via Chelyabinsk to Kuybyshev and then on to the Moscow region. A second outlet runs further north from Surgut over the Urals near Perm to the refineries at Gorkiy, Ryazan, Moscow, Mazeikiai (Lithuania) and the terminal at Ventspils on the Baltic Sea. West Siberian oil is also being distributed to the east, south and south-west. The eastward flow runs from Nizhnevartovsk towards eastern Siberia through Krasnoyarsk to Angarsk, and a pipeline completed in 1986 runs along the bed of the swift Angara river to Irkutsk. A south-westward flow is for the benefit of the Ukraine and runs from Kuybyshev to the refinery at Lisichansk in the Donets basin; an extension reaches another refinery at Kremenchug on the Dnepr river and also brings oil to the tanker terminal in Odessa on the Black Sea. A southward flow of west Siberian crude is piped via Pavlodar and Chimkent to a new refinery which is to begin production at Neftezavodsk near Chardzhou in the Turkmen SSR during the current Plan period.

Other important, though less extensive, pipeline systems are those linking the Caspian and north Caucasus oil deposits to their respective markets. Thus, Baku oil is piped north via Tbilisi to the refinery at Batumi on the Caspian Sea. In Turkmenistan, oil production from the Nebit-Dag field is linked by pipeline to the refinery at Krasnovodsk on the Caspian coast. West of the Caspian, refining centres of the north Caucasus oilfields, such as Dagestan and Groznyy, send oil to the Ukraine and to the Black Sea ports.

Soviet oil from the Arctic is piped from the Ukhta fields in the Komi ASSR south-west to refineries at Yaroslavl, Ryazan and Kirishi, near Leningrad. In the future, the output from the new oilfield on Kolguyev Island and from offshore extensions in the Barents Sea will be piped southwards by a new line now under construction. New deposits that may be discovered in the north of west Siberia will also be served by a pipeline connection. However, experiments have been made to transport oil from this region, including output from the small Kharyakha deposit in the Nenets okrug, by tanker instead of using overland pipelines southward.

In general however, the fall in oil production during the previous Five Year Plan has had an effect upon the construction of new pipelines and refineries. The present plan calls for a larger output of product pipelines than crude oil pipelines, reflecting a growing need for the installation of product pipelines in the major areas of consumption.

Natural Gas (Table 8.2; Fig. 8.1)
There are two contrasting situations in which gas may occur in the earth as 'natural gas' (i.e. not manufactured in gasworks from coal). Natural gas is very often found in association with crude oil within the containing sedimentary rock formations, and is extracted during drilling for the oil. Many of the Soviet oilfields yield large quantities of this oil-well gas. Natural gas may also occur alone as 'dry' or 'lean' gas and not associated with oil. It is this gas that has, together with oil, contributed to the vastly increased use of hydrocarbon fuels in the USSR in recent years.

The oil-well gas or 'wet gas' is commonly encountered first during drilling as it often rests on top of the oil in the containing rocks. The gas can be used in unprocessed form, for instance as fuel in the generation of electricity in power stations, or it may be 'flared' or burnt. Both alternatives are wasteful because the heavier constituents can be utilised in gas-processing plants to produce ethane, the raw material for ethylene manufacture, butane, the bottled liquefied petroleum gas, and natural gasoline.

During the rapid exploitation of the west Siberian oilfields, the recovery of this useful associated gas failed to keep up with the pace of oil extraction, and it was wastefully flared. Thus in the development of the giant Samotlor oilfield more than 8 billion m³ of oil-well gas was wasted by flaring in 1975 because of the absence of gas-processing plants and of gas pipelines of sufficient length to reach such plants. This tremendous waste of gas is now being reduced by the establishment of additional processing plants, as was set out in the tenth Five Year Plan (1976–80). Some of the gas is now transmitted by pipeline to the Kuzbas industrial area of west Siberia, where it is to be used instead of coal in electric power stations, thus reducing urban pollution; it also serves as a basis for the petrochemicals industry of Kemerovo.

The pattern of natural gas exploitation in the Soviet Union has to some extent followed that of oil. As the two forms of fuel are commonly found together in sedimentary rocks, the exploitation of natural gas first occurred in the north Caucasus region. Large-scale production did not begin until

TABLE 8.2 GEOGRAPHICAL DISTRIBUTION OF SOVIET NATURAL GAS PRODUCTION (billion m³)

	1980	1984	1985	1986	1990 Plan
USSR	435	587	643	672	835–850
RSFSR	254	413	462	494	640–650
European Russia*	40	(32)	(31)	(29)	(30)
Komi ASSR	17.5	(16)			
North Caucasus	14	(9)			
Volga	8	(7)			
Urals*	52	(50)	(50)	(50)	(50)
Siberia	162	(331)	(380)	(415)	565–570
Tyumen' Oblast	156	324	(372)		560–565
Urengoy	50	210	(250)		
Yamburg	—	—	—		
Medvezhye	71	(72)	(72)		
Vyngapur	16	(17)	(17)		
Oilfield gas	12	19	21		
Other Siberian	6	(7)	(7)		
Outside RSFSR	181	174	181	178	195–200
Ukraine	52	(43)	(42)	(38)	
Azerbayzhan	14.5	(14)	(14)	(13)	
Kazakh SSR	5	(5)	(6)	(7)	(15)
Uzbekistan	39	(37)	(37)	(38)	
Turkmen SSR	70	75	(82)	83	86

* Bashkir ASSR with one billion m³ in 1985 is included in Urals.
The 1987 total production was 727 billion m³; no further details are available
Sources: as Table 8.1

the later 1950s, as with crude oil. In 1950 it was less than 6 billion m³ rising to 28 billion m³ in 1958. The gas came from the north Stavropol field, one of the first large deposits to go into full production; this, together with gas from other 'dry' gasfields in the Krasnodar area, was supplied by pipeline to Moscow and other cities of the Central Industrial region and also in a southward direction to Transcaucasia. Other dry gasfields utilised at this period were the Dashava field, acquired from Poland after the Second World War, and the Shebelinka field in the Ukraine.

Larger gasfields were brought into production during the 1960s and 1970s. Immense reserves were discovered in Central Asia in both the Uzbek and Turkmen republics, but, as in the case of oil, very long pipelines were necessary to convey the fuel to the consuming areas. By 1965 a new gasfield in the Uzbek republic, Gazli, was supplying fuel to industrial centres in the Urals, such as Chelyabinsk, by means of pipelines more than 2000 km in length; in 1966 the Gazli fields were

yielding over 22 billion m³ of gas, of which more than 80% was piped to Urals industry. The remainder was piped eastward to the cities of Central Asia. A new city, Navoi, was founded in the vicinity of the Gazli field, using the gas for the production of thermal electricity, nitrogen fertilisers and synthetic fibres. Gas from the Turkmen fields of Darvaza and Achak was also piped northwards for consumption in Moscow, Leningrad and the Baltic republics. Additional newer fields brought into the Central Asian supply system were the Uchkyr field near Gazli and the Shatlyk field west of Mary. In the RSFSR the Volga–Urals oilfield began producing gas with the discovery of the giant Orenburg deposit in 1966. The location of this field was favourable for transmission to the western industrial areas of the USSR and for export by pipeline to eastern Europe.

In the later 1960s the huge potential of gas deposits in the north European and west Siberian areas of the RSFSR began to be realised. Development was difficult as the gasfields were located within some of the remotest parts of the Soviet Union where extensive swamps, bogs, forests and permafrost hindered operations, and where no transport facilities, apart from rivers, existed. As an example, equipment for working the giant Medvezhye gasfield of western Siberia was brought down the Ob river to the Gulf of Ob and then taken upstream along the Nadym river to Nadym, which became the base town. This was possible only during the short 3-month ice-free season but during winter, trucks used temporary cross-country roads. Even greater difficulties of access were encountered during the work on the next gasfield to be developed, the Urengoy field, where drilling rigs carried by truck convoys had to be transported for many miles across tundra.

The giant Vuktyl gasfield in the Komi ASSR in northern European USSR became an important contributor to Soviet gas output in the 1970s and had become part of the gas pipeline network in 1968 by a major transmission link called the "Northern Lights," designed also to bring west Siberian gas southwards and westwards towards European USSR along large diameter pipes, with part of the output reaching Leningrad.

At the present time, however, the emphasis in Soviet gas production is upon the supergiant deposits of western Siberia, Urengoy, the largest gasfield, began production in 1978, adding its output to that of the other great field of the Tyumen oblast, the Medvezhye field. The Urengoy field became the focus of development in 1981–5 and reached the designed capacity of 250 billion m³ at the end of that period. Meanwhile, Vyngapur, a smaller deposit south of Urengoy was increasing output in 1980. Gas deposits situated even further north were, however, already scheduled for exploitation, and in 1986 the next supergiant gasfield, the Yamburg, began production. This deposit is located on the eastern coast of the Gulf of Ob in the Yamal–Nenets national okrug, north of the Arctic circle. Difficulties of penetration of permafrost layers in this severe environment were solved by the use of special steam drilling techniques and equipment. Although the Yamburg gasfield was receiving emphasis on development during the twelfth Five Year Plan and was expected to yield 200 billion m³, no permanent settlement will be created as work will be done using Urengoy as a base for a shift workforce to be flown in to the site. A railway extension from Novyy Urengoy will re-

lieve the dependence on water transport of equipment in summer. Later, in the 1990s, the immense reserves of another gasfield north of Urengoy and near the settlement of Tazovskiy, namely the Zapolyarnoye, will be exploited.

This will become part of other gas projects for that period, situated even farther north in the western region of the Yamal peninsula and near the coast of the Ob estuary. Here the Kharasavey, Bovanenkovsk and the Arkticheskoye gasfields will be linked by the railway planned to run northwards from Labytnangi; the latter, being situated on the Ob river and having a rail link across the Urals, will serve as a point of transshipment, with connections southward to the Urengoy and Medvezhye fields. Thus west Siberian gas will play a dominant role in USSR gas production, supplying two-thirds of the total within the twelfth Five Year Plan period and three-quarters by the year 2000.

Part of the gas is used at present to supply gas-fired electricity generating stations in, for example, Surgut and Nizhnevartovsk, two of the major cities which are administrative centres of the industry. But remoter destinations of the gas production are being reached by a great pipeline network aligned westwards to European Russia and beyond to eastern and western Europe. These transcontinental pipes are 142 cm in diameter to carry Yamburg gas as part of a CMEA joint project including a fuel supply to Greece and Turkey. In other parts of Siberia several smaller gasfields produce fuel for local needs: thus Norilsk, the northern metal mining and smelting centre receives gas from two deposits situated to the west of the city. Farther east, in the Yakutsk republic, the Vilyuy river basin contains gas deposits with large reserves; these were formerly considered for a proposal to export gas to Japan and to the United States, which was later abandoned. In the Soviet Far East, gas on Sakhalin is piped to the mainland for the city of Komsomolsk. Completed in 1986, it had an initial capacity of 1.5 billion m³ per year.

Outside the RSFSR the non-Russian republics contributed in 1985 about 28% of the USSR total, equivalent to less than one-half (48%) of the west Siberian output. Of these, the greatest production came from the Turkmen SSR, where the previous Five Year Plan emphasised development of the Sovetabad gasfield, near the Iranian border. This has large reserves and first produced gas in 1982; in 1985 its yield totalled one-fourth of the Turkmen total output. The field will replace the older Shatlyk field, which is now declining, as are other smaller fields in the republic.

In Uzbekistan, the focus is upon the giant Shurtan field, near Karshi, one of a group of gas deposits of the Amu–Darya valley around Bukhara, with Uchkyr and Gazli in the west and in the Karshi district in the south-east. Here, several separate gas deposits yield fuel with a high sulphur content, requiring processing treatment in the centre of Mubarek. The field has large reserves and part of the production is piped to the Tashkent area for thermal electricity generation.

In Kazakhstan, development is centred mainly on the north-western area of the republic on the gas and gas condensate deposit of Karachaganak, lying below salt domes in Permian strata. A pipeline links the deposit to the Orenburg processing complex but the gasfield's own processing installation is being built during the twelfth Five Year Plan; in 1990 an output of 8 billion m³ is expected.

Ukrainian gas, ranking third in output after that of the Central Asian republics of Turkmenistan and Uzbekistan, achieved an output of 40 billion m³ in 1986, but was expected to diminish thereafter. As in the case of oil, it is locally important, being close to consuming centres. The pipelines linking the Shebelinka gasfield in the eastern Ukraine to Ukrainian and Moscow industrial areas were among the earliest to be completed in the USSR. Farther west, natural gas development in the Carpathians centred upon the old Dashava deposit but is now augmented by the larger gasfields to the north around Rudki. Part of the yield is exported by the 'Brotherhood' pipeline system to Czechoslovakia and Austria.

Azerbaydzhan's own output of gas, now much depleted, is maintained by supplies imported from Iranian deposits south of Teheran. The fuel is linked to the Stavropol–Ordzhonikidze pipeline system which extends across the Caucasus range to Tbilisi in Georgia and beyond, thus continuing the north Caucasus supply to central European Russia.

The gas pipeline network

From the above outline of the natural gas resources of the Soviet Union, it will be apparent that the country is now criss-crossed by an elaborate network of gas pipelines, to some extent separate from that which carries oil, yet with similar general directions, and with a concentration in western USSR, the destination of most of the gas. This network expanded in the 1970s at an average rate of 5000 km each year and consisted of large diameter pipes with a marked concentration toward Moscow and the industrialised Centre, leading from the southern gasfields of the north Caucasus region and Central Asia. Linked systems carried the fuel farther west to cities in the Baltic area and onwards to eastern and western Europe, with the Donbas also linked to this main flow, reinforced by gas from the Orenburg field. A sub-system embraced cities of Central Asia such as Tashkent, Frunze and Alma-Ata.

More recently, the large-scale exploitation of the giant gas deposits of western Siberia has required the extension of the pipeline network, which will continue to be augmented during the 1990s, as required by the twelfth Five Year Plan. From the Medvezhye and the Urengoy fields, gas is carried along two basic pipeline routes westwards and southwards. The westward route runs over the Urals to join the pipeline from the Ukhta of the Komi ASSR, feeding fuel to Vologda,

The most northerly gas pipelines in the world are in the USSR: the Mesoyakha-Norilsk line in June, with pipes for doubling the line being brought by truck and trailer

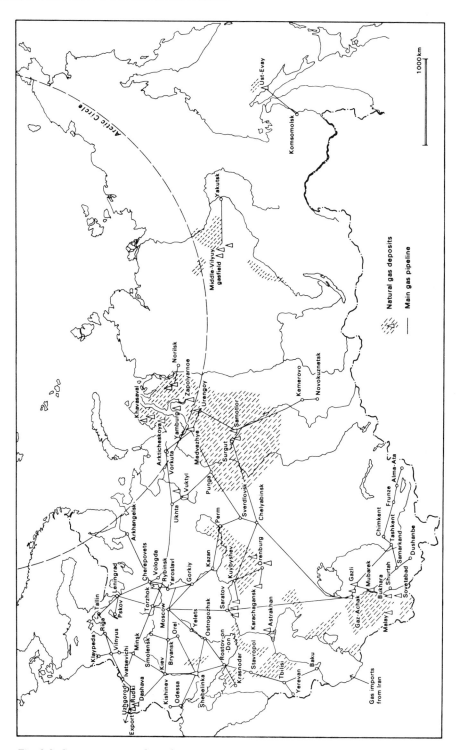

Fig. 8.3 Gas resources and pipelines

Cherepovets and onwards to Leningrad and Tallin and further to eastern Europe. The south-ward route runs through Nadym and supplies gas to the Ural manufacturing areas and onwards to Central Asia.

The immense gas development programme projecting an output of 1000–1100 billion m³ by the year 2000 requires the supply of vast quant-ities of large diameter pipes such as those recently completed for the Urengoy gasfield. These are now being laid to serve the next Arctic gas pro-ject, Yamburg, situated north of Urengoy. This project will also require a similar set of six pipe-lines of 142 cm diameter and having a total length of 23 000 km to convey gas to European USSR for export to eastern and western Europe. Two of these were completed by 1987 linking Yamburg to Yelets, the centre of distribution of gas for central Russia's domestic use. A third, the 'Progress' export pipeline was a joint Comecon project with a completion data in 1988, from Yamburg to Uzhgorod an extension across Bulgaria to Greece and Turkey. The bulk of the large diameter piping is manufactured in Japan and West Germany and is delivered by river and sea routes to the northern gasfield sites. West Siberian gas is also piped southwards to supply fuel to the Kuzbas and other south Siberian destinations.

Smaller and more local developments in the USSR, distinct from the above, include: pipeline extensions improving gas supplies to Norilsk from its local deposits; the completion of gas connections by pipelines made in 1986 in the Far East between the Sakhalin gasfield and Kom-somolsk; additional pipeline links developing the gas network in the Yakut SSR from the Middle Vilyuy field; and the new 150 km pipeline from the Karachaganak gas condensate deposit in west-ern Kazakhstan to the Orenburg processing com-plex. In Central Asia, new pipelines now join the high-sulphur Shirtan gas deposits of Uzbekistan to Mubarek, the processing centre.

Coal

The output of solid fuels has increased during each period of the successive Five Year Plans since the end of the Second World War. Although their relative importance in the national fuel bal-ance was reduced by the rapid growth of the production of oil and gas in the 1960s and the earlier 1970s, the tenth Five Year Plan, for 1975–80 provided for a large increase in the coal

output (Table 8.3), although the projected figure of 775 million tonnes was not achieved. The Twelfth Plan however, raised the target for 1986–90 to 780–800 million tonnes thus main-taining the high status placed on coal production in the USSR fuel balance, and also making the country the world leader in output of this fuel.

The map (Fig. 8.1) shows the distribution of the USSR's coal together with other fossil fuels. The impressively large deposits of eastern Siberia, the Tunguska basin and the Lena basin are inferred but not proven or measured; together the two basins are thought to contain about 3500 billion tonnes of hard or bituminous coal, with brown coal or lignite but the deposits are both inaccess-ible and remotely situated and do not figure in the details of coal production.

Much more realistic are the quantities of solid fuel that have been measured or proved and which form the basis of the Soviet coal industry. The hard or bituminous coals include the deep-mined coking coal required in the production of steel and also for fuelling electric power stations. They are of high calorific value and carbon con-tent. The lignite has much lower heating value for its bulk, yet can be mined cheaply from shallow surface workings. United Nations figures have indicated that the USSR has one-third of the earth's known reserves of lignite and about one-fifth of the coal.

Table 8.3 emphasises the wide dispersal of Soviet coal deposits which are found in geological strata of various ages. As the reserves in the western fields become reduced, more and more of the output is originating in the Asian coalfields of Siberia and Kazakhstan and this gives rise to very long rail hauls of fuel to the consuming centres situated mainly to the west of the Urals. This trend is to increase in the future with the modern plans for extensive development of the eastern deposits of Siberia and Kazakhstan.

The coalfields

The Donets basin (Donbas) is situated mainly in the Ukrainian SSR, and for many years since the middle of the 19th century it has dominated hard coal production except during the German inva-sion of the Soviet Union in the Second World War. It is by far the most conveniently located coalfield, being close to its main markets, the Ukraine and the Central Industrial region. In 1913 during Tsarist times, it provided no less than

TABLE 8.3 GEOGRAPHICAL DISTRIBUTION OF SOVIET COAL PRODUCTION (million tonnes of gross mine output.)

	1980	1984	1985	1986	1990 Plan
				Plan	
USSR	716	712	726	744	780–800
Hard coal	553	556			
Anthracite					
Coking coal					
Steam coal					
Lignite	163	158			
Deep-mined	445	418	421	(419)	420–430
Strip–mined	271	294	305	(315)	360–370
RSFSR	391	(382)	(392)	400	440–445
European Russia*	(85)	(81)	(80)		(75)
Moscow Basin	26	20	19		
Pechora Basin	28.6	29	30		
Donbas (east)**	(31)	(31)	(31)		
Urals*	(41)	(37)	(36)		
Siberia	265	(266)	(276)		(330)
Kuznetsk Basin	145	143	145	(149)	160
Kansk–Achinsk	35	38.4	40.8	42.9	(65)
Neryungri	2.5	7	10		
UKRAINE	197	(191)	(190)	(186)	(186)
Donets Basin**	204	198.3	197.1	(193)	193.3
KAZAKHSTAN	115.4	125.5	130.8	(134)	(146)
Karaganda	48.6	49.6	49.8	(50)	(50)
Ekibastuz	66.8	75.9	80.5	83.6	95.96
CENTRAL ASIA	(11)	(11)	(11)	(11)	(11)
Uzbek SSR	5.7				
Kirgiz SSR	4	4			
Tadzhik SSR	(1)				
GEORGIA	(2)				

* Bashkir ASSR is included in Urals
** Donets Basin, shown entirely under Ukraine, includes part in RSFSR. The 1987 total production was 760 million tonnes; no details available.
Sources: as Table 8.1

87% of the total national output and in 1940 it still yielded one-half of the total national tonnage. Its share of this total was, however, declining, as newer coalfields in western Siberia and Kazakhstan were rapidly increasing production. The importance of the latter two fields became critical in the Second World War when both the Donbas and also part of the Moscow field were in enemy hands.

The coals produced in the Donbas are of high quality, having a high calorific and carbon content and although sulphur is present, they are of great importance for the production of coke for use in blast furnaces, hence the location of large steel-manufacturing enterprises near the coalfield. Several kinds of coal are mined: in addition to coking fuels there are also steam coals, gas coals and anthracite; but many seams are discontinuous and are rather thin, less than about one metre, and lie at considerable depth, so that extraction costs become higher as the deeper seams are worked. This places Donets coal at a disadvantage in competing with that of the Kuznetsk basin where mining conditions are more favourable.

The coalfield extends across two administrative divisions, the eastern Ukraine and the Rostov oblast of the Russian republic. In 1985, more than 84% of the production of the entire field came from the Ukrainian portion, where new, large, deep mines in the main producing areas and also numerous shallow mines in the western part of the basin have been sunk in the Pavlograd district of the Dnepropetrovsk oblast.

Here, new coal towns such as Pershotravensk and Ternovka have appeared, with populations of about 25 000 in the later 1970s. But the deterioration of mining conditions resulting from increasing depths and geological complications has been adversely affecting output in spite of heavy Government investment, so that production has been declining continuously from its peak of 224 million tonnes in 1976 to about 200 million tonnes in 1986.

The western Ukraine also contains a relatively small, yet locally important bituminous coal deposit, the Lvov–Volhynian basin which feeds steam coals to adjacent power stations. These form part of the grid supplying electricity to eastern Europe. Lignite obtained from the Dnepr lignite basin in the Ukraine is converted to briquets for local use.

To the north, lignite is also produced in the Moscow basin, an old field extending south and west of the capital through the urban areas of Tula and Novomoskovsk, which are among the chief mining centres. The output comes from opencast or surface mines and is moved to large thermal-electric power stations which transmit power to the industries of the region. The heavy demand for electricity in the area has necessitated a high output which in former years was over 10% of the Soviet total production, but this is slowly declining as the small reserves become used up and mines close down and are not replaced. In 1986 the total output was 20 million tonnes.

In the Urals, there is a lack of a sufficient supply of good quality coking coal for smelting the rich ore deposits, hence the import of coal over long distances continues to be a feature of the industry. Locally produced coal of suitable quality is thus of great importance and occurs in separate basins whose aggregate production exceeded 10% of the USSR total in former years, reaching a maximum of over 60 million tonnes in the 1960s but declining in the next decade to less than 45 million tonnes. The total production in 1986, including that of the Bashkir ASSR was estimated at 32 million tonnes.

The drastic reduction in coal supplies from the Moscow basin during the last war, and the total loss of production from the Donets basin during the German invasion of 1942–3, impelled the Soviets to open a third coalfield in Europe, the Pechora basin in the remote far north of the Komi ASSR. From this Arctic coalfield coking coal is railed from the town of Vorkuta south-west to the steel manufacturing centre of Cherepovets; part of the output also reaches Lipetsk, in central Russia, and Murmansk in the Kola peninsula. Output in 1986 was 30 million tonnes.

The Asian coalfields

The principal Asian coalfield is the Kuznetsk basin or Kuzbas, situated in south-western Siberia. It occupies a large depression in the valley of the upper Tom river, 300 km long and 160 km wide, flanked by spurs of the southern mountain ranges, the Kuznetsk Alatau in the east and the Salair ridge in the west. Although the output of coal from the Kuzbas (150 million tonnes in 1987) is below that of the Donbas, the very heavy investment in the Kuzbas during the current Five Year Plan period is expected to yield increased output. This will reflect the relative newness of the field and the physical advantages of greater reserves of high quality bituminous hard coal contained in thick undistributed seams close enough to the surface to be extracted from surface workings. These conditions contrast with those of the much older Donbas there mining is virtually all done from very deep mines. In such circumstances, Kuzbas coal is so much cheaper to raise than that of the Donbas that it competes successfully with the latter even after the costs of a long railway haul of 3000 km or more to parts of European Russia are taken into account.

Plans for the Kuzbas envisage an output of 160 million tonnes by 1990, including both coking coal and steam coals. Much of this will be mined from the underdeveloped part of the coalfield, the Yerunakovo district, 40 km north of Novokuznetsk, which contains large reserves suitable for strip-mining operations. An output level of around 220 million tonnes by the year 2000 was predicted in the Long-Term Energy Programme but this may not be achieved as a result of the former lack of investment in the development of this field in comparison with that accorded to the Donbas.

Constraints upon the productivity of this very rich coalfield are threefold; most of the important coking coal is not obtained from surface mines but from sites deeper underground; renewal and extension of such sites is now required for which there has been inadequate preparation. There is also a need for increased supply and modernization of beneficiating equipment. The third limitation is the question of transport of the output, a major proportion of which is destined for the Urals and further west into European Russia and even to the Ukraine, thus imposing an increasing strain upon rail facilities to the west. One proposal to reduce such rail congestion is a coal slurry pipeline such as that now being tried out between Kuzbas and Novosibirsk and designed to carry 3 million tonnes of slurry coal annually over a distance of some 250 km for fuel and power supply to the Siberian urban centre. Due for completion in 1987, this is intended to be the forerunner of a longer transcontinental pipeline.

Between the Kuzbas and Lake Baykal are several separate coal basins; this eastern Siberian group consists of the Minusinsk basin situated in the upper Yenisey valley, and further north, adjacent to the Kuzbas is the Kansk–Achinsk basin; eastward of the latter and north-west of Irkutsk, the Cheremkhovo deposit is found; this formed an important fuel supply late in the last century for locomotives of the Trans-Siberian Railway. The lignite of this field is strip-mined near Tulun.

The largest output from this group of coalfields comes from the Kansk–Achinsk deposit, now the basis of an important lignite working combined with thermal power stations close to the strip mines. This has become a vast project, receiving much emphasis in the twelfth Five Year Plan, with a production that has been expanding to 40 million tonnes of lignite each year, for fuel supply to power stations situated as far as 3500 km distant in Soviet Central Asia. Several very large power stations were planned for transmitting electricity through ultra-high voltage power lines to the Urals and to European Russia. But problems with losses of power in transmission along such UHV lines have resulted in reducing the plans in favour of long-distance transport of the lignite through slurry pipelines, although the feasibility of even this alternative is in question. An experiment for a synthetic fuels industry based on the lignite output was given up in 1987 through shortage of labour and infrastructure in the area. In any case, the wide development of nuclear power in European Russia has reduced the urgent need for low quality fuels from such a distant source as the Kansk–Achinsk project with its attendant problems of logistics.

Another small group of coal deposits east of Lake Baykal includes those of Gusinoozersk, Tugnuy and Kharanor, the latter just north of the Mongolian frontier. The lignite of Gusinoozersk is used for thermal power stations in the Buryat ASSR at the capital, Ulan–Ude, but the supply is to be augmented in the current plan by the development of an additional source, that of the Tugnuy deposit, expected to produce in the future 12 million tonnes of fuel.

In addition to the Siberian coalfields of Soviet Asia, the USSR also obtains a very large output from deposits in northern Kazakhstan, that of Ekibastuz and Karaganda. Of these, Ekibastuz has the distinction of being the USSR's third largest coal basin (after Donbas and Kuzbas). It contains bituminous coal of lower quality than that of the Kuzbas, having a high ash content and inferior calorific value. But the thick seams are easily accessible and cheaply mined by rotary excavator from extensive workings so that production can be maintained at a high level from such sites as the renowned Bogatyr (Hero) mine, the USSR's largest strip mine. It produced 54 million tonnes of fuel in 1986, 4 million tonnes more than its rated capacity. A new working, begun in 1985 is expected to attain an output of 30 million tonnes.

The total output at Ekibastuz in 1986 was 87.5 million tonnes, almost all of which is used only as power station fuel. This is railed so such units in north Kazakhstan and adjacent parts of west Siberia in cities such as Omsk, and to Central Asia, but the largest proportion goes to the Urals. Such wide distribution of huge tonnages of fuel results in overloading of the rail system, adding to that generated by the Kuzbas, and exemplified by the record load that left Ekibastuz one day in February, 1987 with 430 wagons carrying a total of 42 000 tonnes of coal and occupying 6.5 km of track. As up to 3000 trainloads of the fuel leave the coalfield each normal day, the resulting burden on the railway is to be relieved by burning some of the coal in local large power stations on the actual coalfield; the twelfth Five Year Plan requires more of the latter to be constructed. In the same plan, the initial development of the new lignite base of Maykyuben was announced, a project that has been in abeyance since the 1950s and was to be accompanied by the establishment

The Kharanor open-cast coal mine in East Siberia is planned to increase output from 6.5 to 9 million tonnes by 1985. The mine is situated in remote mountainous country about 30 km from the Chinese frontier

of a new urban settlement of Shoptykol, 60 km south of Ekibastuz.

The other major coal deposit of Kazakhstan, the Karaganda basin, resembles to some extent the Donbas in having good bituminous coals, shaft-mined, but with diminishing incremental growth of both coking and steam coals. Their output of some 40 million tonnes in 1985 was supplemented by that of strip-mined lower grade deposits from workings situated north of Karaganda, the Kushoky deposit and the Borly deposit at Molodezhnyy. Another coalfield situated in the semi-desert 300 km west of Kazakhstan is the Shubarkol site, having 2 billion tonnes of reported reserves of low-ash coal for power generation; its location in a remote area almost devoid of settlement requires the provision of infrastructure facilities such as road and rail access and a power-line supply. The projected capacity of the strip mines at Shubarkol is 22 million tonnes.

The four Central Asian republics import coal from the Kuzbas and from Kazakhstan, but small tonnages are produced in their own scattered basins. Several of the best deposits occur in the Fergana valley or in the hilly country around it. The deposit at Angren where lignite is mined has been important in local industrial development in Uzbekistan. Kirgizia also has lignite workings at Kok Yangak, Kyzl Kiya and Sulyukta.

Georgia's small coal basins are in the north Caucasus area at Tkibuli and Tkvarcheli; they range from lignite to hard bituminous quality, and the latter, supplemented by imported coking coal from Donbas, fuels the steel industry of Georgia at Rustavi.

The movement of USSR's coal production eastwards into the Siberian region of Asia has been emphasised by the development of the south Yakutian basin in eastern Siberia. The focus of mining is the town of Neryungri where the output in 1986 reached almost the design capacity of the deposit of 13 million tonnes. Estimated reserves are large. The coalfield owes its importance to its thick seams of high-grade coking and steam coals, extracted by strip mining from near the surface and also from deeper workings. The coking coal is upgraded by beneficiation for export to Japan via the 'Little BAM' rail link to the Pacific coast; steam coal is supplied to local power stations.

The most important coal basins in the Soviet Far East are in the valley of the Bureya river where the reserves are estimated to be large. The mining centre of Urgal is expected to increase its output to 2 million tonnes in 1987, adding to that of the Luzonovskiy mine's similar tonnage from this coalfield. The lower part of the valley, adjacent to the Trans-Siberian Railway contains lignite, mined at the principal centres of Raychikhinsk and Kivdinskiy, for thermal-electric power stations in the area. Farther south, in Primorskiy kray is the large open-cut lignite working at Nadarovka, an important fuel source for the steam-electric power station of Primorsk which supplies current for the electrified local section of the TSR. Among the several small coal basins in the area north of Vladivostok, are Artem and Partizansk; the Artem mine after reconstruction in 1986 doubled its yield of lignite by 2.4 million tonnes.

Slurry pipelines
Pneumatic pipeline transport systems in the USSR depend upon compressed propulsion of trains of cars or containers which carry loads of fine material such as sand, slag or coal slurry. The cars are equipped with wheels equally spaced around the circumference of the pipeline, so that each car is able to follow every turn, climb or descent of the pipeline. Transport costs, including labour, are generally reduced, no cargo losses can occur en route and the system is not affected by weather conditions.

Oil shale
A significant part of the fuel balance of the USSR is derived from oil shale; oil products and petrochemicals can be obtained from the shale after heating, refining and distilling, or it may be burnt as solid fuel for boilers in electric power stations. It has a large ash content and a calorific value similar to the poorer-quality lignites.

Proved reserves of oil shale amount to 6.6 billion tonnes, the bulk of which is found in European USSR in Leningrad oblast and Estonia, and smaller deposits in Kuybyshev oblast along the Volga river. In all these areas, the shale is mined both in deep mines and strip mines. In Estonia the shale-mining industry has undergone rapid expansion, providing fuel for large electric power stations, either directly or as shale-oil and gas, obtained from the shale after a retorting process. Estonia has a surplus of electricity from this source after fulfilling her own requirements, and is able to transmit current to neighbouring Latvia and the Leningrad area. Narva in

Estonia and Slantsy in the Leningrad oblast are the two main centres associated with the industry.

Peat
The USSR has vast quantities of peat, estimated at 60% of the world's total resources. Most of this occurs in the coniferous and southern tundra areas, where the cool, humid summer and cold winter, combined with deficient drainage, provide the conditions for the accumulation of plant residues as peat. Extensive areas of peat are also found in the centre and west of European Russia, in the Baltic republics and Belorussia, where peat occupies basins and hollows which were formerly lakes. This latter area is a main producing and consuming region where peat is used either as fuel or as a soil-conditioning fertiliser for agriculture. About 7% of the total reserves are found here, in such areas as the Ukrainian Polesye and the Meshchera and Balakhna lowlands.

The west Siberian plain has unique conditions for the formation of peat resulting from the excess of precipitation over evaporation, the lower gradient of the rivers and the concave relief. The plain is dominated by a vast peat mire with an area of more than 30 million hectares and peat resources in excess of 100 000 million tonnes, including those contained in the great Vasyugansk bog with its area of 5 million hectares and its 10 000 tonnes of peat.

Peat has the lowest heating value of all the fossil fuels, and provides less than one-third that of bituminous coal; it contains a high proportion of water and it requires to be dried in the open air, a process dependent upon the summer weather. However, within the centre and west of the European USSR where demand for energy is so high and regional fuel resources so meagre, the use of peat as a power station fuel has long been of some importance, and output has remained fairly constant since the end of the last war, averaging about 46 million tonnes annually, but with production levels low in years when drying was affected by bad weather. An increasing amount has been required for agriculture, particularly on the podzolic soils in the non-chernozem zone of the RSFSR, Siberia and the Far East. In these areas, the volume of peat required for the production of peat-based organic fertilisers and other products will steadily increase.

In regions such as Siberia where the conditions differ from those of the European part of the

USSR, large-scale peat extraction will require new techniques, since it is to take place in harmony with industrial and agricultural development, with much of the land being converted to arable use or forestry. However, peatlands contain many different habitats for wildlife, making it necessary for large areas to be conserved in their natural condition.

Various branches of industry are also dependent upon peat resources such as the production of fibre-board, fodder yeast and uses for metallurgical purposes.

At the present time there is renewed emphasis on the use of peat as a source of power because of the shortage of other boiler fuels in western USSR. Expansion of the Shatura station and the construction of other stations in oblasts such as Smolensk and Pskov, based on local peat deposits, are evidence of the continued importance

of this fuel in the USSR. West Siberia has one peat-fired plant at Tobolsk.

NUCLEAR POWER

The trauma of the disaster of Chernobyl in April, 1986, has not impeded the future development of civil nuclear power generation in the USSR. The current Five Year Plan for 1986–90 calls for a continuation of the increase in nuclear power which characterised the preceding two plans. In 1988 the nuclear share of the electricity output was about 11% rising by 1990 to 20%, and by the year 2000, it is expected to attain 30 to 40%. This is in accord with the Energy Programme announced in 1983 which requires a gradual reduction in the use of oil as a power station fuel, but increased use of nuclear power, together with Siberian resources of natural gas and low-grade surface-mined coal.

The map (Fig. 8.4) indicates the present and projected positions of the nuclear stations. A marked feature of their distribution is the almost total concentration in the European area. Only one of the stations is in Siberia, namely Bilibino, situated in the remote north-east. This contrast in distribution is explained by the abundance of fuel and energy sources, oil, gas and hydro-electricity in Siberia but the marked deficiency of these resources in the European area, where the increasing needs arise from the greater concentration of population and industry. Central Asia, like Siberia, also has reserves of natural gas together with increasing hydro-electric facilities, although it is not so well endowed with coal and oil.

The distribution of nuclear stations reflects the development of this industry since the first very small unit was commissioned at Obninsk, near Moscow in 1954. During the 1970s several nuclear stations were built, including those at Beloyarskiy in the Urals and at Novovoronezhskiy in Central Russia, on the River Don. By 1975 new reactor units had been installed at Kola and at Bilibino in the far north, at Shevchenko on the Caspian Sea in Kazakhstan, and near Leningrad. The tenth and eleventh Five Year Plans gave rise to considerable expansion of nuclear power capacity with new units located across almost the whole of European Russia, from the Kola peninsula in the north to Transcaucasia in the south. The twelfth Plan for the period 1986–90 called

TABLE 8.4 GEOGRAPHICAL DISTRIBUTION OF SOVIET ELECTRIC POWER PRODUCTION *(in billion kilowatt hours a year)*

	1980	1984	1985	1987
USSR	1294	1492	1545	1665
Thermal	1037	1144	1169	1258
Hydro	184	203	206	220
Nuclear	73	145	170	187
RSFSR	805	940	963	1047
European Russia*	390			
Urals*	177			
Siberia	238			
Ukraine	236	257	272	282
Moldavia	15.6	17.1	16.8	17.4
Belorussia	34.1	33.2	33.2	37.8
Baltic	35.3	39.0	42.8	46.6
Estonia	18.9	18.3	17.7	17.9
Latvia	4.7	3.8	5.0	5.9
Lithuania	11.7	16.9	20.1	22.8
Transcaucasia	43.2	49.9	49.9	52.6
Armenia	13.5	14.8	14.9	15.2
Azerbaydzhan	15.0	19.8	20.6	22.9
Georgia	14.7	15.3	14.4	14.5
Kazakhstan	61.5	74.6	81.5	88.5
Central Asia	63.4	81.2	83.9	93.3
Kirgiz SSR	9.2	11.2	10.2	9.3
Tadzhik SSR	13.6	15.2	14.8	15.9
Turkmen SSR	6.7	10.2	10.9	13.3
Uzbek SSR	33.9	44.6	48.0	54.8

* Bashkir ASSR is included in Urals.
Sources: as Table 8.1

for the building of entirely new stations and for the installation of additional reactors in existing power stations. Thus in the Volga basin, new nuclear power stations planned or under construction include Kostroma, Gorkiy and Volgograd, and new reactors are planned or are being built for the existing stations at Kalinin and Balakovo. The Ukraine is also to receive further nuclear capacity at the Rovno station and the South Ukrainian station at the city of Konstantinovka as well as at Zaporozhye on the River Dnepr. Chernobyl permanently lost the reactor destroyed by the accident in 1986 and its remains are now entombed in a concrete and metal shield. Of the three remaining reactors, two came back on-line later in 1986 and the third in 1988.

The reactors in all Soviet stations, as elsewhere, generate electricity for the national grid system. Many are similar to those in use at Chernobyl and are the standard Soviet unit, developed from the Obninsk prototype. They are known as RBMK 1000 as they produce 1000 megawatts of electrical power. The reactors employ graphite as a moderator and circulating boiling water as a coolant which extracts the heat generated by fission and radioactive decay. Steam passes through pipes to the power turbines to generate electricity.

Since the Chernobyl tragedy, modifications have been introduced to give greater safety in the operation of these reactors; a second type known as the VVER 1000 which provides all modern safety requirements is also being installed. This is a vessel-type water-cooled and moderated reactor of 1000 megawatts; large scale manufacture of such equipment is in progress at the Atommash Plant, situated at Volgodonsk on the lower River Don which has a planned annual production of eight such units.

The actual location of the stations with respect to their site of construction has been determined by several principles. The possibility of environmental damage by accidental discharge of radioactive material such as occurred at Chernobyl was not given high priority in the choice of a site, so great was Soviet confidence in the safety of their own reactors. Important site considerations included access to a supply of cooling water, nearness to transport facilities and to suitable sites for waste disposal and favourable geological conditions such as impermeable rock to prevent the accidental diffusion of radioactive liquids.

The site chosen for Chernobyl fulfilled some of these conditions and included others considered to be propitious; it was close to the Pripyat river, a tributary of the Dnepr; it was located in a swampy, largely forested region of the Ukraine and neighbouring Belorussia where population density was relatively low; independent power supplies were available from fossil fuels or hydroelectricity and the nearest large city, Kiev, with its population of 2.5 million was situated 100 km to the south.

A new city grew up at the site of the station, called Pripyat, built to accommodate the 35 000 workforce required in the operation of the RBMK reactors, six of which were to be built. The station was named after the town of Chernobyl, an old settlement of 18 000 people situated 20 km to the south.

At the time of the accident in April, 1986, four reactors were on-line and the explosion and fire occurred in reactor No. 4. It caused the death of 31 people, and a total of 237 others were hospitalised suffering from radiation sickness; after treatment most of the latter returned to work.

The geographic consequences have been far-reaching. Radiation fallout was recorded over much of Europe. The total shut-down of the station caused acute power shortages in adjacent parts of the Ukraine and Belorussia, but two reactors were restarted later in 1986. Immediate and complete evacuation of all inhabitants of Chernobyl, Pripyat and neighbouring villages was enforced within a zone of radius 30 km from the reactor; a thorough decontamination of land, vegetation, roads, buildings and water supply was begun and has continued. Beyond the zone a redistribution of population was organised by a mass resettlement of the displaced urban and rural inhabitants; farm personnel of collective farms were transferred to other farms within nearby rayons beyond the contaminated area. As the undamaged reactors came back on-line, technical staff returned to the station, working in shifts from new homes built beyond the zone. Two entirely new settlements have been developed for them: thus Zelenyy Mys (Green Cape) is a new town on the coast of the Kiev reservoir; and a second new town is Slavutich, 50 km from Chernobyl and linked to it by a new road. Many other refugees have been accommodated in Kiev, Chernigov and elsewhere.

Official attitudes to the programme of rapid development of the Soviet civil nuclear industry

Fig. 8.4 Nuclear power stations in the USSR, including major stations under construction or approved for construction. Several of the installations planned were under review in 1989 following the fuller appreciation of the aftermath of the Chernobyl disaster. Some that were planned may never be built.

have not been radically changed by the Chernobyl tragedy. This is because the geographical distribution of organic fuel resources, so much more abundant in Siberia as compared with those of European Russia, urgently requires the latter region to develop nuclear power as the only other adequate source of power available to its vastly greater concentration of industrial activity. But the several very large and powerful nuclear stations which were planned for the future will not now be built. Account will be taken of the lessons of Chernobyl in regard to the choice of sites for the new stations, and these stations will include design features to enhance the operational safety of the reactors.

Natural disasters such as the Armenian earthquake have also increased safety considerations in the building of nuclear power stations. Late in 1988 the Soviet Ministry of Atomic Power announced the suspension of construction work on seven atomic power plants following acute public concern about the safety of Soviet reactors after the earthquake. All were situated within seismically active zones subject to earth tremors; five lie within the belt of active geological faults which crosses the southern part of the USSR, namely Armenia, Azerbaydzhan, Georgia, Odessa and Krasnodar; farther north, those at Minsk in Belorussia, and at Ignalina in Lithuania are also at risk.

The consequent reduction in nuclear electricity generation was compensated by use of other power sources such as coal, oil, gas or hydroelectricity. The USSR has continued to develop atomic energy whilst taking into account the growing concern for the safety of power stations.

Applications of nuclear energy, apart from power station and heat station installation work, have also included the Soviet Union's use of subsurface explosions for canal and reservoir excavations, and also for the development of nuclear-powered ice-breaking ships. The latter, such as the *Lenin*, the *Arktika* and the *Sibir* are required to keep open navigable channels in the ice-bound coastal waters of the USSR.

The geography of uranium resources

Data are not published relating to the geographical distribution of the Soviet Union's basic nuclear fuel supply, uranium. In the absence of information upon the actual distribution of the resource, tentative data have been inferred by Shabad from Soviet publications referring to the development of urban settlements in the USSR. Settlements with high administrative status but relatively small populations which would normally not allow them to qualify for such status, may be assumed to be associated with uranium mining and processing; an additional indication of industry connected with uranium mining is provided by a reference in Soviet publications to a mining site but which omits a discussion of the material produced. The sites of uranium mining on the basis of Shabad's information may be divided into two groups; some are definitely known to exist, having been publicly recognised and long established, in some cases before the Second World War. A second group can be only tentatively named as urban centres probably associated with uranium mining. In some cases they have relatively small populations not compatible with their elevated urban status, and they may not be marked on Soviet maps.

A recognised location of uranium mining in the USSR is the city of Zheltyye Vody (Yellow Waters), situated in the Ukrainian SSR. Here the uranium minerals occur in very ancient (Pre-Cambrian) crystalline rocks associated with the iron minerals of the Krivoy Rog deposit. It is a city of over 50 000 inhabitants, having grown rapidly since the start of uranium activities in 1948. Another well-established uranium centre is Sillamae, situated in Estonia on the Baltic coast near Narva, where uranium minerals occur in the form of phosphatic ores in sedimentary clay formations. This is a small city of 15 500 inhabitants but has a very large municipal area and has the status of a republic-level city, indicating its economic significance. In Central Asia there are several known uranium mining sites: Min-Kush is an urban settlement in the Tyan-Shan mountains which is involved in the manufacture of nuclear fuel. Mill concentrates from Min-Kush are transported for processing 208 km northwards to a plant at Kara-Balta, in the Chu valley. Kara-Balta, a republic-level city with a population of about 45 000, is also a centre for agricultural industries. Other sources of uranium concentrates associated with the processing plant in Kara-Balta are Kadzhi-Say, one of the earliest Soviet uranium mining sites, situated on the south shore of Lake Issyk-Kul in north Kirgizia; and also several mining settlements at the eastern end of the Chu river valley, including Orlovka, which was granted urban status in 1969.

Additionally, a number of recognised uranium sources occur in the west-east ranges of the Tyan-Shan mountains bordering the Fergana valley. Taboshar, near the western end of the valley, originally associated with radioactive minerals, became an official urban settlement in 1937. With the increasing need for advanced processing of nuclear fuel, it grew in importance after the Second World War. Renamed Chkalovsk, it had a population of over 24 000 in 1970. To the northeast is a more recent development, the uranium-fluorspar mine of Naugarzan, a mountain site, with its own urban status granted in 1964, although administered as part of Chkalovsk.

Siberia contains two places which appear to be concerned with uranium mining. Vikhorevka is a settlement in the Irkutsk oblast of significantly rapid development. It is sited where the meta-morphic Pre-Cambrian rocks of the Angara shield, likely to contain titanium-uranium ores, are accessible to mining. It is close to Bratsk and is linked by rail to the Trans-Siberian main line. The second Siberian settlement, Krasnokamensk, is in Chita oblast; it was promoted in 1969 to oblast rank although not previously recorded as an urban place, and in 1977 was made the administrative centre of the rayon of that name. It has a location where the igneous geology may give rise to ores containing uranium. Similarly there are several locations in Kazakhstan likely to be involved in uranium extraction.

Hydro-electricity

Hydro-electricity clearly offers important advantages when compared with other power sources: by utilising precipitation and runoff it exploits a naturally renewed source which can be re-used by locating several units along the same river. It avoids air pollution, will allow flood control and improves navigation; the reservoirs may provide water for cities, industry and irrigation; they often form a recreational resource and may develop useful fisheries. But there are also adverse features such as losses of water by evaporation from large water surfaces, flooding of agricultural or forested land and damage to freshwater eco-systems (see Chapters 4 and 5).

A significant contribution to the total power output from hydro-electricity has occurred during the past half-century in the USSR, which today is

one of the world's leading countries in terms of developed water power, yet only a part of the potential has been realised. More than 80% of the total potential is located in Asian USSR, particularly in Siberia, and less than 20% is in the European areas where the need for power is greatest.

In the early years of the development of the Soviet Union's resources, hydro-electricity contributed about 11 to 14% of the electricity supply. In the last few decades the contribution of hydro-electricity has fallen only slightly in spite of the increased competition in the energy balance of the fossil fuels and nuclear power, and in 1985 was 13% (Table 8.4).

The development of hydro-electric power in the Soviet Union began a few years after the 1917 Revolution, in the European part of the USSR under the direction of GOELRO (The State Commission for the Electrification of Russia). The Volkhov station near Leningrad was completed in 1926 and this was followed by the construction at Dneproges, near Zaporozhe, of the great Dnepr dam in 1932. The power station had a capacity of well over half a million kilowatts and for many years it was the largest water-power project in Europe. Rebuilt and improved after its destruction in the Second World War, this scheme together with more recent projects continues to provide power for Ukrainian industry, particularly power-intensive industries such as the manufacture of aluminium and electric steels. Like other schemes, it is a multi-purpose unit as the reservoir provides water for the irrigation of the semi-arid steppe, and the increased depth of water eliminated the series of rapids that had obstructed navigation.

The exploitation of the potential of the Dnepr river has continued since the 1950s with the addition of five other stations and the expansion of the original station for peak-load requirements. They are relatively small-capacity stations, though their combined capacity is in excess of 2000 MW; they consist of Kakhovka (1956), Kremenchug (1960), Kiev (1964), Dneprodzerzhinsk (1964) and Kanev (1972). The very large Kakhovka reservoir provides irrigation to the dry steppe of the southern Ukraine in partial compensation for the extensive farmland of the Dnepr valley lost by flooding when the reservoirs were formed.

Contemporary developments have also taken place within the Kama–Volga basin; the seasonal

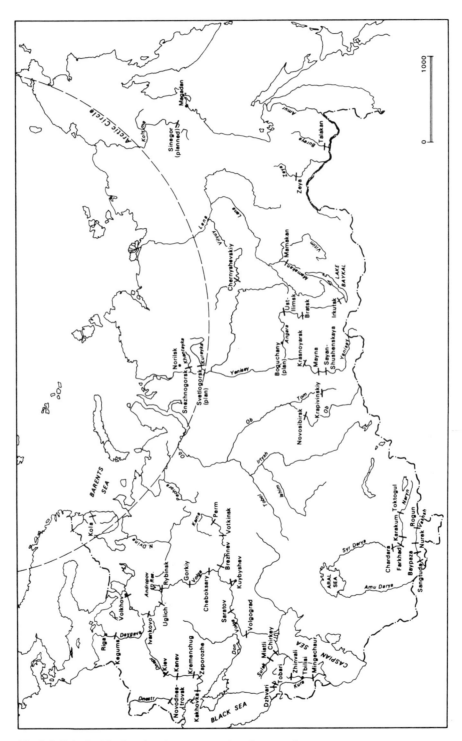

Fig. 8.5 Principal hydro-electric power stations in operation

flooding of the river has been regulated by massive dams which impound very large multi-purpose reservoirs, linked by canals forming a continuous navigable waterway. The canals traverse low watersheds between river basins by a series of locks, allowing inter-basin navigation and water transfer. Thus, a dam on the Volga with a hydro-electric power plant near Ivankovo (built 1937) provided water for the Moscow canal, the navigational link between the Volga and the Moscow river. In 1940 a second dam and hydro-electric unit was added at Uglich, followed the next year by further dams on the upper Volga, creating the Rybinsk (now renamed Andropov) reservoir which at the time was the world's largest man-made lake. The power produced at this dam supplies Moscow and other cities of the area, and the lake provides a fresh-water supply and a recreational amenity. In the post-war years, Volga projects became larger. These units, all based on vast reservoirs on the Volga and Kama, are at Perm (1954–8), Kuybyshev (1955–7), Volgograd (1958–61), Votkinsk (1963) and Balakovo, near Saratov (1970). The Lower Kama project, the third on the Kama, began generating in 1979. All have capacities of over 1000 MW with the exception of that at Perm with 504 MW, and the largest is that at Volgograd with 2540 MW; this takes water from a gigantic lake extending about 640 km upstream. Very high-voltage power lines carry electricity from the Volga stations to Moscow and to the Urals; these have become an integral part of a nation-wide electricity transmission grid linking up the many hydro or thermal stations with their consumers.

The eighth and final stage of the Volga system of hydro-electric projects is at present under construction at Cheboksary, the Volga port 100 km upstream of the confluence with the Kama. Here the new dam will raise the water above its present level, eliminating shallows which create navigational difficulties and provide power for a series of stations yielding 1400 MW, as well as supplying water for irrigation. But the project has required the removal and rehousing of 36 000 people from many small towns and villages.

Other developments in western USSR in process of completion are on the lower Kama river at Brezhnev; and on the River Dnestr at Novodnestrovsk where a project having a designed capacity of 702 MW will supply power for the western Ukraine. In the Latvian SSR there is also a small project on the Daugava river at Kegums, south of the capital, Riga.

The power potential of the Caucasus region with its heavy precipitation and steep slopes is also being realised by new projects. In Dagestan, the Sulak river is harnessed by the 1000 MW Chirkey power system, including several generating units for peak operation to supplement thermal power.

Georgia now has several hydrostations with large reservoirs allowing a constant output for base-loads. The largest of these is on the Inguri river at Dzhvari, where one of the highest dams in the USSR allows a capacity of 1600 MW. Farther upstream is another project, the Tobari station. Entirely new is the Zhinvali hydro-electric scheme on the Aragvi river, begun in 1985 with an ultimate capacity of 130 MW; this is a multipurpose project providing irrigation for river bottomland near Tbilisi.

Another multipurpose scheme is in Azerbaydzhan, utilising the Kura river by the reservoir near Mingechaur for power and irrigation of the Shirvan steppe, and including smaller hydrostations at Varvara and at Shamkhov. In Armenia, a series of hydro stations sited along the Razdan river, the outlet of Lake Sevan, provided much of the republic's power supply; but increasing erosion and falling lake and water-table levels have caused so much concern for the environment that the stations have been closed until conditions may be rectified. Meanwhile, the future of Armenia's power supply is to be based on gas-fired steam electricity together with nuclear power.

Hydro-electricity in the European north was developed early in the Soviet period, based on the harnessing of swift streams flowing from numerous lakes. It is generated mainly in the Kola peninsula, in Murmansk oblast and in the Karelian ASSR by small stations with a combined capacity of about 1600 MW.

The twelfth Five Year Plan emphasised the need for the development in European USSR of pumped storage capacity for the operation of hydro-electric power stations. Such installations would be particularly advantageous in this area where further conventional sites for power development are limited, and where the demand for electricity is so great. These are two-way systems, equipped with reversible turbines which pump water from a lower to an upper reservoir during off-peak periods and then are switched to their normal function during peak times when water from the higher reservoirs is brought into use. After trials with small units, an extended development programme is planned for larger capacity pumped-storage installations at locations such as Kanev on the Dnepr with a capacity of 3600 MW, and Leningrad, with 1600 MW.

In Central Asia, the earliest large hydro-electric schemes were sited on the Syr Darya river and combined power production with the requirements of irrigation; the projects included the Farkhad and the Kayrak-kum dams together with the Chardara station. Much more important were the several major schemes of the 1960s and 1970s, designed to produce power for a high-voltage electricity grid for the whole region, combined with output from large gas-fired power stations. Two of these projects are on the swift-flowing Vakhsh river, the Nurek station, with a capacity of 2700 MW (completed 1972–3), and the Rogun station, upstream with an additional 1.1 million kilowatts. Downstream from Nurek two further stations are planned, the Baipazin and the Main. It is largely due to these plants that Tadzhikistan's aluminium and other enterprises, with their heavy demands on power, have arisen here. The Naryn river in Kirgizia has been similarly put to work by the Toktogul project of 4400 MW, and was completed in 1975, having a function for peak-load output. An earlier unit was the Uch-Kurgan dam at Shamaldy-Say. The latest addition to the Naryn's power output is the Kurpsay station, with a designed capacity of 800 MW, but work is now in progress on the new Task-Kumyr station, started in 1983. These developments have allowed the Kirgiz SSR to become an exporter of hydro-electricity.

The exploitation of the huge hydro-electric potential of the Siberian rivers began in the 1950s with a small unit on the Ob river above the city of Novosibirsk. This was soon to be dwarfed by the development of the immense power prospects of the Yenisey river and of its tributary, the Angara. The latter is particularly suitable for power development as it rises in Lake Baykal which, with its enormous volume, provides an almost constant flow of water; in addition, its constricted, steep-sided valley slopes sharply downstream and is cut into resistant rocks. Hence the ideal conditions of a large head of water combined with solid foundations for dam construction are fulfilled.

The first dam, built near the city of Irkutsk, provided water for the Irkutsk power station, opened in 1958 with a capacity of 600 MW, supplying power for large aluminium reduction plants. Downstream was the newer Bratsk scheme, based on the great Bratsk dam across the Angara gorge; this began generating in 1961 and was upgraded in the 1970s to 4500 MW.

In the meantime, work on other power projects further downstream was in progress. The Ust-Ilimsk project was designed as a twin of the Bratsk scheme and to have a capacity of about 4300 MW, the power being used by a large wood-pulp mill with the surplus being fed into the central Siberian grid. Bratsk and Ust-Ilimsk together represent a significant example of large industrial enterprises established in remote areas of the boreal forest, far to the north of the main line of communications, the Trans-Siberian Railway. Huge areas of forest have been cleared; houses, shops, roads and other infrastructure have been provided for workers and their families.

On the Yenisey river there are even larger projects. These are the Krasnoyarsk station, at Divnogorsk which reached its designed output of 6000 MW in 1971, supplying current to the aluminium works; and also the Sayan-Shushenskaya project upstream from Krasnoyarsk at Sayanogorsk. The latter with its capacity of 6400 MW, reached in 1985, is at present (1989) the most powerful hydro-electric station with a much smaller Mayna station of 321 MW downstream to regulate changes in water levels resulting from withdrawals from the main Sayan reservoir, thus assisting navigation. The twelfth Plan called for an additional project on a site on the middle Yenisey at Abalakova, below the confluence of the Angara with the main river, and near the timber centre of Lesosibirsk. This is to have a capacity of 6000 MW and some preparatory work may have begun during the twelfth Plan period, though completion is not expected until the 1990s.

The large hydro-electric stations on the Yenisey/Angara river system furnish a cheap power source from a heavy initial investment in the construction of the reservoir and the power plant. The abundant electricity is used for timber and pulp products and also for the production of aluminium. This close relationship between power availability and utilisation in the area has resulted in the use of the hydro-electric supply for constant base-load purposes, contrasting with the normal procedure of using it only at peak consumption times, the constant load being borne by thermal stations. However, in dry years the hydro-electricity supply becomes insufficient for base-load work and the latter in future may be supplied by the projected lignite–fuelled power stations of the Kansk-Achinsk basin; such thermal units require to be run continuously as they cannot be shut down at short notice, in contrast with hydro-stations.

On the river Ob, contrasting in flow and gradient with the Yenisey, the opportunities for hydro-electricity generation are few, apart from the one small scheme south of Novosibirsk. However a dam at Yelanda in the Altay mountains, on the Katun tributary, has been proposed but may be cancelled in view of its possible environmental impact.

On the upper Ob basin, the Tom river provides power for a small project of 300 MW situated near Krapivinskiy, in the Kuznetsk coal basin.

Other Siberian projects include that of the Norilsk area on the Khantaika river, a small station of 441 MW supplying power for the nickel, cobalt and other metal industries. In the Yakutsk ASSR the river Vilyuy is harnessed by small projects at Chernyshevskiy and at Svetlyy. The Amur basin contains two sites where the main river's tributaries, the Zeya and the Bureya, provide power. The Zeya project reached its designed capacity of 1290 MW in 1980, a multi-purpose project embracing also flood-control and navigational improvements. To the south-east, the Bureya station at the site of Talakan is designed for 2000 MW and is planned for start-up in 1990.

In north-east Siberia, the Kolyma river is planned to provide power for the gold-mining industry of the Magadan oblast from a hydrostation at Sinegor, near Debin; and another is planned for the Ust-Srednekan station on the Srednekan river.

Energy from solar and tidal sources
The development of these forms of power has not been a high priority in the USSR in view of the availability of fossil fuels, hydro-electricity and nuclear power.

As regards solar power, the physical conditions for its development are not ideal as a result of the high latitude, cloudy conditions and low sun altitude of much of the Soviet territory. However, recent dubiety on future supply of oil has resulted in some increase of interest in the practical applications of solar energy such as for space heating, solar ovens and water pumps. In Central Asia, with the areas's abundance of sunshine, solar hot-water heating systems are operated in some buildings such as hotels. In the desert areas, irrigation water is pumped by solar power obtained by large installations of collecting cells. Distillation of saline water using solar cells has been subject to experimental tests in the desert republics of Turkmen and Uzbek where much of the research on solar applications is being done.

Of some interest is the 5 MW solar plant for research in the Crimea where solar heat is converted into steam-driven turbines. Direct conversion of solar radiation into electric current, as is done on Soviet space stations, is highly prospective. Experiments are in progress for heating buildings in Novosibirsk and even in the far north where the sun does not set for days in the summertime.

In the application of tidal power, several potential sites were surveyed in north European Russia and in northern Siberia, and a 400 kW tidal pilot plant was started in 1969 in a small fiord in the north. However, interest, has faded in view of the difficulty of the rapid changes in generation of electricity in accordance with tidal variations. This problem may be overcome by combining the output of tidal stations with that of conventional fuel stations to eliminate the fluctuations of power inevitable with tidal power.

FERROUS METALS

IRON ORE
The Soviet Union is the world's foremost producer of iron ore with abundant reserve supplies, estimated at 110 billion tonnes. The bulk of the ore is relatively low in iron content as the higher grade ores have been approaching exhaustion and low-grade ore requires concentration or beneficiation before use in blast furnaces. As Table 8.5 indicates, the European part of the USSR dominates iron ore production, yielding over 70% of the total mined in 1985, the remainder originating in Kazakhstan and Siberia.

TABLE 8.5 GEOGRAPHICAL DISTRIBUTION OF SOVIET IRON-ORE PRODUCTION (million tonnes of usable ore)

	1975	1980	1984	1985
USSR.	235.0	244.7	247.1	247.6
RSFSR	88.8	92.4	99.5	104
European Russia	47	50	60	64
Kola	(10)	10.7	11	11
Karelia	—	—	6	10
Kursk	35.6	39.3	41.7	43
Urals	(26)	25	23	23
Siberia	(16)	17	16	16
Kazakh	21.4	25.8	24.0	23.0
Ukraine	123.3	125.5	122.8	120
Azerbaydzhan	1.3	1.1	0.9	(1)

The total production in 1987 was 251 million tonnes; no details available.
Sources: as Table 8.1

The most important European deposit is in the Krivoy Rog basin of the Ukraine, where the ore consists of highly metamorphosed Pre-Cambrian rocks containing magnetite, mined by both shaft and surface excavations. There are several separate mining and concentrating complexes within the basin, converting the low-grade iron quartzite into concentrated material, commonly in the form of pellets.

A major project of the twelfth Five Year Plan has been the installation of a new concentrator in the Krivoy Rog basin, the result of a joint Comecon project, which utilises oxidised wastes resulting from open-pit operations. This will compensate for the gradual depletion of easily accessible or easily beneficiable ores, now requiring extraction at greater depths.

The highest rate of production increase has been in the second great iron deposit west of the Urals, the Kursk Magnetic Anomaly (KMA) in central Russia, north-east of Krivoy Rog. The ancient shield rocks, consisting of magnetites and iron-bearing quartzites are mined from the surface or from pits and concentrated before transport to the nearby Lipetsk iron and steel works or to the three major steel centres in the Urals, Magnitogorsk, Chelyabinsk and Novo-troitsk. The two main mining centres of the KMA, Gubkin and Zheleznogorsk, produced jointly in 1984 nearly 42 million tonnes of iron

quartzite, consisting of 16 million tonnes of rich ore and 25.8 million tonnes of concentrate derived from the low-grade quartzite. The USSR exports a considerable part (about 20%) of the production; the bulk of this is from the KMA and is railed to many of the countries of Europe, including not only the Soviet satellites but also Britain and Italy. The steel industries of the Urals and of the Kuznetsk basin in western Siberia also import iron ore from the KMA.

The Kola peninsula in the north of European Russia, with Karelia, is the third main iron-producing area west of the Urals; low-grade ores are mined at Kovdor and Olenegorsk in the Murmansk oblast and the concentrates are railed to the iron and steel centre of Cherepovets on the Andropov reservoir in northern Russia.

An additional source of iron became available in north European Russia in 1973. An agreement with Finland was made to develop the Kostom-kuksha iron-quartzite deposit in the Karelian ASSR, near the Finnish border, and production began in 1978. Concentrates and pellets, first produced in 1982, are sent to Cherepovets and supplement the concentrates from the Kola peninsula.

In the Urals, the complex of igneous, metamorphic and sedimentary rocks contains a great variety of mineral deposits, of which iron ores were originally among the most abundant and provided the basis of the traditional iron and steel industry of the region. There are several iron-mining centres, although production in them all is declining as the deposits become exhausted. The most important is Magnitogorsk where a large deposit of high-grade ore, the Magnitnaya Gora (Magnet mountain) formed the basis of the Soviet steel industry during the Second World War and in the early post-war years, supplying both Ural and Siberian steel works. Today, as the Ural iron production is insufficient for local needs, additional supplies from the KMA and Kazakhstan are brought in. Modernisation of the Magnitogorsk plant was planned during the twelfth Plan period and includes conversion of the open-hearth process of steel making to the basic oxygen process, with increased utilisation of scrap.

Other deposits of ore are the Kachkanar, a low-grade vanadiferous magnetite, and those of the Nizhniy-Tagil area at Kushva, Vysokaya, Blagodat and Alapayevsk. The Bakal deposit supplies the iron and steel mills at Chelyabinsk.

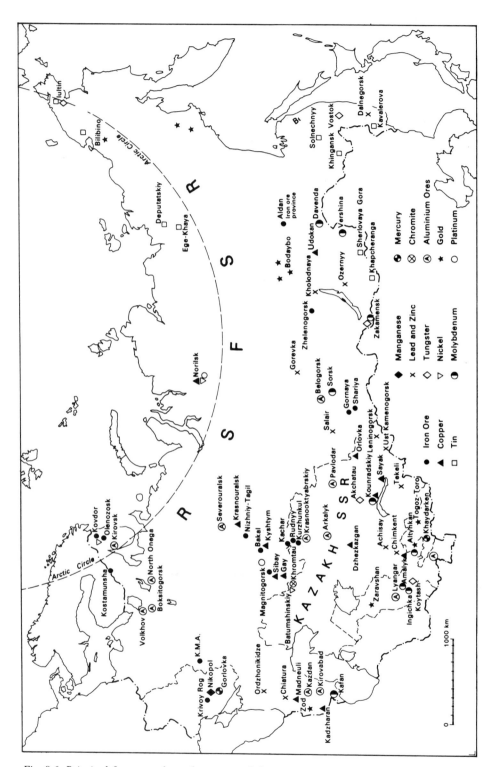

Fig. 8.6 Principal ferrous and non-ferrous metal deposits

East of the Urals, reserves of low-grade iron ore occur in Kustanay oblast in north-west Kazakhstan. Here a very extensive magnetite deposit is mined at Rudnyy, and the concentrated ore is railed to the Urals and east to the Kuznetsk basin. However, the conditions of mining in the Sokolovka-Sarbay deposit at Rudnyy have become more difficult as the depth of the open-pit excavation has increased, resulting in a decline of some three million tonnes of output. The quality of the ore has also fallen as the content of metal has decreased, affecting the output of concentrates. A new mine south of Rudnyy, based on the small Kurzhunkul deposit was opened in 1983, and another new mine, the Kachar open-pit operation west of Rudnyy began production in 1985. These two new sites may compensate for the decline in the output from Kazakhstan's older mines.

In Siberia, the output of iron-ore is below the requirements of industry, and forms only 6% of the all-USSR total; hence a substantial tonnage is railed from producing areas as far west as the KMA of central European Russia to the steel manufacturing centres of the Kuznetsk coal basin. Part of their iron-ore requirements is provided from mines in the Gornaya Shoriya district in the southern rim of the basin, but the principal supply is from the Zheleznogorsk complex in Irkutsk oblast and from mines in southern Kemerovo oblast and southern Krasnoyarsk kray. Rudnogorsk and Tatykhaninskoye are satellite sites of the Zheleznogorsk complex, supplying crude ore to the latter centre's concentrator.

The twelfth Five Year Plan called for the development of new iron-ore deposits in Siberia, situated in the zone to be served by the Baykal–Amur Mainline. These would form the basis of a new integrated iron and steel plant in eastern Siberia using coking coal from the south Yakutsk basin. North-east of the coal centre of Neryungri is the Aldan iron-ore province where various types of iron-ore including magnetite ores are obtained by open-cast working, as exemplified by the Tayezhnoye deposit. Other iron-ore locations in the Aldan plateau are the Chara-Tokko and the Olekma-Amga, situated in the basin of the Olekma river to the west and north-west of Chulman. Magnetite and iron-quartzite ores are both present in these sites with total reserves estimated to be at least ten billion tonnes.

Ores of non-ferrous metals

The ores of all the non-ferrous metals are distinctive in their relatively low percentage of pure metal, commonly less than 5%; hence the need for a preliminary process of concentration prior to smelting. Many are associated with igneous activity during which metallic minerals were formed by the agency of heated gases and vapours, representing a late stage in the intrusion and cooling of a magma.

COPPER

Copper is among the most important of such deposits, because of its value in the electrical industry as a conductor and for electrical machinery. Many other industries need copper, exemplified by the manufacture of motor-cars and trucks; the metal is also extensively used in the production of alloys such as bronze and brass.

The USSR has large deposits of this valuable metal and has been among the chief producing countries for a long period, mining having begun in the Urals in the seventeenth century. In addition to the older Ural centres of Krasnouralsk and Kyshtym, new sources have been exploited at centres such as Sibay, Gay and Uchaly.

As the older deposits in the Urals become exhausted, the greater reserves of Kazakhstan are being increasingly used by the Ural smelters; in central Kazakhstan the great copper complex at Dzhezkazgan has become one of the principal copper centres of the Soviet Union, with steady expansion since the beginning of production in the 1960s. In 1985 the new Akchiy–Spassk deposit was opened up and a new concentrator was added. Copper concentrates are railed to the Urals and to the Balkhash complex farther east (Chapter 9) where the industry is supported also by the Kounradskiy mine and by the newer Sayak mine. Other centres in Kazakhstan are at Orlovka, a new complex in the east, and at Bozshakol where development of a copper-molybdenum deposit was begun during the eleventh Five Year Plan.

Copper has also been derived as a by-product from the extraction of other ores. Important contributions to the USSR copper industry are supplied by Georgia, Uzbekistan and Armenia, Madneuli, in south Georgia, near the town of Kazretiy, is a new non-ferrous metals complex yielding copper, barites, zinc and lead; the first ores were mined here in 1973. The Armenian SSR has a copper-molybdenum site at Kadzharan, near the border with Iran. The Uzbek republic's output of copper is centred upon the Almalyk copper and zinc concentrator. This receives copper-molybdenum ore from the Sarycheku deposit, situated in the Kurama mountains east of Almalyk, exploitation of which began in 1973.

Norilsk, in the far north of the USSR, is being continuously developed as a non-ferrous metals complex, with copper one of its important products. The scale of the operation is indicated by the 'October' mine, one of the largest sites, where there are numerous vertical shafts more than 1000 metres in depth and an extensive series of tunnels. Hydro-electricity and natural gas provide power to the area which is linked by rail to the Yenisey river terminal of Dudinka.

A new and potentially very important source of copper is the Udokan deposit in the northern Chita oblast of eastern Siberia, adjacent to the course of the Baykal–Amur Mainline. The proposed development was to be associated with a new settlement named Naminga. But the project remains at present in abeyance, and no decision has been taken in the Twelfth Plan for its commencement.

LEAD AND ZINC

Minerals containing these metals are commonly found together, often in rocks affected by thermal metamorphism or where heated solutions have deposited veins of the metal in limestones. They are compact deposits, and the lead/zinc ores may also yield gold, bismuth, selenium and other metals. The industrial uses of lead and of its compounds are manifold: for accumulators, for piping, foil, ammunition and for various alloys.

The USSR appears to rank very high in the scale of world producers, probably second after the USA. An early centre of production was Ordzhonikidze, in the Caucasus, where the Terek river supplied hydro power for the plant, which recently acquired new mines and a concentrator. As in the case of copper, Kazakhstan is apparently an important producer, with sites at Leninogorsk in the Altay region, and also at Chimkent, Achisay, Karagayly and Tekeli. A major expansion of the lead/zinc industry took place after the Second World War in eastern Kazakhstan, with the provision of zinc refineries at Ust Kamenogorsk and Leninogorsk (Chapter 9).

Almalyk in Uzbekistan is another important centre.

In eastern Siberia, the twelfth Plan required a start to be made on the development of the Gorevka lead-zinc deposit on the Angara river above its confluence with the Yenisey. This had been delayed by the location of the ore within the flood-plain of the Angara, where a huge open pit for the extraction of the ore would be required. The pit is to be protected by a dyke or dam when the valley is flooded by the new middle Yenisey dam at Abalakova. Farther east, in the lead-zinc development programme of the twelfth Plan, is the Ozernyy project situated in the Buryat ASSR. A railway to this site was reported to be being built in 1980. A further lead-zinc source is the Kholodnaya site at the northern end of Lake Baykal, but the Plan makes no reference to a start to its exploitation, possibly on grounds of its likely environmental impact.

Other deposits are in western Siberia (the Salairskoye group of mines, whose output is refined at Belovo, the zinc processing centre), and also in the Soviet Far East, where the lead-zinc mine and lead smelter of Tetyukhe has been renamed Dalnegorsk ("far mountain").

TIN

Cassiterite, or tin oxide, is found in veins in certain types of granite, often associated with copper pyrites and wolfram. However, most tinstone is recovered as grains from placer washings in weathered soil, or in alluvial silts and gravels of river beds. The metal is important in the engineering and chemical industries and the manufacture of alloys and tinplate.

Tin is one of the nonferrous metals not abundant in the USSR, and supplies are imported from Malaysia and Indonesia. The principal sources in the USSR are in eastern Siberia and the Far East. The oldest worked deposits are in Chita oblast at Khapcheranga and Sherlovaya Gora, but of greater importance at present are the tin sources farther east; Solnechnyy and Khingansk are in Khabarovsk kray of the Far East region, the latter site now having diminished reserves. To the south in Primorskiy (Maritime) kray is another older tin-producing district around the Kavalerovo concentrator, west of the port of Tetyukhe. Farther north, in Magadan oblast are the Valkumey lode and the Krasnoarmeyskiy placer deposit, the latter south-east of Pevek, the port on the East

Siberian Sea. Iultin and the newer deposit of Svetlyy are also in this area, but a much larger deposit is the Deputatskiy source in Yakutia, very inaccessible but named for future development of the lode and concentrator in the twelfth Plan. It would replace an older source to the south-west, at Ege Khaya. Continued working of these newer tin sources gives the USSR apparently the second place in world production after Malaysia.

MANGANESE AND TUNGSTEN

These are two of the key metals, both of importance in the manufacture of alloy steels: tungsten for motor vehicles and for armament steels; manganese for mining equipment and railway tracks. Manganese is abundant in the USSR; it is obtained from the Chiatura deposit in the Georgian Caucasus mountains, a world ranking source; here it is associated with tungsten in the output of the Tyrnyauz mine. However, some 70% of the production comes from Nikopol, in the Ukraine where the ore-bed occurs near the Dnepr river within hollows on deeply weathered Pre-Cambrian gneiss. In Kazakhstan a newer source is in the Zhayrem area where manganese is obtained from an open-pit site and is concentrated at Kentau and Tekeli. An older source is at Dzhezdy in the Dzhezkazgan copper district where a deposit came to be important during the Second World War to replace the loss of the Nikopol basin to the enemy.

Tungsten is derived from wolfram and scheelite, the former associated with tinstone and molybdenum in acid igneous rocks. The USSR is a leading producer of tungsten, and most of the output is mined with tin, as at Iultin in Magadan oblast, and also with molybdenum as at Zakamensk in the Buryat ASSR. Vostok, in the Far East, is a major site. Two other sites are Akchatau, in Kazakhstan, and Koytash in Central Asia.

NICKEL

The main deposits of nickel ore are located in Murmansk, the Urals, the Kazakh SSR and in north Siberia, at Norilsk. The mines at Norilsk, together with the smelters, account for a major part of the USSR production, and this complex also yields copper, iron, gold, silver and platinum.

MOLYBDENUM

Like tungsten, molybdenum is used as a con-

stituent of shock-resisting, heat resistant and high-speed steels, and in metallic form is employed in the electrical and electronic industries. It is mined in association with granite intrusions where copper and tungsten also occur. The mines and related industries are widely dispersed in the USSR over an area extending from Armenia in the west to eastern Siberia. In Armenia there are copper-molybdenum sources around Kafan in the Zangezur mountains; the group in Central Asia includes sites at Amalyk, Ingichka and Lyangar in the Uzbek SSR. In Kazakhstan, molybdenum is mined with copper at Balkhash, and a new source is at Kokentau. Farther east, in Krasnoyarsk kray, the mining and concentrating centre is Sorsk. Further sites are in the Buryat ASSR at Zakamensk and in Chita oblast at Davenda and Vershino.

CHROMITE

Chromite is important in the production of alloys, such as stainless steel and chrome steel. The USSR is the world's chief producer of chromite, and it comes from the Khrom-Tau (chrome mountain) area in north-west Kazakhstan, where it occurs in ultra-basic intrusive rocks.

MERCURY

This is a rather rare metal, though it is found in several sites in the Soviet Union and is required in the chemical, electrical and instrument industries. The principal producing centres are Nikitovka, a suburb of Gorlovka in the Ukraine and Khaydarken, on the southern margins of the Fergana valley in the Kirgiz SSR. Other centres are the Anzob complex, of the Tadzhik SSR, the Shorbulag mine of Azerbaydzhan, the Plamennyy mine in north-east Siberia, and the Aktash deposit in the Altay mountains.

ALUMINIUM

The aluminium industry has undergone rapid development since the end of the Second World War. In the 1930s it was a relatively small operation located in European Russia where bauxite, the principal source of aluminium, was mined near Tikhvin–Boksitogorsk, in the Leningrad region. Hydro-electricity, generated at Volkhov, was used for the process of reduction into aluminium metal. Later, alumina was railed southwards to the new Dnepr hydro-electric plant at Zaporozhye for conversion into aluminium.

Apart from bauxite, aluminium ores include nephelines, alunites, cyanites and sillimanites. The Soviet Union supplements her own resources of bauxite with imported supplies, but also obtains the alumina from nepheline and alunites which are mined at various sites. The industry is therefore widely dispersed, an important locational factor being abundance of electricity for the final process of conversion of the alumina into aluminium metal. The Kola peninsula, Transcaucasia, the Dnepr region, the Urals, Kazakhstan, Siberia and the Leningrad region all contribute to the present increasing production of aluminium in the USSR.

Post-war developments have included the use of nephelite in the production of aluminium; the centre of nepheline and apatite mining is the town of Kirovsk, east of Lake Imandra, in the Kola peninsula, where hydro-electric power is used in the aluminium industries of Kandalaksha and Nadvoitsy. Nepheline from the Korovsk area is also converted to aluminium at Volkhov, the intermediate process of alumina production being carried out in the Boksitogorsk area. A newly discovered deposit of bauxite ore has become available in the European north-west, named the north Onega deposit. It is a high-grade ore body but is in a remote area of waterlogged terrain at the confluence of the Onega and Iksa rivers.

In Transcaucasia, alunite deposits at Zaglik, near Kirovabad in the Kura river valley, are the basis of the alumina plant at Kirovabad; the final process of aluminium production takes place at Yerevan and Sumgait. Yerevan also receives supplies of alumina from Razdan in Georgia.

The aluminium industry in Central Asia utilises the clay material, kaolin, as a source of alumina; the deposit is obtained in the Uzbek and the Tadzhik republics and was planned to support the aluminium reduction plant at Regar, in the Gissar valley west of Dushanbe, where the hydro-electricity of the Nurek project is available. Alumina from the Urals or from Pavlodar in Kazakhstan is also used in the Regar plant.

The German occupation of much of European Russia during the war years, 1941–44, resulted in a significant movement of the Soviet aluminium industry to the east. In the Urals, bauxite and alunite had been worked for several years and production increased at centres such as Kamensk–Uralskiy and Severouralsk, with part of the raw material being railed to Siberia. The

post-war years have seen vast expansion of the industry into the Asian part of the USSR based upon the exploitation of newer sources of the different aluminium-yielding ores, together with the availability of abundant hydro-electricity in the reduction plants at Bratsk and Krasnoyarsk. Thus, bauxite is now mined at Arkalyk, and at Krasnooktyabrskiy, both in Kazakhstan, where an additional new source, the Belinskiy mine in Kustanay oblast, is now in production; Pavlodar has become important, sending alumina to the Novokuznetsk and Shelekhov aluminium plants. In Siberia, new plant has been added at Achinsk, producing alumina from nephelite from Belogorsk, using thermal-electricity derived from the lignites of the Kansk-Achinsk coalfield. But despite the widespread utilisation of non-bauxite materials for aluminium production in the USSR, the industry has been heavily dependent on imported raw materials of both high-grade bauxite and alumina.

GOLD

Gold is abundant in the USSR and the country is second only to South Africa in world production; its reserves of the metal are mainly required for adjustment of its balance of payments, by the sale of gold to the West. Gold mining was one of the earliest industries of the upper Lena river region in Tsarist times, where placer deposits were obtained from washings of alluvial gravels. This method is still used in the middle Yenisey region, operated on a very large scale by huge dredges. Commercial gold mining began in the Urals in 1814 and was followed by the discoveries in Siberia, notably in the Vitim and Aldan plateau areas of the upper Lena. Here, mining continues from the supply centre of Bodaybo, the head of navigation on the Vitim, the Lena tributary. Hydro-electricity is available from a station at Mamakan, to be supplemented by a new station, Telmama, supplying power to a new centre at Sukhoy Log. Other Siberian gold mining sites are in the Magadan oblast, and at Bilibino, in the Chukot autonomous okrug where atomic power is used for the mining operations.

The traditional domination of the Urals and Siberia in Soviet gold production is, however, being changed by newer developments in the south of the country. In Armenia, the Zod deposit is an important lode, mined east of Lake Sevan, and the metal is concentrated at Ararat. Important developments in Uzbekistan, based on lodes, are at Altynkan on the slopes of the Kurama mountains in the Fergana valley; the Kochbulak mine in the same area, and the Zeravshan complex in the central part of the Kyzylkum desert of Uzbekistan. In the Kirgiz SSR is a new gold-lode project, the Togoz-Toro source in the upper Naryn river, near Kazarman. Another site yielding both gold and tin in the area of Yenilchek, has been under development in the twelfth Plan. The Tadzhik SSR also is starting a gold-lode source, that of Toror near Pendzhikent in the Zeravshan mountains south-east of Samarkand.

It thus seems clear that the Central Asian republics named will become some of the principal gold producing areas of the USSR, where the trend of extraction is away from the stream-gravel deposits of placer sites towards the exploitation of deep-lode deposits accompanied by processes of complex milling and other forms of treatment.

Non-metallic minerals

The USSR's resources include significant reserves of several non-metallic minerals, some of which are exotic or strategic: the former include diamonds, and the latter uranium, boron, lithium and beryllium. Uranium was discussed earlier in this chapter.

The USSR suddenly moved to a prominent place in the world production of diamonds with the discovery in Yakutia of diamond-bearing kimberlite, the ultra-basic igneous rock, in the 1950s. Soon afterwards, the new town of Mirnyy became a centre of production near the deposits, with an output of both gem and industrial diamonds. In the next decade, two additional diamond centres were established at sites much farther north, in the Arctic region at Aikhal and Udachnyy.

Siberia also provides materials required in the manufacture of nuclear and aerospace equipment, notably, boron, lithium and beryllium. At Dalnegorsk in the Soviet Far East, the boron industry is reported to employ a work-force of nearly 6500. Lithium and beryllium are mined in eastern Siberia, probably in the Chita oblast.

Asbestos is abundant in the USSR and is produced at several centres, Asbest and Kiyembay, both in the Urals, and at Ak-Dovurak in the Tuva ASSR of Siberia; but a second major Siberian deposit still awaits a start to development. This is the very extensive Molodezhnyy deposit which lies in the northern part of Chita oblast and apparently has no equal in the world in its reserve quantities of textile-grade long-staple fibres. It is within the zone of the BAM near the line's future Taksimo station.

Apatite

Apatite is an important source of phosphorus and is a constituent of most kinds of igneous rock; it is commonly associated with alkaline igneous intrusions of pegmatite. The first deposit of this kind to be developed in the USSR was at Kirovsk in the Kola peninsula. Concentrated as superphosphate it is railed as fertiliser to many parts of the Soviet Union. A by-product of apatite production is nephelite concentrate, used as a non-bauxite material for commercial aluminium production. A second major deposit of phosphate to be exploited was in the Karatau mountains of Kazakhstan. The twelfth Plan emphasised the development of the Seligdar apatite, in eastern Siberia near Aldan; this, though lower in phosphate content than that of the Kola deposit, has large reserves and a potential output of 40 million tonnes of crude apatite per year with a concentrated yield of 4–5 million tonnes. Also in eastern Siberia is the Oshurkovo deposit at Ulan Ude but the reserves are smaller than those of Seligdar, the ore is poorer in quality and the extraction of the material is considered to produce effects detrimental to the sensitive environment of Lake Baykal from pollution of the Selinga river which flows into the lake.

BIBLIOGRAPHY

Dewdney, J. C. (1976), *The USSR, studies in industrial geography*, Dawson, Folkestone.
Dewdney, J. C. (1982), *The USSR in maps*, London
Dienes, L. (1977), 'Basic industries and regional economic growth: the Soviet South,' *Tijdschrift voor Economische en Sociale Geografie*, **77**, No. 1, pp. 2–15.
Dienes, L. (1981), in Jt. Economic Committee US, *Energy in Soviet policy*, Washington, 101–19.
Dienes, L. (1987), 'The Soviet Oil Industry in the 12th

Five Year Plan,' *Soviet Geography*, **28**, pp. 617–655.

Dienes, L. and Shabad, T. (1979), *The Soviet energy system*, Wiley, New York.

Goldman, J. I. (1980), *The enigma of Soviet petroleum: half full or half empty*, London.

Kaser, M. (1983), The Soviet gold mining industry, in *Soviet natural resources in the world economy*, ed. R. G. Jensen *et al.*, Univ. of Chicago Press.

Legasov, V. (1987), 'A Soviet expert discusses Chernobyl', *Bulletin of the atomic scientists*, **43** No. 6. pp. 32–34, Chicago.

Markov, V. D. and Khoroshev, P. I. (1986), The peat resources of the USSR and prospects for their utilization, *International Peat Journal*, **1**, 41–47, Helsinki.

North, R. (1987), 'Transport and communications,' in Wood, (ed.) 130–157.

Osleeb, J. P. and Zumbrunnen, C. (1984), *The Soviet iron and steel industry*, Totowa, N. Jersey.

Pralnikov, A. (1988), Chernobyl' today, *Soviet Union*, No. 9 (462).

Pryde, P. R. (1978), 'Nuclear energy development in the Soviet Union,' *Soviet Geography*, **19**, pp. 75–83.

Pryde, P. R. (1984), 'The Soviet development of solar energy', *Soviet Geography*, **25**, pp. 24–33.

Shabad, T. (1969), *Basic industrial resources of the USSR*. New York.

Shabad, T. (1986), 'General fuel production trends,' *Soviet Geography*, **27**, pp. 248–279.

Shabad, T. (1987), 'Geographic aspects of the 1986–90 new Soviet five-year plan,' *Soviet Geography*, **27**, pp. 1–16.

Shabad, T. (1987), 'Geographic aspects of Chernobyl' nuclear accident', *Soviet Geography*, **28**, pp. 504–526.

Shcherbak, I. (1989) *Chernobyl, a documentary story*, Moscow; English translation, University of Alberta, Edmonton.

Tochenov, V. V. *et al.* (eds) (1983), *Atlas of the USSR*, Moscow.

Wilson, D. (1983), *The demand for energy in the Soviet Union*, London.

Wilson, D. (1987), 'Oil and gas resources', in Wood, (ed.), 96–129.

Wood, A. (ed.) (1987), *Siberia, problems and prospects for regional development*, London.

Ziegler, C. E. (1987), *Environmental policy in the USSR*, Printer, London.

9 Industry

Industrial development in the Soviet Union has had to overcome difficulties which have been no less formidable than those which faced agriculture. It is true that the climate does not pose nearly so many problems for industrialists, though it does impede some industrial activities, such as mineral extraction and movement of goods and labour, and the water resources are very unevenly distributed. Industrial development also is not hindered, generally, by soil variations. On the other hand, the great extent of the country, which has conferred reserves of land on Soviet agriculture, has meant great transport problems for industry, both in the assembly of raw materials and labour at suitable locations, and in the distribution of finished or semi-finished goods. Strategic issues have also posed problems for Soviet planners, with a constant need to balance development in the European parts which have the greatest population and, therefore, demand, against the wish to locate as much industry as possible in the less accessible regions where they are safer from invasion and aerial bombardment. Economic issues have likewise presented problems in whether to locate industries near the raw materials, many of which are in the eastern regions, or in the areas of main demand, where raw materials are fewer. In either case, specialised regions of industry mean long transport hauls, whereas a more even distribution of industry throughout all populated regions means less dependence on centralised or specialised industrial areas and more regional security.

The problem of ensuring a sufficient supply of labour and of, in particular, skilled labour, has also always been paramount in Soviet industry. Although after the emancipation of the serfs there was considerable migration from the farms to the towns, the workforce was not particularly well adapted or sufficiently well educated for skilled work. Later generations have benefited from being brought up in an urban and industrial environment, but the newcomers reinforcing the industrial ranks have almost always been from rural sources and in need of training. In later years the flow of migrants from the rural areas has slackened as the reserves of manpower have become less, and industrial planning has had increasingly to think in terms of coping with actual labour shortages, particularly in the less attractive regions. It has been increasingly necessary for labour reasons, as well as being politically desirable, for industrial development to be speeded up in Central Asia rather than in Siberia, the increase among the traditionally Moslem peoples of the Central Asian republics being the most rapid in the Soviet Union, whereas population has increased only slowly in Siberia, with much outward migration to areas of better climate.

Against the advantages of location in either Siberia or Central Asia have been the attractions of greater economies of operating manufacturing industries in the western or European areas, where skills and training facilities are best developed and contacts most readily established with the Soviet Union's partners in COMECON, as well as with the capitalist countries in western Europe and North America, on whom the USSR has depended for much of the expertise required to develop modern industries.

In 1917 the new government inherited a considerable range of industries from tsarist times but industrial development had depended very much on foreign engineers and foreign capital. For example, British capital developed textile mills (which employed managers and overseers from Lancashire), east of Moscow and at Narva in Estonia. German capital was invested in engineering and in the chemicals industry and was particularly important in the Baltic region, where there was a large German minority. Ukrainian heavy industry was financed by British, French and Belgian money, and the city of Donetsk was founded by a Welsh ironmaster, Hughes, being originally named after him (Yuzovka). French money was also liberally invested in railway development and Swedish money in telephones, while the Nobel millions were made in the Baku oilfields, to name but a few examples. The tsarist authorities gave the greatest encouragement to industries which might help Russia strategically and St. Petersburg, then the capital, grew as an industrial centre supplying the Imperial armed forces. As noted in Chapter 1, manufacturing industry was relatively well developed around Moscow, the traditional commercial capital of the country, while the coal and iron ore of the Ukraine attracted heavy industry. Added to the industries of the tsarist empire were also the textile manufacturing of Lodz and the coal and iron-making of Dabrowa, for much of Poland was then under Russian control.

Nevertheless, in 1914, two-thirds of the manufactured goods required by Russia had to be imported and over three-quarters of manufacturing industry was controlled by foreign capital. Many simple everyday needs were still supplied to the market by peasant industry and much commerce continued to pass through the traditional great fairs, like that of Nizhniy Novgorod (now Gorkiy).

SOVIET INDUSTRIAL DEVELOPMENT

When the Bolsheviks seized power in 1917 they proclaimed workers' control of industry but in the confused conditions of the Civil War (1918–21) this was curtailed, central planning was developed and by mid-1918 all major industries had been nationalised. After this period of 'War Communism' Lenin permitted more freedom in what became known as the New Economic Policy. After his death in 1924 no clear programme emerged until Stalin took control and initiated an ambitious industrialisation drive in the first of the Five Year Plans (1928–33). His aim to create 'socialism in one country' resulted in almost total centralisation of economic planning with the emphasis on basic industries such as coal, steel and heavy engineering.

Since the Soviet leaders were not prepared to permit industrial development by private entrepreneurs there was no possibility of letting individuals make the decisions on capital investment and location of industry and then suffering if the decisions turned out to be bad ones. With the State as the only entrepreneur there was need for an immense and complex bureaucracy to make the decisions and guide development. The principal organ for industrial development was GOSPLAN, the State organisation for industrial planning, itself guided by the leaders in the Communist Party.

Gosplan laid down directives proposing the spread of industry more evenly over the country by development of the eastern regions; the development of regional specialisation at the same time as maximum overall development in the major planning regions; and encouragement for any development which reduced the burden on transport. In the early years, planning suffered from a gigantomania, illustrated by the concept of the Ural–Kuzbas Kombinat. This was a vast interregional link-up between industries in the Ural region and in western Siberia. Kuzbas, in west Siberia, was to ship its coking coal by rail to the Ural region, where suitable fuel was lacking, with return loads of iron ore. The idea was not entirely new, a similar project using water transport having been suggested by a Commission in 1916, but it was inadequacy of the railways to handle the bulk freights demanded that weakened the scheme. From the early 1930s also, supplies of Karaganda coal began to reduce the Ural's dependence on Kuzbas output while discovery of iron ore in western Siberia reduced the Kuzbas need for ore from the Urals. By the mid-1930s, the Ural region emerged as the second great metallurgical producer after the Donbas as its large new iron and steel works went into operation.

Much publicity was given to the development of the eastern regions but some of the greatest but least advertised achievements were in industrial construction in European Russia, notably in the Moscow and Gorkiy districts, while new industries began to appear in Baykalia and in the Far East in response to the strategic threat created by Japanese control of Manchuria. However, some regional economic development in the 1930s had strong political undertones. For example, the Moslem lands of Central Asia, of uncertain allegiance, were tied more tightly to the rest of the Soviet Union by increasing economic interdependence. Land in Central Asia was turned over increasingly from food to industrial crops (e.g. cotton), urgently needed for industries in other parts of the country, while Central Asian food deficiencies were made good by Siberian wheat sent along the Turksib railway, completed in 1931. Central Asian manufacturing industry was also made heavily dependent on raw metal and chemical supplies sent from the Ural region and Siberia, or even from European Russia. It was not until the wartime emergency that iron and steel making was established in Central Asia (Begovat/Bekabad) to use locally available scrap, and the region was kept continually dependent on pig iron and alloy metals from Siberia and the Ural region.

As the world situation deteriorated in the late 1930s, developments with obvious strategic implications were planned, such as the opening of the Pechora coalfield, the working of metallic minerals on the lower Yenisey and exploitation of the Ural–Volga oilfields, while further progress was made in industrialisation in the Ural region and Siberia. Of these developments, few were anywhere near completion by the time of the German invasion. As much equipment as possible was removed from the path of the advancing German armies and re-erected east of the Volga, notably in the Ural region and in Siberia. Central Asia acquired food and textile plants among others. Many of the evacuated plants left the machinery in their 'temporary' sites when they returned to European Russia where they were re-equipped after the German defeat. This accounted in part for the surprising growth of machine tools and transport equipment in the Ural region and western Siberia after the Second World War. The 'eastern regions' also benefited from new plants erected there after having been dismantled as reparations in Germany: this was marked, for example, in the chemicals industry of western Siberia.

Postwar planning

After the war, reconstruction was achieved under the strict centralised control favoured by Stalin, with first priority being continued for the producer goods industries. Khrushchev introduced the first major reforms in the late 1950s. His decentralisation of management to regional economic councils (*sovnarkhozy*) (Chapter 12) did not lead to the greater production he had hoped for and after his removal from power the central ministries were again given full control over the most important branches of the economy. Nevertheless, under Khrushchev the economy became somewhat better balanced in terms of production of consumer goods, the reconstruction of the chemicals industry was seriously begun, the motor vehicle industry was expanded and the production of petroleum products substantially increased.

Under Brezhnev's conservative leadership centralisation was again favoured, with a trend towards grouping of enterprises. In 1973 it was decreed that associations (*ob'edineniya*) would become standard in basic industry in order to achieve economies of scale and enhance technical progress. Production associations (*proizvodstvenninye ob'edineniya*) linked enterprises horizontally or vertically, while industrial associations (*promyshlennye ob'edineniya*) executed ministerial requirements at all-Union or republic level in selected industries such as fuel and power. As with individual enterprises, associations had to accept limited economic accountability (*khozraschet*) in an attempt to reduce waste of resources but progress was minimal in this period (now known in the USSR as 'the period of stagnation').

Integration was pursued also in the creation of territorial production complexes. Much stress was placed during the 1970s on these areal forms of organisation, intended to facilitate both specialisation of individual enterprises and integration of overall regional production. The planning for such complexes embraced both long-established industrial regions such as the Donbas and pioneer areas like those being built in association with the Baykal–Amur railway.

Official literature stresses the part played by the planning bodies of each republic, oblast and rayon in the construction of plans by the Gosplan organisation. The Gosplan of each republic is supposed to take account of all territorial requirements and it is presumably through these processes that individual enterprise managers bid for their own places in the national plans, in terms of both production and supplies. It is clear, however, that national interests are supreme in the formulation of each five-year plan and the longer-term plans that are accepted as essential for full development of the economy.

Perestroika

When Mikhail Gorbachev came to power in March 1985 it was widely recognised that it was necessary to do more than tinker with the industrial structure of the country, which was clearly falling further and further behind the advanced Western economies in practically every respect. There were different opinions on how this widening gap should be reduced but Gorbachev decided to go all out for what many considered to be reforms of a capitalistic nature and therefore highly contentious in the context of Soviet revolutionary evolution. Gorbachev stressed the necessity of making profitability the key to the future allocation of resources. From 1 January 1988 a wide range of enterprises became responsible for their own financing and their continued existence was supposed to become dependent on their making sufficient profits to plough back into their firms. In practice there had to be limitations on this process because major industries could not be allowed to collapse if their directors, totally inexperienced in managing their affairs to this extent, could not make the transition quickly to the new conditions. For smaller firms, however, there were obviously new opportunities to move quickly towards greater attention to market requirements—quality control and ultimate profitability.

It soon became clear that restructuring was going to take many years. Apart from the resistance from conservative and entrenched interests, difficulties were caused by the varying rates of change in different branches of industry. Thus, in 1988 a clothing firm in Latvia reported rapid advances in product design and manufacturing but was hindered by still being totally dependent on the state organisation for supply of materials. Probably the type of organisation most able to take advantage of the new freedom was the small family unit or partnership specialising in small scale production or personal and repair services. For the large national firm or association there were many new rules on allocation of profits to the various funds, including state reserves,

cultural and social provisions and housing accommodation, for which responsibility for its own employees still lay largely with the firm. Further stages in *khozraschet* became operative from 1 January 1989. Prices were, however, still regulated by the state to avoid undue disruption, profiteering and inflation. Price reform is scheduled to take place gradually with an acceleration in the early 1990s.

REGIONAL INDUSTRIAL DEVELOPMENT

Postwar planning has generally favoured industrial development in the better placed areas and the relatively weak regions have hardly improved their position. Thus, approximately half the total national investment has remained in European Russia, while the rate of industrial growth in Transcaucasia and Central Asia has generally been below the national average. In Kazakhstan and in Belorussia, however, it has been above the national level, and investment per head of population has been greatest in Siberia, Kazakhstan and north-west European Russia. Major mining, metallurgical and railway developments in southern Siberia, and giant new hydro-electric barrages providing cheap power for large new electro-metallurgical and electro-chemicals industries provide the basis for widespread developments.

Despite the lack of satisfactory statistics, there is every indication that European Russia (including the Ural region) remains the main location of industrial output, contributing well over half the total output. As much as one-fifth of total output may come from the Ukraine; another fifth from the Central Industrial region; over one-tenth from the Ural region and a little less may now be contributed by the Volga lands. Siberia and the Far East together contribute probably between 10 and 15%. The geographical pattern of industry represents a product of the struggle between the centripetal tendencies to locate plants in the most economic sites and the centrifugal tendencies to spread industry as widely as possible throughout the country for strategic reasons.

The attractions of locating the maximum industry practicable in the eastern regions is apparent, for then there is minimum transport of energy and minerals to the point of manufacture or processing. As the markets are mainly in the west there is then, however, the necessity to transport finished products. Furthermore, labour supply is limited in Siberia and the Far East, though plentiful in Central Asia, which is the only major region in which the total labour supply can be expected to increase significantly in the coming decades. Hence the advantages of each location for each industrial development must be carefully balanced by Gosplan before a decision is taken. There are clear advantages in locating in the east the industries catering for local consumer needs, as in the cases noted above. Less obvious in transport terms are some of the others developed in the east but these are commonly located there in response to some specific overwhelming advantage, such as cheap power for the aluminium and electrochemical industries in Siberia and Central Asia.

Thus, there can be no general rule about the economic desirability of an eastward spread of industry as a whole and many branches will continue to be developed predominantly in the west irrespective of the riches of the east. Strategically also there is clash of interests between emphasis on development in the east and in the west. There is powerful motivation to encourage Soviet people to move into the regions bordering China, with its 1100 million population posing a threat, at least in Soviet eyes, to the Siberian and Far East lands, the legal possession of some of the border territories being, in any case, in dispute. On the other hand the integration of the Eastern European economies with that of the USSR and the need for trade with 'the West' make the European regions more attractive for investment. The policies of *perestroika* encourage more investment in the west, but selective investment in the east is continuing.

THE MAJOR INDUSTRIAL GROUPS

Each of the main groups of manufacturing industry will now be reviewed, followed by a résumé of the principal industrial regions.

The metal production industries

The comparatively rich endowment of the Soviet Union with minerals, detailed in Chapter 8, has been a considerable aid to industrial growth. Exploitation of deposits in the harsh physical environments of northern Siberia and Central Asia has posed many problems and a dilemma. The choice is whether to save on transport by installing expensive refining and purifying equipment at the mines, or to save such investment but pay heavier transport costs by evacuating raw ore for refining and treating elsewhere, carrying away considerable quantities of what is ultimately 'waste'. Solution of the equation is difficult since factors such as nature, size and location of the deposits, the relative priority of development of different minerals and ease of provision of transport, as well as labour supply, are all involved.

Fuel minerals are well represented. Coal has ceased to be the principal source of energy, having been displaced by petroleum and being now rivalled by natural gas (Table 9.1). Coal deposits of varying quality are widely scattered. Most significant are those suitable for metallurgical coking coal, the continuing supply of which does give rise, however, to some anxiety. An important influence on industrial development has been the great improvement in availability of oil and natural gas as detailed in Chapter 8, while a wider

TABLE 9.1 PRODUCTION OF FUELS BY TYPE (CALCULATED IN STANDARD UNITS)

Year	All	Petroleum	Natural Gas	Coal	Peat	Shales	Wood
1940	100	18.7	1.8	59.1	5.7	0.3	14.4
1950	100	17.4	2.3	66.1	4.8	0.4	9.0
1960	100	30.5	7.9	53.9	2.9	0.7	4.1
1970	100	41.1	19.1	35.4	1.5	0.7	2.2
1980	100	45.3	27.1	25.2	0.4	0.6	1.2
1987	100	40.0	37.7	20.6	0.2	0.4	1.1

Source: Narodnoye khozyaystvo SSSR, various years.

spread territorially of deposits has eased the supply problem.

Iron ore, generally in quality better than deposits currently used in the Western world, has also been shown to be available in generous proportions and there is a rich endowment with most of the more important alloy metals, especially manganese. The wide spread of deposits is again an important locational influence. The situation for non-ferrous metals is a little less happy, notably because of the adverse physical environment in which several important 'deficiency' metals (e.g. tin) are found, but also because some metals are poorly represented. A special office of COMECON deals with intra-communist bloc use of these metals and recovery of non-ferrous scrap.

It should also be borne in mind that the Soviet Union is the major supplier of raw materials to the COMECON countries of eastern Europe, and has encouraged these countries' dependence on Soviet supplies for political reasons. Soviet regional economic policy, with its emphasis on a high degree of regional self-sufficiency and reduction of the burden on transport, has encouraged development of deposits, even where these have been unpromising in quantity and quality and are consequently expensive to exploit.

A special note should also be made of electricity, which occupies a notable place in Marxist–Leninist dogma: Lenin said 'Communism is Soviet power plus electrification of the whole country'. The greater part of the electricity generated is by thermal power stations. A little over an eighth of the current is generated by hydro-electric stations. Because of the remoteness and great physical difficulties, it is unlikely that more than a small proportion of the immense water power potential of Siberia will be harnessed in the foreseeable future. The creation of a national grid for electricity distribution presents problems because of the great distances involved, but this may be overcome by Soviet advances in high tension transmission technology, allowing greater economical distances for electricity 'transport'. Meanwhile, several regional grids have been constructed. It has been argued that availability of conventional fuels and water power make the necessity to develop nuclear powered generating stations less urgent in the Soviet Union than in western Europe, but the Soviet Union has, in fact, continued to develop nuclear power stations, as described in Chapter 8.

IRON AND STEEL MAKING

The major iron and steel producer is the Dnepr–Donbas district. Although of lower relative importance than formerly, it still produces about one-third of all Soviet steel. Heavy metallurgy consumes about 40% of the Donbas coal output. A particularly important group of works lies on the western flank of the Donbas coalfield, obtaining iron ore from Krivoy Rog and Kursk, and comprising plants at Donetsk and Makeyevka, which together smelt half the output of pig iron in the Donbas, and also Yenakiyevo, while Gorlovka is a large coking centre. At Kramatorsk there is electric steel production. The plants of the eastern Donbas concentrate on specialised work such as pipes at Voroshilovgrad, complicated sections at Kommunarsk and special pig iron at Almaznaya, while Taganrog is important for tubes, pipes and boilers. Gorlovka coke, Yelenovka limestone and a mixture of Kerch and Krivoy Rog ores are used at Zhdanov, where there are some of the largest rolling mills in Europe. The highly phosphoric Kerch ore is suitable for tube and welding steels. Although an ore producer, Krivoy Rog imports Donbas coal and coke to smelt its own fragile ores that do not stand transporting elsewhere. An intermediate location in the bend of the Dnepr marks large plants at Dneprodzerzhinsk and Dnepropetrovsk, while Zaporozhye, using cheap Dneproges electricity, makes electric steels and cold rolls, and has a ferro-alloy plant. Nikopol, with manganese mines, is also a rolling mill centre. The Dnepr area has a big advantage in its water supply, which contrasts with the water-deficient Donbas.

A blast furnance of the Karaganda iron and steel works at Temirtau

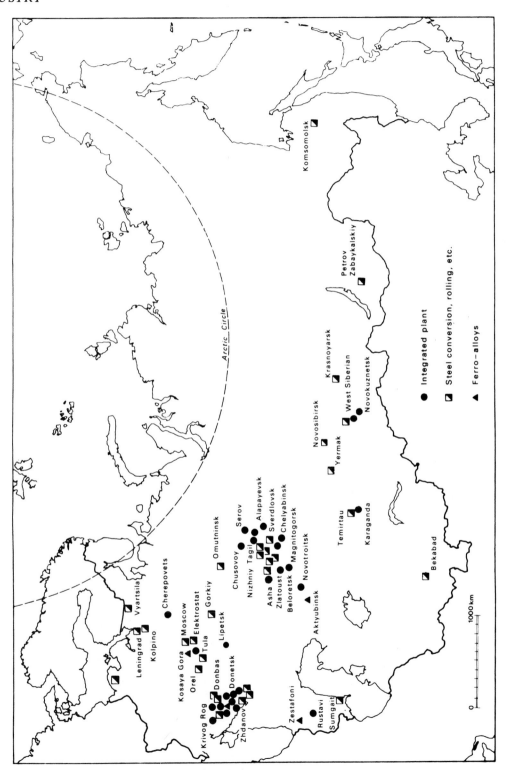

Fig. 9.1 Distribution of major iron and steel plants

The second main centre, producing about 30% of the total, is the Ural region. The low cost of Ural ore offsets the high costs of coke and coal imported from western Siberia and Kazakhstan, but it is now insufficient and ore is imported from as far afield as Kursk. Unlike the Donbas, there is not such a difficulty in the Ural region in obtaining the large quantities of water needed for processing, except in the southern areas. Manganese is often used generously in place of flux. The main plants lie mostly on the eastern slope of the Ural ranges along a belt of intense mineralisation which supplies their raw material. Nizhniy Tagil, one of the largest works, is supplied from nearby mines such as Vysokaya Gora and Blagodat or most recently from Kachkanar, with coal and coke mostly from the Kuzbas. There are also some old charcoal smelting plants in the area. The Sverdlovsk district depends mostly on converting engineering scrap into steel for electrical engineering, while Polevskoy has one of the largest Soviet tinplate works. Since 1943, quality and alloy steel has been made at Chelyabinsk, using Bakal ores and Karaganda and Kuzbas coal and coke. Zlatoust makes instrument steels. One of the largest of all Soviet plants is Magnitogorsk, which uses local ores from the Magnetic Mountain (Magnitnaya Gora) and recently from Kazakh deposits, with coal and coke brought from the Kuzbas and Karaganda. A serious problem has been the supply of water for processing in the dry steppe, largely overcome by a barrage across the Ural river near the town. In the southern Ural, Novotroitsk has a large modern works, using the chrome-nickel-iron ore from Khalilovo and imported coal and coke. The small works on the western flanks of the Ural mostly concentrate on rolling, plating or drawing.

Western Siberia was developed first as a steel producer in association with the Ural region as part of the Ural–Kuzbas *Kombinat* of the early Five Year Plans. One of the largest integrated iron and steel works in the USSR is sited at Novokuznetsk, where a second large plant is being developed, and where there is also a big ferro-alloy plant. Novokuznetsk has one of the world's largest continuous sheet mills. Coal and coke are available locally and ore comes from Tashtagol, Shalym and Abakan, and from the more distant Angara, Ilim and Pit basins, with coal from a new field at Chulman in southern Yakutia. Plans have been put forward for a steelworks at Tayshet and

modernisation of the old Petrovsk–Zabaykalskiy works.

The Central Industrial region around Moscow has traditionally had high production costs offset by low transport charges to local consumers, but its future has been transformed by the KMA development. Coal and coke come from the Donbas, while ore is drawn from Krivoy Rog and other areas, but some ore is also mined locally. Tula and Lipetsk are the main smelters, while the latter has large rolling mills and a continuous strip mill, as well as a large diameter tube plant and a cold rolling mill. Both Kosaya Gora and Lipetsk produce very high quality steels for special purposes. Some steel conversion from scrap is carried on in Moscow at large engineering works. The upper Volga has some small rolling mills (e.g. Gorkiy). The Moscow and Leningrad regions are also now supplied from Cherepovets (near the Rybinsk reservoir) which uses Vorkuta coal and Karelian ore. Leningrad converts steels for local engineering plants and Vyartsilya in Karelia makes electric steel.

Using Donbas and Caucasian coal and Dashkesan iron ore, the Transcaucasian Rustavi iron and steel works near Tbilisi, opened in 1955, supplies the Sumgait tube works near Baku. At Zestafoni, using local ores and cheap hydro-electric current, is one of the main Soviet ferro-alloy plants. In Kazakhstan, based on Atasu ores and Karaganda coals, there are new iron and steel works at Temirtau and Karaganda and a large

Tractors coming off the production line at a Tashkent factory

ferro-alloy plant is being built at Yermak (Pavlodar), while the Aktyubinsk ferro-alloy works uses ore from Khrom-Tau. To use available scrap and convert imported Siberian pig iron, a steelworks was opened in 1943 at Bekabad (Begovat) near the Farkhad dam in Uzbekistan. In the Far East, Amurstal at Komsomolsk serves a similar purpose, and it also has important tinplate works.

Non-ferrous metallurgy

Some of the important locations of the non-ferrous metals industry have already been mentioned in connection with the steel industry, and others in describing the occurrence of important ore bodies in Chapter 8. It is necessary, however, to stress that this is a vast and vital branch of industry, of no less importance than the steel industry, to the functioning of the modern complex economy, and to note the main plant locations.

COPPER

The Ural region has supported copper smelting since the early part of the eighteenth century. Relocation has occurred as local deposits have become exhausted, with the southern area becoming dominant as new metal deposits have been discovered. Verkhnyaya Pyshma, near Sverdlovsk, is the largest refining centre. Concentrates are shipped in from other ore-rich regions including Kazakhstan. Within the Kazakh SSR, the Balkhash plant, near Lake Balkhash, refines concentrates from the several ore bodies of the area. A major new East Kazakhstan copper refinery is being developed to use the Nikolayevsk deposit and concentrates from other ore bodies. Armenia has one of the oldest copper smelters at Alaverdi. This plant has been rebuilt in Soviet times and is now a major works. Copper smelting has also been developed in Uzbekistan, using Uzbek and Tadzhik ores.

NICKEL

Nickel is worked along with copper in several plants, notably at Monchegorsk and Nikel in the Kola peninsula and the Norilsk area in northern Siberia. In copper and nickel working the emphasis is increasingly on very large smelters, which achieve marked economies of scale and enable valuable by-products to be recovered. This is, in turn, facilitated by modern concentration methods which reduce the amount of waste material transported from the mining areas.

ALUMINIUM

Production involves even more transport than does the movement of concentrated copper and nickel ores to smelters, because the reduction of alumina to aluminium metal requires immense inputs of electric power for electrolysis, about 18 000 kilowatt-hours per tonne of metal produced, and hence the most favoured locations are in areas of major hydro-electric stations. On the other hand, ore is not usually available conveniently nearby and in any case has first to be smelted. Fortunately, the alumina is suitable for long distance transport to the reduction works.

The first Soviet aluminium reduction plants were built at Volkhov (1932), near Leningrad, and Zaporozhye on the Dnepr (1933), the former location being influenced by the availability of bauxite in the Leningrad area. Bauxite deposits also influenced the location in the Ural region of the third and fourth plants. All post-World War II plants, however, have been located in the regions of cheaper power—Novokuznetsk, Volgograd, Sumgait (Azerbaydzhan), Yerevan (Armenia) and, in eastern Siberia, at Shelekhov near Irkutsk, Krasnoyarsk and Bratsk, all three in the 1960s. Over one-half of the Soviet output of aluminium is now produced in Siberia. Aluminium plants have, however, also been constructed in Karelia and Kazakhstan, and Central Asia has also been selected for development of this industry because of its hydro-electric resources. All the locations in the east have become more attractive with the use of nephelines and other ores as alternatives to bauxite (Chapter 8).

MAGNESIUM AND TITANIUM

Like aluminium these are of great strategic importance through their application in the aerospace, nuclear and other advanced technologies. For the early applications of magnesium, the need was met from the plants at Zaporozhye (1935), on the Dnepr, and the two plants in the Urals, Solikamsk (1936) and Berezniki (1943). Production of titanium began only in 1954 and the two metals are now produced together at Podolsk, south of Moscow, Zaporozhye, Berezniki and Ust–Kamenogorsk, a large new plant in Kazakhstan dating from 1965, and at a similar large plant at Kalush in the western Ukraine. Titanium production requires a very high consumption of electricity, greater than that for aluminium, so availability of power is crucial in location. In the ores, however, titanium is found in association with vanadium, tantalum and other minerals and integrated processing plants make for the greatest economy of extraction of all these light metals and associated elements.

LEAD AND ZINC

Smelters for these metals were formerly located in coalfield areas as the retorts used large quantities of coal and coke. Old smelters of this kind are still found in the Urals. Ukraine and Kuzbas. More modern methods of handling the polymetallic ores, however, have now been adopted. Mining areas now have concentrating plants, the concentrates being treated in integrated plants typically using natural gas for smelting lead, which results in production of sulphuric acid as a by-product, which is then used in the electrolytic refining of the zinc concentrates. For the latter processes, large quantities of electricity are needed, which, together with the location of ore bodies, has resulted in an emphasis on these industries in the eastern regions. The largest centre is Ust–Kamenogorsk in Kazakhstan, dating from 1947 but since much expanded. It is fairly close to large hydro-electric stations on the Irtysh as well as the Zyryanovsk metal deposit. It produces lead, zinc, cadmium, copper and other metals together with sulphuric acid and zinc sulphate. The titanium-magnesium plant already mentioned is located nearby. Other major refineries have been developed at Leninogorsk, not far away, and at Almalyk in Uzbekistan. Also in Uzbekistan, Chimkent smelts lead while zinc concentrates from the area are railed to electrolytic refineries. In the Urals, only Chelyabinsk produces zinc. Belovo in the Kuzbas is similarly the only zinc producer in Siberia, while lead is smelted at Tetyukhe in the Far East. Otherwise, ores from the areas east of Lake Baykal are shipped as concentrates to other regions. In the North Caucasus, the Ordzhonikidze zinc plant, dating from tsarist times, has been developed into a major integrated plant for lead and zinc, while another old plant at Konstantinovka in the Donbas, has been modernised and operates wholly on concentrates from distant areas.

TIN

As tin deposits are relatively small and scattered (Chapter 8), no major centre for smelting has yet emerged near the ore bodies. Concentrates are shipped to the existing smelters at Podolsk, south of Moscow, and at Novosibirsk. Hence, there is a large transport input. Promising new fields are being developed in the Far East and processing is beginning there.

It will be seen that when the basic metal processing industries, whether iron and steel, base metals, polymetallic or light metals, are examined, certain regions recur repeatedly in the analysis. Although these clusters were indicated in the brief regional introduction to this chapter, such a cursory survey could not stress the importance of the modern complex plants that have developed. Nor does the traditional emphasis on the iron and steel industry adequately indicate the breadth and variety of the plants that produce the metals for the engineering, constructional and other industries that in turn produce and transport the goods required for further processing, (and, indeed, the processing of the minerals and metals themselves), for transport services and for all other forms of consumption. It will now, however, be clear that in addition to the older industrial regions (the Centre, the Donbas–Dnepr area and the Urals) that of western Siberia and Kazakhstan, in particular, has received an immense amount of investment in basic production facilities in recent decades. Though water supply remains a major problem it has become a formidable addition to Soviet industrial regions. Central Asia and Transcaucasia are also emerging as considerable industrial regions in their own right through the availability of mineral resources and hydro-electric power. These trends will be confirmed in examination of other industries, including those producing chemicals, but first, the main users of metal, the engineering industries, will be reviewed.

Engineering

The use of metals in all branches of the economy depends on metal-cutting and shaping equipment and the machine-tool industry is thus of fundamental importance. Over a hundred enterprises manufacture major metal-working machines, forges and presses, lathes and tools of all kinds. More than half of the total are classed as automatic or semi-automatic, many having digital programme control. The Soviet Union claims to be the birthplace of the most advanced machines for electrophysical and electrochemical working of metals.

The Moscow area and Leningrad, traditional centres of the machine-tool industry, remain dominant in this form of engineering, but many plants have been built in the Urals, especially in Orenburg and Chelyabinsk, the Volga area centred on Saratov, the Lower Don and the North Caucasus, the Ukraine, Belorussia, especially Vitebsk and Minsk, Lithuania and Transcaucasia (Fig. 9.2). Novosibirsk is the main centre in Siberia, while in Central Asia, Kirgizia was first to develop a machine-tool industry. Design and labour skills, markets and the availability of steel are important in location of these plants. Instrument-making factories have expanded particularly rapidly with the accelerating demand for all forms of control units, calculators and computers, electronic equipment and photographic apparatus. High labour input and demand for special skills cause these factories to be located mainly in Moscow, Leningrad, Kiev, Lvov and other large urban centres. Availability of female labour is a particular advantage, both in terms of numbers and the traditional dexterity of feminine fingers.

The construction of heavy machines, providing plant for ferrous and non-ferrous metallurgy, mining, power stations, cement and chemical plants, however, is more often located in metal-producing regions because of the large and heavy content of metals in the products. The Donbas, Dnepr and Ural areas are leaders in these branches. Other major works are sited in the Kuzbas, Novosibirsk, Petropavlovsk in North Kazakhstan, Irkutsk, Krasnoyarsk and Alma Ata, the last three being especially concerned with mining equipment for their areas. Turbines, generators and steam turbines are produced mainly in Leningrad, Kharkov and Sverdlovsk.

Machinery for the textile, food and other light industries is produced in great quantities and varieties in Leningrad, with the Moscow area also important. The old textile centres of Ivanovo and Kostroma are prominent in textile machinery. Like the food industry itself, the manufacture of equipment for processing food is extremely widespread. The greatest number of plants is located in the Ukraine, with the Central, Volga and Transcaucasian areas also important, but the industry spreads all the way across to the Far East where there is specialisation in equipment for the fishing industry. Other regional specialisations include the manufacture of special dredges and mining equipment in eastern Siberia where gold and diamonds are mined, and oil drilling equipment in Transcaucasia, near the first Russian oil-producing area.

Prior to the revolution some railway vehicles were built in Russia but most were imported. Among the now numerous railway works are the old but reconstructed plants at Voroshilovgrad (formerly Lugansk), Kharkov and Kolomna, which build a range of vehicles. Electric locomotives are constructed at Novocherkassk, Riga and Tbilisi and diesel locomotives at Gorkiy and Ulan Ude in Siberia. Moscow's underground and suburban lines are equipped largely from the Mytishchi works. The heavy demand for railway wagons in the industrial districts of the Donbas and Ural regions is covered largely from Dneprodzerzhinsk and Nizhniy Tagil respectively, those of northern and central Russia from Kalinin and Kolomna. There is a considerable amount of specialisation, for example, tank wagons at Zhdanov and refrigerator vehicles at Bryansk. The only constructional plant in the eastern regions is at Novoaltaysk. There is now considerable interchange with the other members of COMECON.

The Soviet Union has greatly increased production of motor vehicles in recent years. Cars accounted for only 20% of the 363 000 road vehicles of all kinds completed in 1950, but in 1987 the output of cars exceeded 1 300 000. Buses and lorries are, however, still produced in much greater numbers. The largest plant for cars is that at Tolyatti on the Volga, named after the Italian Communist because the plant was built under contract by the Fiat Company. Other plants producing small cars are at Moscow and Zaporozhye while larger cars are made at Gorkiy. Cars are also produced at Ulyanovsk and Miass in the Volga and Ural areas respectively.

The most important lorry producing plants are at Minsk and Zhodino, both in Belorussia, and at Naberezhniye Chelny on the Kama river in the Tatar ASSR and there are numerous smaller plants, some of them associated with car factories. Some of the buses are also produced in these factories, notably in Moscow and Gorkiy, but

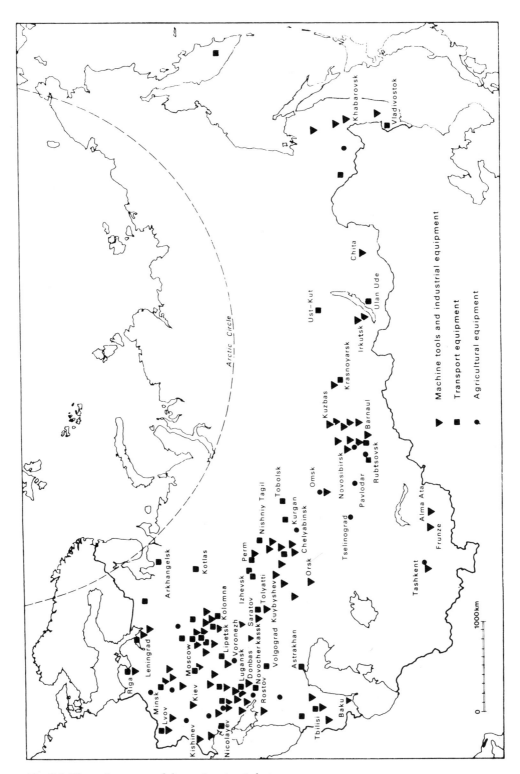

Fig. 9.2 The main centres of the engineering industry

there are also specialised enterprises in Lvov, Likino (near Moscow), Pavlov (near Gorkiy), Riga and Kurgan.

There are now also specialised works for motor vehicle engines, the first being at Yaroslavl and others at Ulyanovsk, Kharkov and Barnaul in Siberia. It will thus be seen that altogether the Soviet motor industry is now of major importance with plants widely distributed, although with the majority still in the European regions.

The construction of both military and civil aircraft is organised by design bureaux which tend to specialise in particular types of aircraft, for example Tupolev, involved mainly with large transport aircraft, bombers and long-range reconnaissance types, often based on the same fundamental designs. There is no hard and fast rule, however, and the Yakovlev designs range from light aircraft through fighters to transports and Antonov from agricultural biplanes to some of the largest transport aircraft in the world. Most of the design bureaux and research facilities are in Moscow but there are newer establishments at Kuybyshev, Kazan, Perm, Novosibirsk, Omsk, Sverdlovsk and Ufa, i.e. mainly in the east for strategic reasons. Production factories are not directly linked with particular design bureaux but more often with the production of particular classes of aircraft—transports, strike aircraft, helicopters, etc. Most were rebuilt in the eastern regions during the war and these have been further developed though there are new plants also in European Russia, which has regained its former importance.

Building of river vessels occurs on the main rivers; for example, on the Volga at Rybinsk, Gorkiy, Krasnoarmeysk and Astrakhan, with Perm and Votkinsk on the Kama; while the Dnepr is supplied from Kiev and Nikopol. Kotlas is the main yard for the Northern Dvina and Tobolsk supplies the Ob–Irtysh basin, while the Yenisey receives its ships from Krasnoyarsk and the Lena from Ust–Kut. Sea-going vessels (mostly naval) are built at Nikolayev and Leningrad, but major repair facilities are available on other seas—Arkhangelsk and Murmansk in the north and Vladivostok for the Pacific waters. Astrakhan and Kaspiysk build Caspian vessels. Many ships are bought from foreign yards.

Widespread agricultural engineering usually reflects local needs, with works both in engineering centres and in the countryside. Numerous large tractor works are scattered across the country: Kharkov, Volgograd and Chelyabinsk are most important, but Lipetsk is also a significant plant and Rubtsovsk supplies Siberia. Grain combines, first built in 1930, are made mostly in grain growing regions—Zaporozhye, Omsk, and Barnaul have important works and the Rostselmash plant at Rostov-na-Donu builds one-fifth of all Soviet farm machinery output. Tula makes potato harvesters and sorters; Bezhetsk, flax processing machinery; in Transcaucasia (Batumi, Tbilisi, Poti) production is mainly of machinery for citrus fruit cultivation and tea growing; machinery for Central Asian oasis cultivation comes from Tashkent, Chirchik and Frunze. Tselinograd and Pavlodar supply machinery, notably to the 'virgin lands' area of Kazakhstan and southern Siberia, while Perm makes silage cutters and threshers and Kurgan, dairy machinery. There are many other large works.

The chemicals industry (Fig. 9.3)

A large and diverse chemicals industry is essential for a modern industrial state, and its importance is likely to increase further with the development of new processes and materials, both in the heavy ('industrial') and light (pharmaceuticals, cosmetics) sectors. The Soviet chemicals industry has been developed intensively since the late 1950s. Raw material supply tends to be a most significant locational factor for heavy chemicals production, so that plants are tied closely to mineral deposits or to other industries whose waste or by-products form raw materials. Light chemicals tend to be sited near to consumers and in centres of skill and research.

The emphasis put on agriculture since the mid-1950s has brought corresponding development of artificial fertiliser manufacture, for which abundant raw materials are available, though often in inconvenient locations. There is a marked agglomeration of nitrogen plants in the Central Industrial and Black Earth regions of European Russia, while the Donbas produces sulphate of ammonia from coal by-products, which is also made at cokeries in the Ural region and western Siberia. Several large phosphate plants lie in European Russia (e.g. Voskresensk, Sumy, Zhdanov, Leningrad), in the Ural region, southern Kazakhstan and in Central Asia, where there are also nitrogen plants. Transcaucasia covers its own needs for artificial fertilisers. Large potash

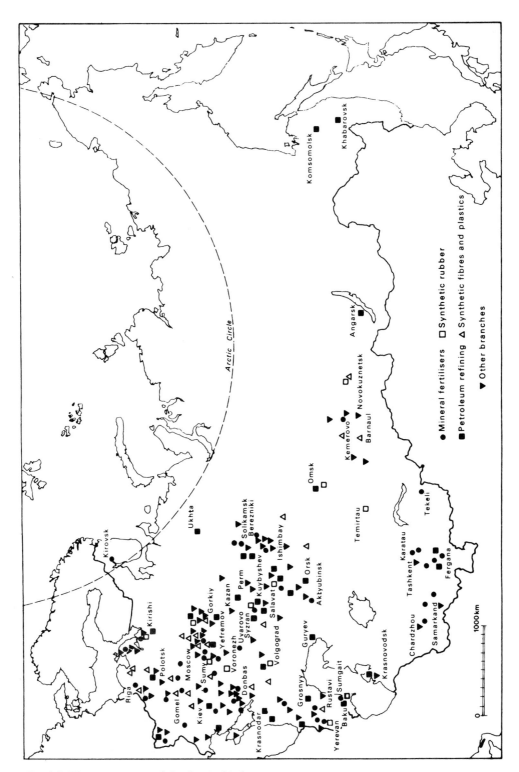

Fig. 9.3 The main centres of the chemical industry

deposits are used at Solikamsk in the Kama basin as well as in Belorussia and Transcaucasia. Location of plants is usually near raw materials.

Sulphuric acid, a basic raw material in chemicals production, is made in many widely scattered centres from 'secondary' sources of sulphur (by-products, waste) but near to consumers since it is not easy to transport. Among the largest plants are Voskresensk and Novomoskovsk in the centre of European Russia, Leningrad, Solikamsk and Berezniki in the Kama basin and Aktyubinsk in Kazakhstan, as well as others in the Donbas coalfield and at Tekeli in Central Asia. Crude oil is an increasingly important source, especially in the Volga–Ural area. Soda production is chiefly from the Donbas, the Urals (Berezniki and Sterlitamak), Siberia and the Kara Bogaz Gol.

To reduce its dependence on imported rubber, the Soviet Union has developed artificial rubbers, though some natural latex comes from the *Sagyz* plants native to Central Asia. Synthetic rubber manufacture expanded rapidly after 1945, using German reparations equipment and processes. Plants at Yaroslavl, Voronezh, Yefremov and Kazan use ethylene derived from potatoes; calcium carbide is used at Yerevan and Temirtau (Kazakhstan); but production increasingly uses by-products of petroleum refining with the Volga area most important.

Petrochemicals and chemicals derived from coal have been developed into a major industry with a wide variety of products including plastics. The main groups of plants are at Moscow and elsewhere in the centre of European Russia, in the Donbas and on the Volga, but several major plants have been built in other areas in Belorussia, the Ukraine, Transcaucasia, Central Asia, Siberia and the Far East. Synthetic fibres are widely produced—the Central Industrial region, the Volga region, the Ukraine and the Baltic region all have plants. Cosmetics, pharmaceuticals and films are dominated by Moscow and its environs, though towns such as Kiev and Kharkov are also significant. Growth of this industrial branch is being helped by the purchase of foreign technology and plant, with British ICI and the German Krupp concern notably important.

As noted in connection with the metal industries there has been a marked spread of chemical plants away from the old centres of industry. Thus, Moscow and the Central region, Leningrad and Kiev areas have notable and long-established concentrations of chemical plants, relying mainly on imported raw materials. The Donbas–Dnepr area is another important chemical-industry region, better endowed with natural resources, except for water, which is in rather short supply. The western Ukraine has also developed chemical plants on the basis of local minerals (sulphur, rock salt and petroleum) but the expansion further north in Belorussia and the Baltic republics is based on imported materials. The Caucasus–Transcaucasus regions have notable developments based on oil and other local resources, as have the Volga–Ural areas. Natural resources are also the basis of most of the current developments in Central Asia and the Far East.

Timber, wood chemical and building industries

A marked overlap in these branches of industry occurs because of the role of timber in the production of building materials, but timber is also important in the chemical industry and pulp and paper production. Also, although much reduced by more advanced methods of harnessing energy, timber is still widely used as a fuel in forested and mountainous areas. Siberia and the Far East contain almost 80% of the Soviet Union's timber resources so in European areas there is now urgent need for conservation of timber stands.

The first stage of exploitation, that of logging, is carried out by *lespromkhoz* units which are also responsible for storing and some processing of timber, and for replanting. Large plants with powerful machinery and drying kilns have largely replaced small sawmills in the most productive areas. Factories, producing packaging materials, furniture and prefabricated building items, plywood and chipboard are commonly integrated with the timber milling plants. Several such complexes, extending to the manufacture of pulp and paper, are being developed in Siberia, at Bratsk and in the Ob, Yenisey and Amur river basins of Siberia and at Arkhangelsk and Kotlas in the European North. Most of the plywood enterprises are, however, located in the European northwest, especially Belorussia, the Centre and the western parts of the Urals, 95% of the plywood output being from the European forest areas, with the eastward shift a relatively new feature. A number of sawmills are found at the mouths of northern rivers, catering for the export trade which is mainly in sawn timber.

Paper factory timber yards in Siberia—one of the industrial plants using power from the Bratsk hydro-electric scheme

The pulp and paper branch of industry is dominated now by large plants, notably those at Solikamsk on the Kama river, where there are many smaller plants, while Balakhna on the northern reaches of the Volga is another important location. Most of the older plants are in the European areas, but, as already noted, major new complexes are being built in the eastern areas. Associated with these new integrated developments, in particular, are wood-hydrolysis plants which produce a variety of chemical products such as the familiar resins, turpentine and acetic acid and the newer technical oils, formalin, acetate phenols, special glues and pharmaceutical and other products in ever-increasing diversity.

The manufacture of furniture has also been developing in the eastern regions but traditionally it has been a market-oriented industry, with most factories in Moscow and Leningrad, with other cities, notably the capitals of the Union Republics in the European areas, also important. Valuable beech, oak, hornbeam and other hardwood resources have ensured the North Caucasus region an especially important role in this industry.

Timber is one of the Soviet Union's most valuable resources and even more valuable because it is renewable. As yet only the European areas, a relatively narrow strip of forest along the Trans-Siberian Railway and areas around mining and other settlements in Siberia have been exploited. Future development must be increasingly in the east, but even more important in the long term is the improvement of utilisation practices. Timber is a renewable resource, but only if conservation is constantly kept to the fore alongside exploitation will these resources be available for future generations. The USSR, like most countries, has in the past allowed extravagant use of timber, careless extraction, unnecessary burning and waste through bad transport and processing methods. It has yet to be seen how efficient the timber enterprises have become in an age which recognises the need for, but all too rarely implements, conservation and restoration of natural resources.

One aspect of the economy in which the use of timber has declined, at least relatively, with the development of alternative materials and methods, is the building industry. While timber still provides the preferred material for many purposes, such as window and door frames and some kinds of flooring, and almost everything deep in the forest zones, cement and concrete

have become the basis of most large-scale construction, including the ubiquitous Soviet blocks
of residential flats. The USSR claims to be the
leading world producer of cement, asbestos-
cement and pre-cast concrete. Formerly, large-scale
cement production was based on the few deposits
of good quality marls near Novorossiysk on the
Black Sea coast and in the Volga area near Volsk.
Now, however, there is a widespread industry
based on limestones, blast-furnace slags and other
waste materials. Many other materials as well as
the familiar bricks enter into the building
industry's requirements and, as far as possible,
factories are distributed to minimise transport
and hence are near large centres of population,
subject to availability of raw materials. Glass is a
case in point, with most works in the Donbas, the
Centre and Belorussia, where materials are available reasonably close to demand. The largest
plant is, however, at Saratov, with emphasis on
economies of scale.

Light industries

Although the classification of industries into
heavy and light is not entirely satisfactory, partly
because there is overlap from the timber, chemicals and other branches into the consumer goods
industries, such as furniture, domestic plasticware
and toiletries and cosmetics, the industries traditionally classed as light still tend to have a
freedom in location not found in heavy industry.
This is because light industry is mainly concerned
with the utilisation of raw materials which can be
readily transported so that they can be processed
near to markets, or use more ubiquitous materials
such as some agricultural products. The finished
products represent high value-added items, as in
the case of textiles and clothing, which can with-
stand distribution costs of fairly high level, but
which in any case present no particular problems
in transport, or which, like processed foodstuffs,
present no problems if distribution is kept within
moderate bounds, as in the case of bread and
milk. Excessive distances in these cases do impose
problems so plants are generally distributed in
some degree comparable with the pattern of
population. Hard and fast rules cannot, however,
be laid down. The pattern of location varies according to the particular product, the sophisti-
cation of the transport and distribution network,
etc. but, as there are not normally very heavy
loads, noxious processes or severely limited

origins of raw materials, at least a substantial
degree of flexibility is possible.

TEXTILES

The cotton industry is the most important of the
textile group of industries. Spinning mills are
mainly in the Central region, which accounts for
about 75% of output. This is mainly explained by
historical circumstances, the raw cotton being
transported from Central Asia or, originally, the
importing ports. The advantages of the Central
region lay in the availability of an adaptable
workforce, accessibility for the foreign interests
that provided much of the capital, and availability of markets, as well as of suitable skills for
maintenance of machinery, and ease of providing
the required degree of humidity. These conditions
no longer restrict location and large cotton mills
have been developed in Central Asia, especially
Tashkent, near the source of the raw material, in
the Ukraine, Estonia and western Siberia, with
lesser developments in other regions, showing the
flexibility that is possible although most of the
cotton will continue to come from Central Asia.

Wool manufacturing has also spread widely.
Early centres, using imported wool, were developed in the Baltic cities and in the Central region.
As internal supplies of wool became more important the industry was expanded in the Ukraine,
Belorussia and the North Caucasus. The large
number of sheep in the Caucasus, Kazakhstan
and Central Asia led to increased emphasis on
processing in these regions, and a number of mills
have been built in Siberia. Again, flexibility is
possible and a very large number of centres are
now engaged in processing local and transported
wools and in making garments. The traditional
carpet-making craft industry remains in Central
Asia but large mechanised enterprises using both
wool and artificial fibres have been developed in
Transcaucasia and other regions.

The linen industry, with a more restricted
demand for its products, has remained in its
traditional location in the Central and north-west
areas, with expansion also in Belorussia and the
Baltic republics, close to the flax-growing areas
and also to the main markets.

Of the silk mills, most are close to the supply of
raw material in Central Asia and Transcaucasia,
but with weaving and finishing in the Central and
other European areas. The true silks remain important commodities in the USSR but, as in other

Tadzhik, Uzbek and Russian shoppers in Dushanbe, capital of Tadzhikistan.

countries, rayon and synthetic fibres have become increasingly important. They are manufactured mainly in the Centre and other European areas and, in the case of fibres derived from the petro-chemical industries, western Siberia, and other areas with major oil resources, appear likely to become increasingly important.

The clothing and knitwear industries are eminently suited to locations with no special advantages except an appropriate labour force so are often used to initiate industrial development in small and relatively isolated towns. There are, however, numerous such establishments in Moscow, Leningrad and other major cities so it can be said that this is an industry of which the distribution accords closely with that of the population, though with some specialisation.

The leather and footwear industries, though of a handicraft nature originally and correspondingly widely distributed, have become more concentrated in response to mechanisation of production and are located mainly in the European areas, but with some expansion in other areas to make for a more even distribution. The

fur industry, associated with one of the oldest sources of wealth in Russia, is based predominantly on the forests of the European and Ural areas with a major centre at Slobodskoy, where an ancient trade route from Siberia crossed from the Ural mountains to the Vyatka river. Other historic trading centres such as Kazan and Chita are among the other locations of the fur industry.

St. Petersburg was the original home of the Russian china industry, where the market of the court and wealthy citizens of the former capital provided ample stimulus. Leningrad today is an important centre of the china and pottery industry which has also been long developed in the Novgorod–Volhynsky area, the Moscow area and the Ukraine. The output includes porcelain and ceramic goods for industrial and building uses, glazed and wall tiles and domestic goods. Factories have more recently been built in the Central Asian and other areas to provide local supplies of such goods as well as employment.

The manufacture of consumer goods generally is widespread but with the main centres in the European regions, especially the Moscow and

Leningrad areas, where technological expertise and labour are both available. A substantial portion of the output of factories producing photographic equipment, watches, radios, sporting guns, etc. is earmarked for export but all such light and valuable items can be distributed also to all parts of the Soviet Union at costs low in relation to their value.

FOOD INDUSTRIES

Food manufacturing plants are scattered widely over the country in relatively small units. The aim has been to use local produce to keep transport to a minimum and there is a relatively low level of development of such modern processes as freezing, dehydration and even of canning. Factories that do carry out such processes are, however, situated in areas producing the vegetables, fruits, etc. involved, for ease of supply of fresh products. Similarly, sugar-beet processing, and the production of wines, butter, cheese and other products from perishable commodities are located in the primary producing areas. In the case of town milk supply the producers are taken, so to speak, to the consumers, the farms being established near the cities. This is true also of horticulture for table vegetables to be fresh. Meat processing plants are being increasingly concentrated in areas of specialisation on beef cattle and pig production east of the Volga and there are large fish processing factories in the Far East where the major fishing fleets are based. Poultry processing is another branch that has become characterised by large units, particularly in the European areas. Whereas cattle can be transported to processing plants (although costly), this is not practicable on a large scale with poultry.

A certain amount of relocation has been attempted on a planned basis in some of the major food industries. On the principle that flour is more expensive than grain to transport and more susceptible to deterioration, flour mills have been built widely outside the main grain growing areas and within about 50 km of consuming cities. This was done on a modest scale in the 1930s in Transcaucasia, Belorussia and the eastern regions, and in the 1960s larger plants were built in Central Asia, Siberia and the Far East. Bread and confectionery are essentially localised enterprises located near consumers. Similarly, while wineries must be located in the grape growing areas because ripe fruit cannot be transported, bulk wines

The Hall of Cosmonauts, commemorating Soviet achievement in space, is a popular pavilion at the Exhibition of Scientific and Economic Development in Moscow

can be transported and it is more economic to bottle the wines in the areas of consumption. Production of spirits is also traditionally linked with the wine areas and the potato growing areas of the central European USSR but new distilleries and bottling plants were built in the 1960s in large cities of the Urals, Siberia and the Far East to reduce the transport of bottled vodka and other spirits. In virtually all branches of the food industries the aim is to replace small, inefficient and sometimes unhygienic units with modern factories located in accordance with principles of transporting the commodity which is easiest or least costly to transport in the stages from farm to consumer.

Although not a part of industry in the ordinary sense, the retail establishments provide an essential part of the distribution network. The Soviet organisation of retailing has lagged far behind the development of the production side of the economy and most shops are small and very traditional in their methods of handling and selling goods. Combined with unsophisticated and often crude forms of packaging this results in shops looking very unattractive to western eyes, while purchasing is a slow and complex procedure. There is, however, one advantage resulting from this situation, the Soviet Union has little of the rubbish problem that affects both town and countryside in the west, most packaging is simple and easily degradable and there is relatively little of it, while disposal is facilitated by education of children from an early age not to scatter litter and by an army of otherwise perhaps unemployable persons to sweep and clean public places.

THE MAJOR INDUSTRIAL REGIONS

To draw together the diverse patterns of the various industries which have been described, the principal regions within which Soviet industry is concentrated will now be considered briefly (see Fig. 9.4). The regions of spatial concentrations here described overlap administrative divisions of all levels, as do the agricultural regions described in Chapter 7. The distribution of industry and of agriculture in each of the formally designated Major Economic Regions, used for large-scale planning purposes in the USSR, will be dealt with in Chapter 12. The more detailed treatment of industrial aspects of the regions has been reserved for Chapter 12 because it is for the official Major Economic Regions that comparable statistics are made available from year to year from Soviet sources.

The Industrial South
The European Industrial South is formed by the sub-regions of the Donbas, the Dnepr Bend and Krivoy Rog and is closely linked to the great industrial towns of Kharkov and Kiev. The resource base is the coal (much of coking quality) of the Donbas, the iron ore of Krivoy Rog, Kursk and Kerch, Yelenovka limestone, Nikopol manganese, salt, imported petroleum and natural gas. Hydro-electric current from the Volga and Dnepr barrages is augmented by local thermal generators. A locational problem of the dry south is

finding adequate water supplies for industry. It is the most important region for iron and steel making, notably on the coalfield at centres such as Donetsk and Makeyevka, on the orefield at Krivoy Rog and in the Dnepr Bend (notably ferro-alloys). Coking by-products and local salts plus imported raw materials are the basis of a heavy chemicals industry, notably on the coalfield. Raw metal and a large local market attracted heavy engineering both to the coalfield and to nearby centres. Well served by railways, its products are distributed widely throughout the country.

The Central Industrial Region
The Central industrial region includes Moscow and its environs, the textile towns of the Klyazma basin, and the Tula lignite field. Lignite and peat are used for electricity generation, but many thermal stations formerly using imported coal have turned to oil or natural gas supplied by pipelines. Some electric current is obtained from the Volga region. Nearly all raw materials have to be imported and the area is a major focus of the Soviet railway system, with 11 main lines converging on Moscow. A large local market and soft water were important factors in attracting textile manufacture, despite having to import raw cotton and wool, and this remains the principal Soviet centre of the industry. The chemicals industry depends on imported raw materials plus local lignite and phosphates, but there is also a large pharmaceuticals sector based on the Moscow market and research facilities. Engineering includes a wide range of branches, particularly the more sophisticated ones such as aerospace and electronics, and those requiring a large market, as well as good design and development facilities and a pool of highly skilled labour, so that some of the principal Soviet plants are found in or around Moscow.

The resource base of the Centre was transformed by the discovery of the vast iron ore deposits of the Kursk Magnetic Anomaly (KMA) in the 1950s. These are now claimed to be the largest known deposits in the world, with reserves of at least 30 000 million tonnes of rich ores, more than the total of all other Soviet deposits. New works at Novo-Lipetsk have been added to the basic production facilities at the old centres, such as Tula and Gorkiy, and the northerly Cherepovets integrated steelworks built in the

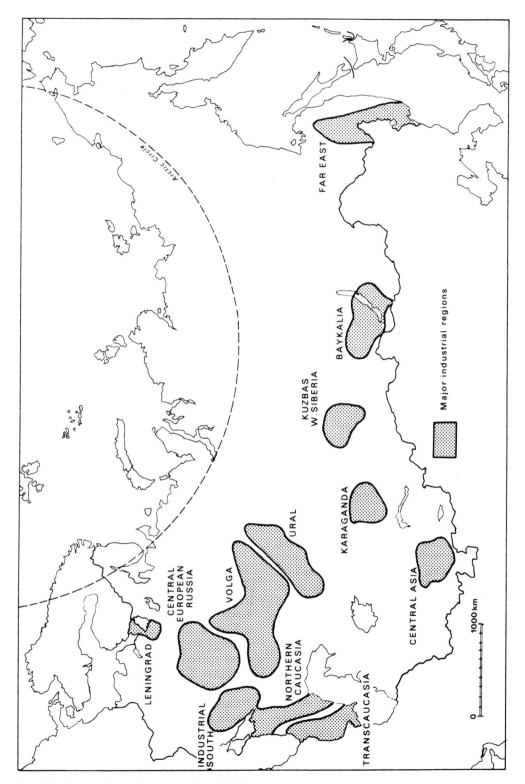

Fig. 9.4 Major industrial regions of the USSR

1950s. Other developments are projected, as a result of which this area is now being looked on as the third metallurgical base, a term previously applied to western Siberia. The development of the KMA will facilitate increases in production in the Ural and in the South, with which the Centre will virtually merge.

The Ural Region

The Ural region comprises a number of industrial clusters separated by forest and mountain country. The main metallurgical plants which form the basis of Ural industry lie on the eastern slope of the mountains where iron ore and other metallic ores are found. Chemicals associated with local salts or petroleum are typical of the western flanks where there is some coal around Kizel. The Ural region is, however, generally deficient in fuel, and coal and coke are imported from western Siberia and Kazakhstan, though local fuels—coal, peat and wood—are also used to generate electricity. The physical nature of the Ural mountains makes them unsuitable for hydro-electric development. Petroleum and natural gas are imported from the Ural–Volga oilfields, Central Asia and from western Siberia. The production of iron and steel, non-ferrous metals and heavy chemicals from the Ural region is widely distributed by railway to other regions, while, since the Second World War, the production of heavy engineering and transport goods has begun on a considerable scale. This region, however, is still regarded primarily as an iron and steel producer, the 'second metallurgical base'.

Western Siberia and North Kazakhstan

The Kuzbas coalfield is the focus of industrial western Siberia, which includes several large towns lying at some distance from the coalfield, such as the engineering centres of Novosibirsk and Rubtsovsk and the textile town of Barnaul. Kuzbas coal and local iron ore are used by the iron and steel industry (notably at Novokuznetsk), while there is also smelting and refining of non-ferrous ores. By-products of these industries, plus wood from the tayga, form the basis of chemicals production. Development of vast resources of water power will provide cheap electric current and lead to an emphasis on electro-chemicals and electro-metallurgy in the future. This would be the core for possible development of a larger 'metallurgical base' of southern

Siberia. For transport, the region depends on the Trans-Siberian Railway and supporting rail routes and the airways. North Kazakhstan is virtually an extension of this region. It has a new integrated iron and steel works and a new ferro-alloys plant but industrial development is hindered by a shortage of water only partly overcome by the Irtysh–Karaganda canal.

The Middle Volga Region

A newly emergent industrial region is the middle Volga, using its own natural gas, petroleum and salts, while timber is easily obtained from the northern forests and coal from the Donbas, which along with the Ural and Centre regions can also supply raw metal. It has developed electro-chemicals and electro-metallurgy, consuming current from its own large hydro-electric barrages, while petro-chemicals have developed in new refineries. Engineering developed during the Second World War and a vast car assembly plant has been built at Tolyatti, between Ulyanovsk and Kuybyshev. It is a major cement-producing region. Crossed by several railways and served by the Volga–Kama and associated waterways, it enjoys a nodal position in the Soviet transport system.

Other industrial regions

Other regions are of much less importance. Leningrad shows a notable dependence on manufacturing from materials imported from home or abroad. Textiles, engineering and chemicals are all represented and ship-building is important. Transcaucasia has local resources of ferrous and non-ferrous metals, petroleum and coal. It is a major producer of ferro-alloys and synthetic rubber and has a new integrated steelworks at Rustavi. Novorossiysk in northern Caucasia is one of the largest cement producers in the USSR. There is a specialised engineering industry for the Caucasian oilfields. Central Asia has a rich assortment of minerals and fuels augmented by supplies of raw materials from Siberia. There is engineering (textile machinery) in Tashkent, chemicals—notably fertilisers—at several centres, and Tashkent and Frunze have textile mills using local cotton. New hydro-electric stations offer great scope for industries with high power needs. Industry in the Far East has a strong strategic character and primarily serves local needs.

The USSR's achievements have failed to supply many manufactured items, especially consumer goods which are therefore in short supply in the Soviet Union and many bottlenecks are reported in the production process. The complexity of modern society has defeated centralised planning. It was to remedy this situation that Mikhail Gorbachev introduced *perestroika* involving private enterprise and acceptance of the profit motive.

Progress with *Perestroika*

As noted above, the reconstruction programme faced great difficulties and even after several years it had made only limited impact in the major industries. Complaints about rising prices, continuing lack of supplies both of producer and consumer goods and poor quality were more common than reports of significant improvements. Nevertheless, progress was made and should not be underrated. Thus, the Vaz motor car plant at Tolyatti on the Volga reported that self-financing methods resulted in a significant improvement in the quality of the vehicles produced, which compete in international markets, a claim supported by foreign companies selling Soviet cars abroad. A split of profits brought about equally between the plant and the state and a rising percentage of wages being linked to quality and productivity were ingredients of the recipe for success.

In many industries much reliance has been placed on the 'brigade' method of organisation. Brigades, or teams, responsible for particular stages of the production, or of operations supporting manufacture, have been allocated resources and given freedom to improve their methods and derive financial benefits if successful. Clearly, however, if a brigade fails to meet the requirements of the factory it cannot be allowed to bring down production elsewhere and this kind of situation has resulted in many problems. It is especially difficult to allocate blame if the brigade can point to shortages of materials and components arising from failures in other production units or transport agencies.

At the Nineteenth All-Union conference of the Communist Party called in June 1988, especially to authorise reform, Mikhail Gorbachev condemned the excessive central control over the production of individual plants with state orders that resulted in production of goods not in demand and promised measures to reduce such control by the state planning committee. The need to satisfy consumers was given special emphasis and the congress was followed by instructions to the Ministry of Light Industry to plan for production increases in such goods as clothes and shoes and for other enterprises to increase sharply output of household appliances, television sets, video recorders, cars and other consumer goods. These measures were to bring immediate results increasing in effect in the thirteenth Five Year Plan, running from 1990 to 1995, in which there is less emphasis on planning and more on individual and corporate enterprise, with, for example, enterprises able to conclude contracts with each other free of detailed instructions from the state.

BIBLIOGRAPHY

Abouchar, A. (1979), 'Regional industrial policies in the USSR, the 1970s,' *Regional development in the USSR*, NATO colloquium, Oriental Research Partners, Newtonville, pp. 93–103.

Bater, J. H. (1976), *St. Petersburg: industrialisation and change*, Edward Arnold, London.

Berliner, J. S. (1988), *Soviet industry, from Stalin to Gorbachev, essays on management and innovation*, Edward Elgar, Aldershot.

Bond, A. R. (1987), 'Spatial dimensions of Gorbachev's economic strategy,' *Soviet Geography*, **28**, pp 490–523.

Cole, J. and German, F. (1970), *A geography of the USSR*, Butterworth, London.

Conolly, V. (1967), *Beyond the Urals*, Oxford University Press, Oxford.

Conolly, V. (1975), *Siberia today and tomorrow*, Collins, London.

Current Digest of the Soviet Press (1949–), Joint Committee on Slavic Studies, Washington, DC.

Dewdney, J. C. (1976), *The USSR, Studies in Industrial Geography*, Westview Press, Boulder, Colorado.

Dienes, L. and Shabad, T. (1979), *The Soviet energy system*, Wiley, New York.

Dobb, M. (1966), *Soviet economic development since 1917*, Routledge and Kegan Paul, London.

Economist, London, various dates; *Financial Times*, London, various dates

Hamilton, F. E. I. and Linge, G. J. R. (1979), *Spatial*

analysis, industry and the industrial environment, I—Industrial systems, Wiley, Chichester.

Hewett, E. A. (1988), *Reforming the Soviet economy*, Brookings Instn., Washington.

Hutchings, R. (1971), *Seasonal influences in Soviet industry*, Royal Institute of International Affairs, London.

Kalesnik, S. V. and Pavlenko, V. F. (eds.) (1976), *Soviet Union; a geographical survey*, Progress Publishers, Moscow.

Kistanov, V. A. and Epshteyn, A. S. (1972), 'Problems of optimal location of an industrial complex,' *Soviet Geography*, **13**, pp. 141–52.

Lavrishchev, A. (1968), *Economic geography of the USSR*, Progress Publishers, Moscow.

Linge, G. J. R., Karaska, G. J. and Hamilton, F. E. I. (1978), 'Appraisal of Soviet TPC concept,' *Soviet Geography*, **19**, pp. 681–97.

Lonsdale, R. E. (1977), 'Regional inequity and Soviet concern for rural and small-town industrialisation,' *Soviet Geography*, **18**, pp. 590–602.

Lydolph, P. E. (1979), *Geography of the USSR, topical analysis*, Misty Valley, Elkhart Lake, Wisconsin.

Mathieson, R. S. (1975), *The Soviet Union; an economic geography*, Heinemann, London.

Matrusov, N. D. (1970), 'Geographical problems in the development of machine-building in the Ob-Irtysh complex,' *Soviet Geography*, **11**, pp. 464–71.

Narodnoye khozyaystvo SSSR v . . . (various years), *Statisticheskiy ezhegodnik,* Tsentralnoye Statisticheskoye Upravleniye SSSR, Moscow,

NATO Economic Directorate (1979), *Regional development in the USSR*, NATO Colloquium, Oriental Research Partners, Newtonville, Mass.

Nove, A. (1977), *The Soviet economic system*, Allen and Unwin, London.

Pallot, J. and Shaw, D. J. B. (1981), *Planning in the Soviet Union*, Croom Helm, London.

Pryde, P. R. (1968), 'The areal deconcentration of the Soviet cotton-textile industry,' *Geographical Review,* **58**, pp. 575–92.

Shabad, T. (1969), *Basic industrial resources of the USSR*, Columbia University Press, New York.

Shabad, T. (1987), 'Siberian development under Gorbachev,' in *Soviet Geography studies in our time*, eds. Holzner, L. and Knapp, J. M., Milwaukee, pp. 163–74

Shabad, T. and Sagers, M. J. (1987), 'The chlor-alkali industries in the USSR,' *Soviet Geography*, **28**, pp. 434–455

10 Urban and Rural Settlement

In response to the creation of a powerful industrial structure in the Soviet Union since the late 1930s, there has been a vast movement of people from country districts both into established towns (which have grown substantially) and into new towns founded in the process of economic development. Although the population of the USSR rose from 147 million in 1926 to 281.7 million in 1987, growth in town population was even more impressive since the number of country dwellers fell from 120.7 million in 1926 to 95.7 million in 1987. Marxist–Leninist views on social, economic and ideological organisation have made the town the focus of attention. Although great changes have been made in towns, many long-standing features of the Russian town remain. Great changes have also been brought to the village—the outcome of Communist reorganisation of agriculture and the need to raise the standard of living of country dwellers to the level of the townspeople, while many experiments have been made to develop new forms of rural settlement adjusted to the contemporary economic and social conditions.

At the tsarist census of 1897, within the boundaries of the Russian Empire of the time, only 16% of the people were living in towns, but the most strongly urbanised parts of the empire—Russian Poland and the Baltic countries—were lost as a result of the First World War. During the Revolution and subsequent civil war, several million people left the starving and disrupted towns to find shelter and sustenance in the countryside, from which many had come a decade or two earlier. In 1926, the Soviet authorities, believing

TABLE 10.1 POPULATION GROWTH IN TOWN AND COUNTRY

	Urban	Rural	Total
		Millions	
1897 census	20.0	108.1	128.2
1926 census	26.3	120.7	147.0
1979 census	163.6	98.8	263.4
Change 1926–1979	+137.3	−21.9	+111.4
1987	186.0	95.7	281.7

Source: Soviet statistical handbooks.

some measure of stability to have been reached, made an exceptionally detailed census that showed that some 82% of the people were still resident in rural districts. But this proportion was to begin to change rapidly two years later when the first of the new economic plans was put into operation and the transformation of the economy started.

The 1939 census showed that urban population had risen to 33%, but shortly afterwards the destruction and disruption of the German invasion again drove people out of the towns of the occupied areas and large numbers fled into the Volga region and further east, to work in factories which had been evacuated from the path of the German armies. An official post-war estimate of 197.9 million was issued in 1956 and showed that 45% of the people were living in towns—a higher proportion than many people had expected. The first post-war census of 1959 showed that of the 208.8 million people, 48% were now urban dwellers and by 1961 it was estimated that 50% of the population of 216.2 million were in urban areas. In 1979, of the 263.4 million people, 62% were living in towns. The figure for 1987 was given as 66% and it has been estimated that the urban population will comprise 75% of a population of 333 million by the year 2000.

TOWN POPULATION AND REGIONAL VARIATIONS

Migration to towns

An important aspect of the increase in town population and its relationship to changes in the rural population is the process of population growth. One of the main mechanisms has been migration, usually into comparatively nearby towns from rural communities. This occurs most commonly within European Russia, but also often from both towns and country districts in European Russia to towns in the eastern regions. Towns in European Russia, notably in the western parts, have been important suppliers of settlers to the towns of Siberia. Movement has commonly been a voluntary choice, in response to encouragement given by the authorities in the form of better housing, pay incentives and 'fringe benefits', and has been marked by a high proportion of young single people or young married couples, though in some remote and inhospitable environments (such as Magadan in the Far East) there has been difficulty in recruiting sufficient women to build a balanced sex-ratio. Forced migration cannot be omitted, since it played a part in tsarist times (when the dreaded prison settlement, the *katorga,* was the destination of many exiles) and even after the Revolution, notably in the Stalinist period. Many towns in the Soviet Arctic were peopled at their inception mostly by politically unreliable elements and even by criminals, as, for example, Norilsk in the late 1930s when it was a wooden shanty town. Tsarist and Soviet governments have often forced migration of whole communities to carry out their population policy. Some Ural towns, for example, were founded by enforced migration of serfs in the eighteenth century.

Town population has also grown by natural increase, which accounted for a fifth of the growth in town dwellers between 1939 and 1959, whereas migration into towns comprised over half the total increase in numbers. The balance is accounted for mainly by boundary changes or by changes in status. The difference between the urban birth and death rate (both lower than in the countryside) has been less than in the country, with a consequently lower rate of natural increase. Families are generally larger in the countryside, despite the proportion of women of childbearing age being higher in towns. Although the rate of natural increase in the countryside has been markedly higher than in the towns, the drift from the country to town has been so strong that it has caused an actual fall in numbers of rural population. The number of people classed as urban has also increased by administrative measures; first, by giving town status to settlements as soon as they have reached the legal threshold for such designation, and, second, by expanding the boundaries of towns to encompass adjacent suburbs or rural districts.

Regional variations in urbanisation

The considerable regional variation in the proportion of town dwellers in the total population is shown in Fig. 10.1. It can be seen that this ranges from around 20% in the oasis areas of Central Asia such as Kashka–Darya, Surkhan–Darya and Khorezm to about 80% in Magadan oblast of the Far East and to 90% in Murmansk oblast in northern European Russia. The proportion of urban dwellers is generally below the national average (66%) in the farming districts of western and central European Russia, northern Caucasia and the Ukraine, all areas comparatively densely settled by Soviet standards. The proportion of urban dwellers is well above average in much of Siberia, the Ural region and the Karaganda district of Kazakhstan: these are all areas where industrial development has been pushed ahead since the early 1930s. It suggests that such lands have, in Soviet times, been settled primarily for industrial or transport development and that migration has been focused into towns and workers' settlements—the rural settlement of the eastern regions has been much less important, though some change came with the inception of the virgin lands scheme in the mid-1950s. Evidence shows that many towns in Siberia and the Far East, for example, lie in virtually uninhabited and undeveloped country. In Magadan oblast, 1 199 100 km^2 in area, 445 000 people (81%) out of a total

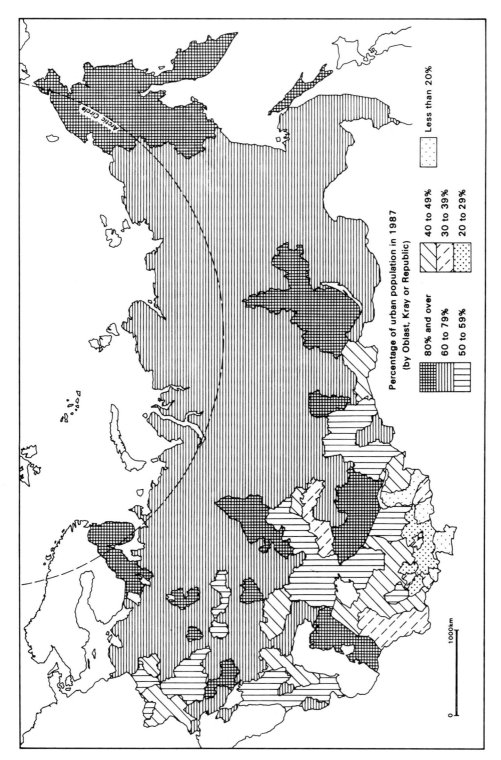

Percentage of urban population in 1987
(by Oblast, Kray or Republic)

80% and over

60 to 79%

50 to 59%

40 to 49%

30 to 39%

20 to 29%

Less than 20%

Arctic Circle

1000km

Fig. 10.1 Percentage of urban dwellers in the total population. With some 66% of the national population living in towns, the average is substantially exceeded over most of Siberia, whereas the richer agricultural areas of European Russia and the native societies of Central Asia and Transcaucasia show below-average levels of urbanisation.

population of 550 000 are classified as urban dwellers—148 000 urbanites are found in Magadan city alone—so that the density for the whole oblast is 0.46 persons per km², while for rural districts it is under 0.1 persons per km².

Numbers of towns

In January 1987, there were in the Soviet Union some 2176 towns and 3992 'settlements of town type', falling into the size categories shown in Table 10.2. Between 1926 and 1987, the number of towns had risen by 1467 and the 'settlements of town type' by 2776.

Definition of a town

What criteria are used in the Soviet Union to define a town and a 'settlement of town type'? Fundamentally, the definition is strictly an economic concept, based on a minimum adult population and a minimum level of employment outside agriculture. The requirements vary somewhat among the 15 Union Republics in response to local conditions, and are generally least exacting in the less developed republics, and some similar concessions are made even in the less developed parts of the RSFSR. Conditions for selected republics are shown in Table 10.3. Once a settlement has the status of a town, its progression up the urban administrative ladder is related to the

level of local government to which it is subordinated—whether to the rayon, the oblast (or kray) or to only the republic (as in the case of, say, a capital of a republic). If fortune is adverse, however, a settlement may be downgraded—several small market towns in the western territories, incorporated into the Soviet Union in 1945, suffered the fate of being reduced to village status. A 'workers' settlement', the first state on the road to urban development, may even be dissolved and disappear should the reason for its existence, such as a large construction project or a mine, fail or be abandoned. Nevertheless, many villages may be designated as a workers' settlement or even as a 'health resort' earning them the classification of 'settlements of town type' and may utlimately be given the status of town (*gorod*). Such elevation in status takes place also as the result of industrial development or where an important transport function emerges (as on a new railway or at a road-rail junction), or even through development of a significant resort.

The town plays an influential part in the political geography of the Soviet Union, since Marxist–Leninist theory has regarded it as an excellent milieu for the evolution of a Communist society and the 'victory of the proletariat'. It has, therefore, been seen as a vital focal point from which to disseminate Marxism–Leninism to the

TABLE 10.2 SIZE DISTRIBUTION OF URBAN SETTLEMENTS IN THE USSR

Number of inhabitants	Number				Population (millions)			
	1926	1959	1975	1987	1926	1959	1975	1987
TOWNS								
Less than 5000	141	205	175	124	0.5	0.7	0.4	0.3
5–20 000	350	726	825	837	3.8	8.6	10.2	10.6
20–100 000	187	600	803	927	7.9	25.1	33.5	38.8
100–500 000	28	123	201	236	4.1	24.2	43.1	50.8
More than 500 000	3	25	39	55	4.1	24.2	44.6	61.6
	(2)	(3)	(13)	(23)	(3.6)	(9.49)	(26.2)	(40.8)
TOTAL	709	1679	2043	2176	21.7	83.0	131.8	162.1
SETTLEMENTS OF TOWN TYPE								
Less than 5000	927	1542	2015	2086	2.0	4.5	5.7	5.8
5–20 000	281	1368	1673	1846	2.4	11.8	14.2	16.3
20–50 000	8	30	51	58	0.2	0.7	1.4	1.6
More than 50 000	—	—		2	—	—		0.2
TOTAL	1216	2940	3739	3992	4.6	17.0	21.3	23.9

Figures in brackets refer to cities of one million or more inhabitants.
Source: Narodnoye khozyaystvo SSSR, various years.

TABLE 10.3 SETTLEMENTS OF TOWN TYPE: CRITERIA FOR DEFINITION IN SELECTED REPUBLICS

| Republic | EMPLOYMENT STRUCTURE (% WORKERS IN INDUSTRIAL SECTORS AND MEMBERS OF THEIR FAMILIES) | | | MINIMUM TOTAL POPULATION IN THOUSANDS | | |
	Workers' settlements	Settlements of town type	Towns	Workers' settlements	Settlements of town type	Towns
RSFSR	not less than 85%	non-existent	not less than 85%	3	non-existent	12
Ukraine	preferably majority	over 60% (50% for rayon centres)	preferably majority	0.5	2	10
Georgia	non-existent	not less than 75%	not less than 75%	non-existent	2	5
Moldavia	preferably majority	not less than 70% (not less than 60% for rayon centres)	preferably majority	0.5	2	10

Source: Khorev, B. Gorodskiy poseleniya SSSR, Moscow, 1968.

A boulevard in Tbilisi. People are gathered around a kvass cart, kvass being fermented fruit juice.

countryside, for so long markedly resistant to Communist views. The towns form a series of nodes on which the territorial-administrative framework is hung, since each unit in the system—rayon, oblast or kray, or republic—must have a suitable town as its centre. The concept of the basic territorial-administrative unit—the oblast (or kray in less developed areas)—is seen as the hinterland of its major 'proletarian' centre, and such a view is found to pervade the whole spectrum of Soviet regional planning. It is thus perhaps easier to understand why urban status itself is conceived in economic terms rather than in functional, morphological or demographic measures.

Historical aspects

Towns within the Soviet Union date from most ancient times to the most recent. Not all owe their origin and character to Slav peoples—in the Baltic region and the western frontier districts, German, Polish and Rumanian influences have been strong; in Transcaucasia and Central Asia, Iranian, Indian and even Chinese influences have been found. Some towns in Central Asia, Transcaucasia and the Black Sea littoral may be traced back to pre-Classical and Classical times, though they have not always been on exactly the same site. Mary, in the Turkmen SSR for example, is believed to be one of the oldest town foundations in the world, while Samarkand (as Maracanda and Afrosiab) dates from the third millenium BC (Fig. 10.2). In the Baltic littoral, German influence—first introduced in early medieval times by missionaries, traders and knightly orders—remained influential until the latter part of the nineteenth century, as may still be seen in Riga and Tartu (Dorpat). There are also other towns

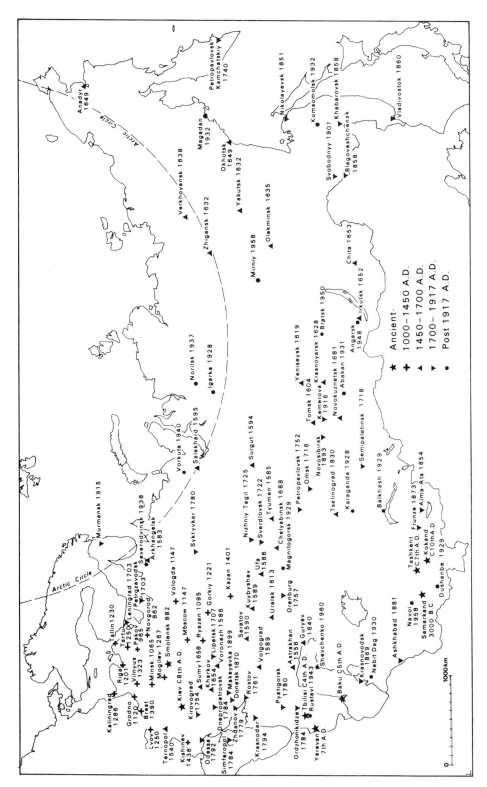

Fig. 10.2 Dates of foundation of towns. The early dates of town foundation in Central Asian oasis lands and parts of European Russia contrast with the much later spread into the Trans-Volga lands and Siberia, though in all regions major foundations of the Soviet period are to be found.

which lay for long periods outside the Russian state but which have been absorbed since 1945— for example, Polish Lwów (Austrian Lemberg— now Lvov), Uzhgorod, Mukachevo and Chernovtsy (Rumanian Cernauti), while Königsberg, a typical German Baltic town until 1945, when it was badly damaged, has been rebuilt as a characteristic Soviet town and renamed Kaliningrad.

The early town in the Russian lands was a close association of defensive, administrative and trading functions. The administrative functions were usually exercised from within a walled enclosure (*gorod*) within which lay a castle or *kreml* (in northern Russia, *detinets*). There was also frequently a cathedral or monastery, though in some parts of northern Russia, the ecclesiastical functions completely replaced the administrative and defensive roles. Outside but also walled was the *posad*—the traders' and artisans' quarters. Beyond this there was in some cases a peasant settlement, the *sloboda* outside the walls or forming part of the *posad*. The kreml (still seen in Moscow, Gorkiy, Tula and many other towns) usually lay on high ground or on a bluff above a river, while the merchants' and traders' quarters stood on lower ground, not uncommonly some distance away. The early Russian towns date from the three or four centuries before the catastrophic Tatar invasion (1240) which laid the majority of them in ruins. These towns lay in areas where reasonable farming land and good fishing were available, especially at break-of-bulk points on river portages (of which Smolensk is the 'classical' example), and were thus scattered through the wooded steppe, the better parts of the mixed forest or the fringe of the boreal coniferous forests. A few venturesome towns lay towards the steppe fringe, though here they suffered most from interference by raiding steppe nomads. Everywhere wood was the common building material, though from the thirteenth century stone began to be used for churches and fortifications— Vladimir was already renowned far and wide by the mid-twelfth century for its stone buildings.

EXAMPLES OF SOVIET TOWNS

Moscow

Moscow, capital of Russia, except in the period 1708–1918, is one of the world's largest towns,

with a population of 8.8 million in 1987. It commands the 'mesopotamia' between the main river routes and portages of the centre of the plain of European Russia. On the east, it was protected by the marshes of the Klyazma and Oka basins, though to the south it lay open to the wooded steppe and was defended by a number of fortified monasteries. The first record of Moscow is in 1147 AD when there was a wooden kreml on a bluff above the Moscow river, protected on either flank by small streams. Around this nucleus, the town grew. In the fourteenth century, a solid wall was built round the fortress and, separated from it across the Red Square, arose a traders' quarter, the *Kitaygorod* (from the Tatar meaning 'fortified town'), which in 1534 also received a substantial wall. In 1520, Moscow reputedly had 45 000 houses and was claimed to be larger than London. As the town grew, there developed the 'White Town', the home of courtiers and rich citizens, around the kreml: in 1586, it also received a wall. Beyond it newer suburbs began to emerge, the 'Earthen Town', named from the clay wall around it, and settlement spread across the Moscow river. Beyond the Earthen Town, a number of suburbs for foreigners grew up, like the German suburb—*Nemetskaya Sloboda*—in the north-east.

The kreml ceased to be mainly a defensive point in the seventeenth century and was then much ornamented and elaborated (mostly to foreign designs). In the early eighteenth century, the outermost suburbs were surrounded by a wall with 14 customs barriers (*zastavy*). Despite the ornate churches, the brick kreml and stuccoed wooden palaces, most streets remained unpaved and low timber houses, set in large gardens with high wooden fences and many outhouses, gave the town the appearance of a vast village. Even though the houses and palaces of the aristocracy and the rich began to be finished in central European style, basically they remained of wooden construction. The Napoleonic attack on Russia was marked by the firing of Moscow (1812) and three-quarters of its buildings were destroyed. But even though the capital had been moved to St. Petersburg, the town quickly recovered and began to increase its commercial and manufacturing functions.

The industrialisation of late nineteenth century Russia was essentially focused upon Moscow. New and squalid industrial slums appeared, and

A view of Moscow from the river showing some of the most desirable residential flats and a new high-rise administrative block

the old walls were pulled down and boulevards made in their place—the wall of the Earthen Town, for example, was replaced by a park-like ring, the *Sadovaya*. The line of the old walls can still be seen as broad green circles in the contemporary plan of Moscow, and some of the old barrier names remain (e.g. the Kaluzhskaya Zastava). Industry had gathered particularly in the south near the river harbour or in the east. The growth of railways created a ring of terminal stations roughly along the line of the old *zastavy*. Brick, stone and concrete have been increasingly used, but even now many wooden buildings remain, though in the early 1960s a determined effort was begun to remove them.

The return of the capital to Moscow after the Revolution brought a new significance and a change in the town's function, with the meteoric rise in the bureaucratic superstructure of the Soviet state. The population of 600 000 in 1897 had risen to 1.6 million by 1912 and to 2.03 million in 1926, while the 1939 census enumerated 4.14 million people. Industry was expanding, using the pool of skilled labour available, and workers for the new bureaucratic machine flowed in. In the early 1930s an elaborate development plan

was drawn up, with the widening of streets and removal of the last remains of the old walls, as well as the laying out of new parks. In the central districts extensive changes were made in the street pattern and many new official buildings erected, though some of the more grandiose architectural concepts were never realised. Great improvements were made in public transport: construction of an underground railway began and suburban railways were electrified, besides expansion of the bus, trolley bus and tram networks. To increase rail mobility, a ring railway was built (to be augmented by an outer ring line after the Second World War) and in 1937 the Moscow–Volga Canal provided a shorter route to the Volga than the winding course of the Moscow and Oka rivers. The southern river port at Nogatino was now augmented by installations in the northern part of the town at Khimki.

During the inter-war period, building of homes received a relatively low priority, but from the early 1950s expansion of residential accommodation has produced vast new housing quarters (mostly large blocks of flats with supporting services) around the city. The growth of motor traffic, even though still below Western levels, has led to

The tourist industry is one of the main sources of foreign currency as well as employing many people—here Soviet and foreign tourists stream into Petrodvorets, the summer palace of Peter the Great near Leningrad.

the provision of widened streets and appropriate junctions, while during the mid-1960s a ring motorway was built. Many new buildings have been added, including the erection of skyscrapers. Nevertheless, the side streets remain poorly paved and wooden houses are still common. Moscow is a mixture of architectural styles: beautiful old structures such as the Kremlin or the former palaces; dismal, late-nineteenth century buildings; a few buildings in a clean functional style of the early 1930s and a vast assortment in the heavy, ornate, almost bizarre style of the Stalinist period. A return to a more functional style has been made since the death of Stalin, with much use of concrete and glass, reflected in the striking headquarters of COMECON, several hotels in the central area and a variety of other buildings.

Leningrad

Leningrad (St. Petersburg or Petrograd) in its elegance and farsighted central layout is in great contrast to Moscow. It owes its origin to Peter the Great, an admirer of western Europe, who built it in Russia's then most-westerly territory with access to ice-free waters through which he hoped European ideas would find a way into the backward and inward-looking Russia of the early eighteenth century. The town was founded in 1703 on the flat, marshy delta of the Neva, subject to autumn and spring floods, especially when onshore winds ponded back the waters of the river. The town lies at the base of a narrow isthmus between the Gulf of Finland and Lake Ladoga, and was protected by its outer fortresses, like Petrokrepost on Lake Ladoga and Kronstadt on Kotlin Island in the Gulf of Finland. St. Petersburg was built by some 40 000 peasants recruited in the countryside around, steadily replaced by new recruits as their numbers were depleted by accidents and illness, and directed by Italian and other foreign artisans. In the soft substratum, all the larger buildings had to be supported on piles and the problem of their weight tended to restrict their height—the low, even skyline, broken by fine, tall and spindly spires is characteristic of the older part of Leningrad. The generous and roomy appearance of the central districts is given by wide radiating streets, broad arms of the delta and drainage canals and a fortunate balance between open space and buildings. In 1987 the population was 4.95 millions.

The focus of the town was the low-walled fortress of Peter–Paul, but around the eighteenth century core, there is a broad belt of less attractive buildings of the nineteenth and twentieth centuries, mostly flats and factories, in style similar to other Russian towns. Along the coast lie, however, a number of small and pleasant holiday resorts, while the southern side of the delta is the site of Leninport which, although closed by ice in winter, remains one of the largest and busiest Soviet harbours.

Kiev

Kiev, capital of the Ukraine, lies in the marchlands between the northern forests and the southern steppe alongside a major north—south routeway, the Dnepr river. A settlement appears to have existed on the high right bank of the Dnepr before the ninth century, from which the later

A new development of flats and shops in the suburbs of Leningrad with a rapid transit tramway in the foreground.

nucleus developed. It was an easily defensible position on bluffs overlooking the river, and isolated by deeply incised valleys to either flank. To the south lay the later royal village of Berestovo, site of the Pechersk Monastery, one of the main centres for the dissemination of Christianity to the Russian lands, especially under Vladimir in the tenth century. Internal dissension and attacks from steppe nomads brought the decline of Kiev in 1230, and its fortunes were seldom good until it passed back into the Russian state in the seventeenth century. With the expansion of the Russian hold throughout the Black Sea littoral and the Ukraine, Kiev began to revive. Houses spread along the high river bank and in the deeply incised valleys that dissected it. A broad open site by the confluence of the Dnepr and Pochayna river, known as Podol, became a centre of artisans and the site of the river port, changing in the nineteenth century into Kiev's industrial district. During the late 1930s and in the post-war period, development on the eastern (the low meadow) bank included the large industrial suburb of Darnitsa, while in the 1960s an underground rainway was built to link the two sides of the river. In 1987 the population was 2.5 million.

Novgorod

Novgorod is an example of the oldest Russian towns and has been restored as an historical 'monument' since extensive damage in the Second World War. Lying in poor agricultural country, but enjoying a commanding position on the Volkhov river, part of the important early medieval trade route from the Baltic to the Black Sea, at a point to which portages across the Valday Hills from the Volga basin led, Novgorod was one of the earliest great trading centres in the Russian lands. On a low but steep bluff on the western side of the river lay the kreml whose walls also enclosed the cathedral, and around this but separated from it by a broad square was the administrative and ecclesiastical quarter, the *Sofiyskaya Storona,* itself surrounded by a wall and a moat. Across the river but joined to the western bank at an early date by a bridge, lay the traders' and artisans' quarter—the *Torgovaya Storona—* behind its own wall, where a large square served as the main trading point. To Novgorod came merchants from the Hansa and from Visby as well as Russian and Asiatic traders. The town was largely built of wood and was renowned for the beauty of its buildings, but it was razed by Ivan the Terrible, who reputedly slew 60 000 citizens, and never fully recovered its former glory. Much of the present town is composed of the dreary early post-war style of Soviet architecture.

Siberian towns

Town foundations in the Volga lands and the steppe in the sixteenth and seventeenth centuries and in Central Asia in the eighteenth and early nineteenth centuries were a process of consolidating hard-won territory, and most began life as fortified posts. The Siberian towns were fortified wooden trading posts commonly situated to take advantage of relief for defence and to use better patches of ground where some cultivation might be undertaken. As the rivers and portages were main routeways, most towns were situated along rivers, particularly at points where portages began or where there were ferries. Wooden stockades surrounded low wooden houses set behind high fences, brightly painted wooden churches and administrative buildings, with white-washed stucco (but wooden-frames), that lined wood-paved streets. Deep open drains were often framed over with wood to make a pavement or sidewalk; these still exist in the older parts of Siberian towns. In the nineteenth century, some of the more important towns like Irkutsk became rich and could afford better buildings, parks, museums and other signs of municipal affluence. With a fund of exiled intellectuals to draw on, Siberian towns were distinguished by their museums and literary and philosophical societies.

The wooden town is still being built in Siberia: Divnogorsk, near Krasnoyarsk, on the banks of the Yenisey, is a new town of large wooden blocks of dwellings set amid the tayga, with roughly graded roads and simple facilities.

The coming of the railway in the late nineteenth century brought a new wealth and purpose to towns that lay at points where railway and river crossed and many of these came to outshine towns on the old *trakt,* the former main highway across Siberia. In this way, Novonikolayevsk, a mere insignificant village in the early 1890s, grew into the city of Novosibirsk, with its large theatre, administrative buildings, industrial plants, and the now world-famous 'Academic City' (Akademgorodok) nearby: it now greatly outshines the formerly more important town of Tomsk, with its university which was established

in 1880—Novosibirsk has 1.4 million people and Tomsk about one third of that figure.

There are in Siberia many boom towns: apart from the foundations of the interwar years like Komsomolsk, postwar foundations have included Angarsk in 1948, now a town of 262 000 people, while in the west Siberian oilfields, Surgut rose from 6000 people in 1959 to 227 000 in 1986, while Nizhnevartovsk rocketed from 16 000 in 1970 to 212 000 in 1987.

Soviet Central Asia and Transcaucasia

Distinctive native towns are found in Soviet Central Asia and in Transcaucasia. These lands, which passed to the Tsarist empire in the nineteenth century, already had their own strongly developed native society with a moderate scatter of towns much influenced by Indian or Persian culture. The native towns of Central Asia, for example, were composed of a mass of small alleyways along which lay low clay courtyard houses with flat roofs and few, if any, windows facing on to the streets. They were dominated by mosques and minarets, often beautifully decorated with coloured tiles, if somewhat dilapidated. A central

Flats in Ordzhonikidze in the north Caucasus region, the design being similar to those found all over the Soviet Union.

market place (*registan*) was surrounded by bazaars composed of small shops and one-roomed workshops. These towns usually lay within a towered wall, also made of sun-dried mud bricks, and some were commanded by an old fortress. Adjacent to many of these ancient towns lies the Russian foundation—originally a garrison town—with regular tree-lined streets, small Russian-style houses and public buildings also in the style of European Russia. The Russian barracks usually took up a generous part of the town and, where the railway existed, a large area was occupied by copious sidings and a station enclosure. In the Soviet period, these towns have been changing—the old and less salubrious parts of the native towns have been slowly cleared and replaced by multi-storied blocks of flats or by more open spaces. Nevertheless, in places, the older native buildings of interest have been retained as historical monuments. The contemporary architectural style, while distinctively 'Soviet', has sought to incorporate a local motif as well as designs on the basis of the slogan 'national in style, socialist in content'. Reports from Central Asia suggest, however, that the native people have commonly been reluctant to move away from traditional homes into Russian-style houses or blocks of flats.

Towns 'of socialist realism'

Latest among the Soviet towns are the towns of 'socialist realism', mostly foundations of the late 1940s and thereafter. These are often associated with major industrial developments, though to this group may also be added many of the towns built in the early prewar Five Year Plans (e.g. Magnitogorsk, Karaganda, Komsomolsk) whose growth has allowed the incorporation of the current concepts. The towns focus on their main plant, though they are often separated from it by a green belt, lake (reservoir for industrial water) or a recreational area. Within the built-up area is a careful evaluation of a balance between population, employment, services and amenities. The design is usually one of distinct 'neighbourhood units', with all supporting services. The striking feature, as in so many Soviet towns, is the fewness of shops, but it should be recalled that each shop has a very carefully estimated population to serve. Some of these towns, for example, Karaganda, comprise within one administrative area, a series of industrial enterprises or mines, each with its

own settlement linked to a central administrative and service core, so that the whole pattern of territorial occupancy by buildings is loose and diffuse. These new towns have been criticised for failings in their planning concepts: for example, too many were conceived with too great a compactness in building intended to give the 'big city effect', so that there is monotony and overcrowding of buildings that are too high and line over-wide streets. Many plans have been criticised for being too geometric, especially in focusing too forcefully on the main industrial plant.

In the growth and planning of towns, several problems face the Soviet authorities. Housing for too long received only a low priority, but the demands for improved standards of living, together with the hygienic and social problems arising from overcrowding, have forced attention to the large deficit in most towns. Large new housing and neighbourhood units of prefabricated blocks of flats have been erected, usually accompanied by all essential services. Some of these schemes have been markedly successful, as for example in Kharkov, where the new districts stand amid attractive wooded gullies. In some towns, these new buildings are erected in the gardens of old wooden houses whose residents are moved into the new homes on completion. An attempt has been made to enforce a vigorous policy of issuing 'residence permits' in the towns with the most serious housing deficits. The vigorous migration from country to town creates difficulties in educating rural people to town life and in catering for the countryman's need of such things as a small allotment garden to cultivate and to keep a few domestic animals such as rabbits. These allotment gardens are, nevertheless, a useful contribution to the problems of distributing food and nourishing urban populations, since the distributive industries appear to have been an intractable problem under central planning concepts.

The growth of towns in inhospitable environments—the arid lands of Central Asia or the intensely cold Arctic Siberia—has presented questions of design and form. One of the major problems in both areas has been water supply: in arid areas, reservoirs need to be built to collect water (e.g. at Magnitogorsk) or aqueducts built to carry water from other areas. An aqueduct is presently being built from the Irtysh to Karaganda and Temirtau in Kazakhstan. The oil-mining town of

Shevchenko in the Mangyshlak peninsula, developed in the 1960s, has an elaborate atomic-powered water desalinisation plant, and Krasnovodsk receives fresh water by tanker across the Caspian Sea. Over much of Siberia, permafrost, a distinctive civil engineering hazard requiring special methods of construction, makes water supply and sewage disposal difficult (Fig. 10.3). Wooden frame buildings are preferred in many places because they adjust more readily with the movement of the sub-stratum permafrost caused by an upset of the thermal balance. The importance of shade and greenery in arid lands as an amelioration of the micro-climatic environment is also stressed: in western Turkmenistan, in the largest settlements there are 3–4 m^2 of greenery for each inhabitant, but the Soviet authorities maintain that this must be increased to 25 m^2.

The problems of assembling labour and building materials in remote and virgin country present many transport headaches: the building of Bratsk demanded a temporary workers' settlement and special electricity supply across 700 km of virgin and empty country. Little could be achieved until a railway branch line of several hundred kilometres had been built to the site. Once a settlement is established, it has to be fed and this adds a further burden on transport. It is common to develop glasshouses and special installations to provide fresh food—Norilsk and Igarka are supplied from a state farm on a sandy island in the middle of the Yenisey, whose massive volume of warm water creates a local positive temperature anomaly and consequently keeps the island free of permafrost. Reports suggest that the mining settlement of Deputatskiy in north-eastern Siberia has been built under a gigantic roof supported in the permafrost by adjustable supports and with the buildings, partly subterranean, suspended from this roof. It should not be forgotten that in the worst Siberian blizzards, it is almost impossible to move about outside, even in towns.

Optimum size of a town

A long discussion has taken place in the Soviet Union on the optimum size of the town so that the best amenities may be provided and the best return may be had on investment while running costs are kept to a minimum. Various proposals have been made—in the early thirties the concept of the vast multi-million town was accepted in the then-prevailing 'gigantomania', but since then

views have changed several times. One view will not accept communities of less than 250 000 or more than 500 000 as within the optimum limits in relation to local circumstances. In the mid-1950s, Soviet planners appeared to reject the view that any city should exceed 400 000 people, but the continued growth and the difficulty in controlling growth of large cities has led to estimated populations being exceeded anything up to a decade earlier than forecast. Examples of rapid urban growth have already been quoted for Siberia, but others include the new motor vehicle manufacturing town of Naberezhnye Chelny on the Kama which grew from 16 000 in 1959 to 346 000 in 1981, or the nearby Nizhnekamsk that rose from 49 000 in 1970 to 183 000 seventeen years later. Another vehicle manufacturing centre, Tolyatti, also rose from 72 000 in 1959 to 627 000 in 1987, perhaps the best example of the Volga basin's boom towns. In the commuting sphere around Moscow, Zelenograd rose from 7000 in 1959 to some 148 000 in twenty five years, while Staryy Oskol in the newly developing Kursk iron ore district moved from 27 000 in 1959 to 167 000 in 1987 and will continue to grow substantially as the planned new metallurgical complex is further developed.

It is suggested that the type of industry basic to a new town is an important guide to the eventual likely size to which the town should be expanded. In some instances, however, estimates for ultimate size have had to be revised steeply upwards in the light of real growth—the chemical town of Volzhskiy on the Volga, with 257 000 people (1987), was planned originally as unlikely to exceed 50 000, but further projections suggest that 300 000 should be the upper limit. Soviet planners maintain that the tendency has been to design for too *few* rather than too *many* potential inhabitants. Plans for Navoi in the Uzbek SSR changed from 50–70 000 to 250–300 000, though it had reached only 110 000 inhabitants by 1987. Angarsk, founded in 1948, was planned on too small a scale, with the subsequent problems of adjustment to a much greater population of 262 000 by 1987.

The geographer Pokshishevskiy, speaking to the press in 1970, said, 'practical experience has shown that the further expansion of giant cities has become very undesirable. Vigorous efforts have already been made to limit their growth and their share in the total will presumably diminish

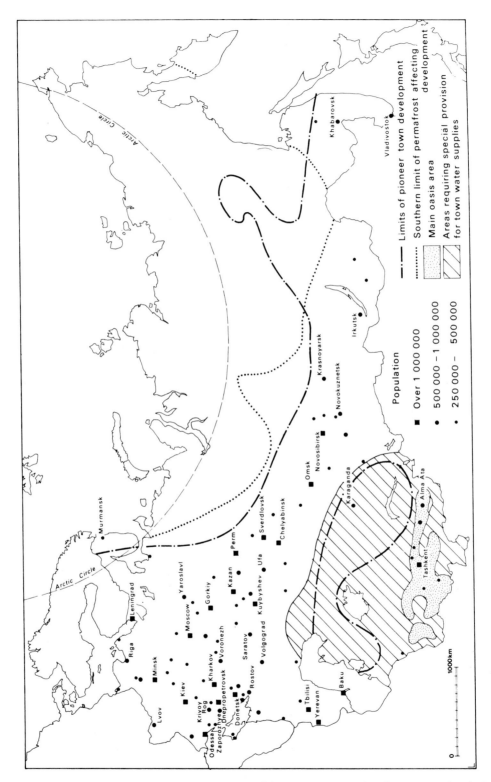

Fig. 10.3 *The largest contemporary towns. The distribution of the largest towns within the main populated triangle with its base between the Baltic and Black Seas and its apex in western Siberia is impressive. Beyond this triangle, factors of* **the physical environment** *play a significant role in town foundation*

considerably by the year 2000'. It has been calculated that the 45 major cities (all over 500 000), which account for 20% of total population and 32% of urban population, will drop to 13% of total population and 17% of urban population by 2000 AD, if growth of population for the group does not exceed 10–15%. Pokshishevskiy then says, 'alternative types of communities—small towns and urban-type communities—are unable to provide their residents with a sufficiently high level of services . . . It is much less economical to provide amenities of various kinds in small urban communities. That is why Soviet town planning theory recommends avoidance of unduly small urban communities while restricting the growth of giant cities. A part of the existing small communities will presumably have grown into medium-sized towns by the year 2000 and the setting up of new communities will depend on economic requirements. In 2000 the bulk of the urban population will probably be living in well-appointed, very healthy and exceedingly economical medium and large communities but not giant cities'. However, by no means all Soviet planners agree with these views and many argue the 'progressive' nature of urban agglomerations.

THE VILLAGE

The Russian lands were traditionally lands of villages, while the village was the common unit of settlement in the sedentarily settled areas of Central Asia and Transcaucasia as well as in the non-Russian parts of European Russia. Moreover, wherever Russian settlers went, they took the concept of the village with them. In form, size

TABLE 10.4 NUMBER OF VILLAGE SOVIETS IN THE USSR

Year	Soviets
1928	72 997
1950	74 863
1964	39 623
1968	40 558
1977	41 249
1981	41 511
1987	42 599

Source: Norodnoy khozyaystvo SSSR, various years Moscow.

and house type, there were, however, regional variations (Fig. 10.4).

In European Russia, the village played a vital part in the emergence of Russian society—particularly through the village council, the *mir*—and was distinguished in tsarist times according to whether it had a church (*selo*)—and was, therefore, a focal point of the countryside—or whether it was without (*derevnya*). Such a distinction no longer exists in the Soviet view and now villages are important if they are the seat of a collective or state farm or a village Soviet.

In northern European Russia, villages tend to be small—many are of a dozen or so houses only—in widely scattered positions on drier ground such as sandy mounds or river terraces. In the central Great Russian districts and in the north-west and Belorussia, villages are larger with several dozen houses, though the type of house varies considerably in size and richness of its appearance from district to district: especially poor and meagre are the villages of the wet lands such as Polesye. The most common Russian village type is the street village, a simple line of houses scattered along either side of a broad and unpaved street, but small nucleations do now more frequently occur as the result of growth around farm centres, where educational, cultural and technical facilities are concentrated. Several villages may belong to one collective or state farm and some have a little handicraft industry.

The steppe villages tend to seek shelter from the cold winter winds that sweep across the open steppe: thus villages are commonly sited in valleys or erosion gullies, particularly where water is available. The villages are usually large and several may be strung out along a gully or a valley, forming an uninterrupted line of buildings for several kilometres, though each community can usually be distinguished by its village pond. Particularly large chessboard villages of a rectangular street plan are common in the newer settled areas of the steppe in the Ukraine and notably in northern Caucasia, as, for example, in the Cossack village type (*stanitsa*) of the Kuban. The villages of southern Siberia on the steppe fringe are often large and prosperous looking and make a generous use of space.

Other settlement types

Non-Slav peoples show other settlement types. In the Baltic republics, villages still tend to be small,

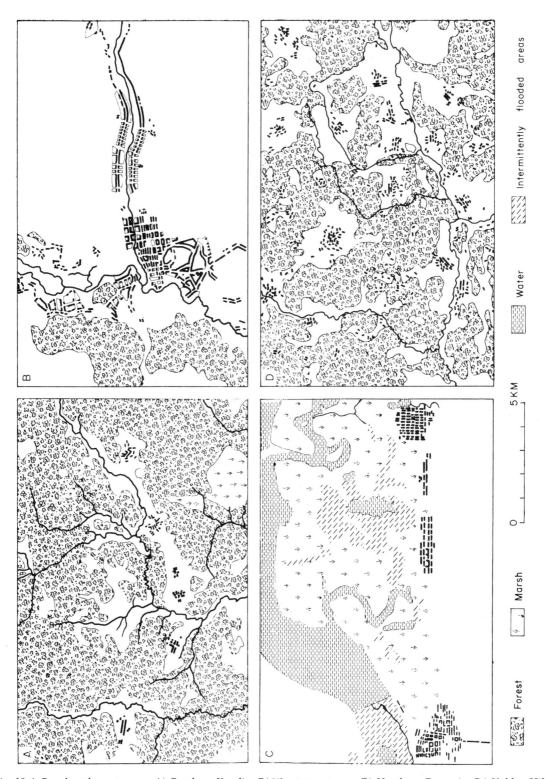

Fig. 10.4 Rural settlement types: A) Southern Karelia; B) Ukrainian steppe; C) Northern Caucasia; D) Valday Hills

comprising 20 houses or so, and the land reform of the nineteenth and twentieth centuries led to the development of scattered individual farms. Houses were often long and low with a mansard roof. Since re-incorporation in the Soviet Union, collectivisation has tended to encourage nucleation. Small hamlets and villages or even scattered farms, distinguished by large houses with barns, are also found in Karelia. In the Volga lands, Tatar settlement areas have been identified by their irregularly shaped villages, usually close to a river or spring, though they are not found on the best lands from which they were displaced by early Great Russian colonisation. The Bashkirs, who only abandoned nomadism in the nineteenth century, live in villages of the street type, and until recently, with simple housing of mud bricks and plaited wattle. Air-dried bricks or wattle frame houses clad by mud or plaster and colour-washed are common in the southern Ukraine and Moldavia. These houses often have long verandahs.

The Caucasian mountain villages are irregular in their plan but closely nucleated. Low stone houses, usually with flat roofs, stand on easily defended slopes amid guard towers. In the lower hill lands, villages are commonly large and houses have spacious verandahs or balconies. Some Caucasian houses show strong Turkish influence of stone basements with overhanging upper floors in wood. Central Asian oasis villages are large and commonly closely nucleated, surrounded by a low wall. The flat-roofed courtyard houses crowd along narrow alleys (with open sewers), but in many places the Russian-style house, and more recently, houses and flats of bricks or concrete, have begun to appear, especially where nomads have recently become sedentary. In nomadic lands, settlements are mobile and consist of tents and *yurts* of various kinds and shapes, but in the Arctic and sub-Arctic in Siberia, native winter houses are partly below ground, built of wood, sods and stone; in summer they are deserted for tents or wooden huts, in some cases raised above ground on stilts. Nevertheless, even in these regions, the Russian wooden house—the *izba*—has been widely adopted.

Modern ideas on rural settlement

Modern Soviet ideas on rural settlement have been centred on ways to bring rural life into line with life in towns. A major deficiency has been the amenities of village life: a large proportion of villages have until recently had no piped water and no electricity and complaints about the state of rural roads have been common. Encouragement has been given to nucleation—outlying

Listrenichnoye, a typical Siberian village on the shores of Lake Baykal.

settlements have been encouraged to move into central collective and state farm villages where better amenities can be provided. A concept re-introduced by Khrushchev was the *agrogorod*—the 'agricultural town'. The first was tried in the 1930s on a state farm in northern Caucasia and this form appeared again on the immense state farms of the virgin lands of northern Kazakhstan. It can offer all the amenities of the small town and provides the authorities with a better supervision of the community. Each day workers are taken to the fields in lorries or buses, as commuters are taken to work in a town. Plans formulated in the 1960s earmarked about 16% of the existing villages for development, while the remainder were to stagnate or even die away, but such radical changes are now regarded generally as unrealistic though grouping of villages for more efficient provision of services continues widely.

The Soviet town provides the urban geographer with a study of the problems of massive urbanisation in a rapidly industrialising economy set in a generally harsh physical environment, especially in arid Central Asia and Arctic Siberia. The application of Marxist–Leninist concepts in the framework of a rigorous central planning also introduces influences in the evolution and growth of towns not experienced in the western world.

The remodelling of Soviet society and economy set in pace by Gorbachev has just begun to show in one of the most sensitive areas—living conditions in towns and villages. Already, the pressure for better accommodation has been recognised, with plans to build larger apartments, construct less "regimented" buildings, replacing high-rise blocks by three or four-storey blocks of flats. Individuals or cooperatives are to be helped to build their own homes. Aspirations for greater freedom of choice and more consumer goods may well soon be reflected in a revolution in the provision and quality of shops and restaurants. Yet, however fast change may be allowed, the image of the Soviet town and village will long bear the imprint of seventy years of central planning and a modest priority for the needs of the individual citizen. At the same time, the great historical traditions of the town in the Soviet lands should not be overlooked, while the nature of the village, in its varying forms, from which so many towns have evolved, has been an important factor in the emergence of the Soviet society.

BIBLIOGRAPHY

Adams, R. B. (1977), 'The Soviet metropolitan hierarchy: regionalisation and comparison with the United States,' *Soviet Geography*, **18**, pp. 313–328.

Baburin, V. L, Gorlov, V. N. and Shuralov, V. Ye. (1987), 'Economic-geographic issues in the development of the Moscow Region,' *Soviet Geography*, **28**, 63–70.

Bater, J. H. (1980), *The Soviet city*, Edward Arnold, London.

Bater, J. H. (1989), *The Soviet scene, a geographical perspective*, Edward Arnold.

Bugromenko, U. N. (1979), 'Complex planning of a city and the study of its spatial structure,' *Soviet Geography*, **20**, pp. 160–169.

Burlachenko, G. F. (1979), 'Problems and prospects of development of rural nonfarm places in the USSR,' *Soviet Geography*, **20**, pp. 305–309.

Demko, G. J. *et al* (1987) Restructuring the settlement system in the USSR, five papers in *Soviet Geography*, **28**, pp. 707–776.

French, R. A. and Hamilton, F. E. I. (1979), *The socialist city*, John Wiley, Chichester.

Hamilton, F. E. I. (1976), *The Moscow city region*, (*Problem Regions of Europe*), Oxford UP, Oxford.

Hamm, M. (ed.). (1976), *The city in Russian history*, UP of Kentucky, Lexington.

Harris, C. D. (1970), *Cities of the Soviet Union— Studies in their functions, size, density and growth*, Rand McNally, Chicago.

Hausladen, G. (1987), 'Recent trends in Siberian urban growth', *Soviet Geography*, **28**, pp. 71–89.

Holzner, L. and Knapp, J. M. (1987), *Soviet geography: studies in our time*, Milwaukee.

Hooson, D. J. M. (1969), *The growth of cities in pre-Soviet Russia*, University of California Slavic and East European Series, Berkeley.

Khorev, B. S. (1975), *Problemy gorodov*, Mysl', Moscow.

Kochetkov, A. V. and Listengurt, F. M. (1977), 'A strategy for the distribution of settlement in the USSR: aims, problems and solutions.' *Soviet Geography*, **18**, pp. 660–674.

Konstantinov, O. A. (1977), 'Types of urbanisation in the USSR.' *Soviet Geography*, **18**, pp. 715–728.

Kravchuk, Ya. T. (1973), *Formirovaniye novykh gorodov*. Izd. Literatury po Stroitelstvu, Moscow.

Kitovka, O. P. (1980), 'Urbanisation in the USSR: problems of spatial differentiation,' *Soviet Geography*, **21**, pp. 30–36.

Lappo, G., Chikishev, A. and Bekker, A. (1976), *Moscow, capital of the Soviet Union*, Progress, Moscow.

Mellor, R. E. H. (1963), 'The Soviet town.' *Town and Country Planning*, **31**, pp. 90–94.

Mellor, R. E. H. (1976), 'Sowjetunion. IV'– *Bevölkerungsverteilung, und ethnische Zusammensetzung*, Harms Erdkunde, Verlag Munich.

Pallot, J. and Shaw, D. J. B. (1981), *Planning in the Soviet Union*, Croom Helm, London.

Petrov, N. V. (1987), 'Settlement in large cities of the USSR: an analysis of effectiveness,' *Soviet Geography*, **28**, pp. 135–157.

Saushkin, Yu. G. (1966), *Moscow-geographical charac-teristics*, (Trans.: T. Kapustin), Progress, Moscow.

Shaw, D. J. B. (1978), Planning Leningrad, *Geographical Review*, **48**, pp. 183–200.

Smirnov, N. V. (1979), 'Stages in the development of the demographic structure of large cities,' *Soviet Geography*, **20**, pp. 219–224.

Tikhomirov, M. (1959), *Towns of ancient Russia*. Foreign Languages Publishing House, Moscow.

Underhill, J. A. (1976), *Soviet new towns—Housing and national urban growth policy*, US Government Printing Office, Washington.

Wein, N. (1987), 'Bratsk—pioneering city in the taiga; *Soviet Geography*, **28**, pp. 171–194.

11 Transport

The contemporary Soviet transport system has evolved entirely under the state direction. Railways in the Russian empire were developed mainly by the state and were thus readily assimilated into the Socialist system after the Revolution. The introduction of the all-important airways and of motor transport began under state direction in the 1920s, though they remained of minor significance until the late thirties. Under Gorbachev's reforms transport managers were set new economic targets in the late 1980s but transition to full economic accountability appeared likely to proceed at only a very slow pace. Only in private car ownership is there likely to be scope for independence from government operations and in 1989 this was still much more restricted than in any other highly developed country.

A modern industrial state depends on an adequate transport system to assemble raw materials for its factories, to carry food from the country or ports to its townspeople, and to distribute the products of its industry, as well as to move people for recreational and employment reasons. The selection of the means of transport to do these tasks depends on the nature of the traffics generated in relation to distance and volume and in relation to the physical environment through which they will have to move.

In the immense and diverse continental environment of the Soviet Union, development of the transport system has been a key to the feverish creation of a strong economic structure based on large-scale industry since the Revolution. Policy has demanded an even spread of development among the regions, despite a clearly uneven distribution of natural resources and conditions, and has thus accentuated the significance of transport. National self-sufficiency has demanded provision of transport to remote mining areas far beyond the existing limits of sedentary settlement. Yet at the same time it has required the burden on transport to be kept as low as possible, in order to reduce to the minimum level investment of scarce national resources in transport rather than in more productive growth sectors of the economy. Transport has, therefore, been an almost decisive factor in such aspects as the evolution of high-cost arctic agriculture (rather than carry cheap food from southern producers) or the relation between plant location and raw material supply among inter-linked areas like the Ural, western Siberia and Karaganda.

TRANSPORT IN A SETTING OF THE PHYSICAL ENVIRONMENT

A basic Soviet transport problem is distance. The USSR, the world's largest compact political unit, is three times the area of the United States and 90 times the area of the United Kingdom, while its east to west extent is so broad as to give a ten-hour difference in time. Moscow is some 9200 km by rail from Vladivostok and there is a seven-hour time difference: the two towns are further apart than London and New York. Moscow is also 370 km *further* by rail from Tashkent than it is from Paris.

The influence of the climate

The strongly continental climate has an important influence on transport. The anomalous cold, with winter considerably exceeding summer in duration over most of the country, and the related phenomenon of permafrost (reputedly affecting 47% of the country's area) are a hindrance over wide areas to construction of roads and railways or even buildings. Intense cold also can cause brittleness in metals, affect ferro-concrete and the working tolerance of machinery or reduce the fluidity of oils and increase thermal loss from heat engines. Frost heaving breaks up roads and upsets the alignment of railway tracks, besides affecting the stability of bridges, but it does at least provide a hard surface over which to move, so that winter has been traditionally a period of movement, when even bogs, almost impassable barriers in spring and summer, and some rivers, can be used by temporary 'winter roads'.

Throughout the greater part of the country, rivers, lakes and even coastal waters suffer ice hazards for 100 days or more each year (Fig. 11.1). Ice makes access to Russia's coasts difficult, except for a few favoured ports seldom seriously troubled or which may be kept open easily by ice breaker, so it is not surprising that the Soviet Union has contributed much to the design of ice breakers and of merchant ships for navigation in ice. In some waters, notably on the Pacific coast and in the Arctic, mist and fog even in the open-water period are also a major hazard; dust storms in spring and summer in the Black Sea can seriously reduce visibility, while along the Caucasian Black Sea coast very strong föhn winds off the shore can make entry to some ports (e.g. Novorossiysk) awkward.

The spring thaw, traditionally the *rasputitsa* (the 'roadless' season), results in bad roads, because the lower layers of the soil remain frozen and the thawed surface layer produces large areas of standing water and floods, turning the ground to a quagmire. The water level in rivers rises and currents increase, while ice floes endanger shipping and bridge supports or pile up in bends and constrictions to form temporary dams whose collapse releases masses of water to sweep downstream as destructive floods. Low-lying areas of the 'meadow banks' (usually the left bank) are regularly flooded and avoided by settlement, and shipping is forced to lie in winter harbours until the period is past. In Siberia, the upper southern reaches of the rivers thaw first and water flows north on to still-frozen lower sections, turning areas like the west Siberian lowlands into huge shallow swampy lakes that are gradually reduced to vast swamps by summer. In late summer, many rivers suffer from low water, forcing navigation to a halt, while the heavily silt-laden streams of Central Asia are plagued by evershifting shoals and banks, making them of little navigational value. In the spring quagmire, it is often only railways, raised on low embankments, which can keep moving; while sudden vicious summer thunderstorms can bring traffic to a halt and disrupt urban trams by damage to overhead lines. Thunderstorms help to settle the dust, frequently a visibility hazard, and stop drifting dust or sand that can block roads and railways.

Physical obstacles to the development of the transport network

The relief map of the Soviet Union suggests few obstacles to easy transport: vast plains and moderately dissected plateaus form large areas of the heart of the country (Fig. 11.2). Even the parallel Ural ranges that lie across the main east-west transport arteries are low and marked by clearly defined through-ways, particularly in their central section. High mountain terrain lies mainly in the southern and eastern peripheries, areas of underdevelopment until recently, and border areas where international transport has not been encouraged for strategic reasons. It is perhaps paradoxical that minor features of relief commonly form more troublesome obstacles to the development of the transport network than do the major features.

In the northern part of the great plains, because of low temperatures (insufficient for evaporation to exceed even the meagre precipitation) and the exceptionally gentle gradients which impede drainage, the land suffers from too much water and is characteristically wet and marshy. Drainage of roads and railbeds is a critical problem since saturation causes bad running and frost heaving in winter. Railbeds are often laid on several feet of sand, and are usually raised on low embankments, which also prevent snow drift in winter and inundation in spring. The softness of the substratum and its poor bearing capacity are also a hindrance to railways, demanding additional or unusually long rail sleepers (ties) or imposing severe weight and axle-load restrictions.

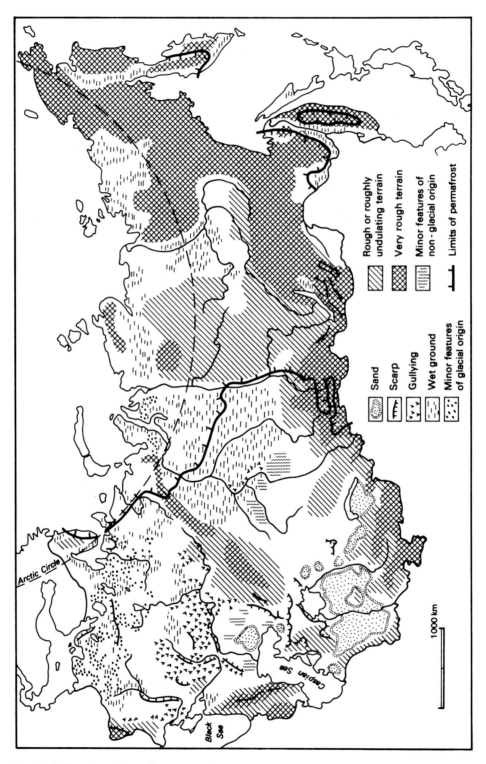

Fig. 11.1 Terrain problems for transport

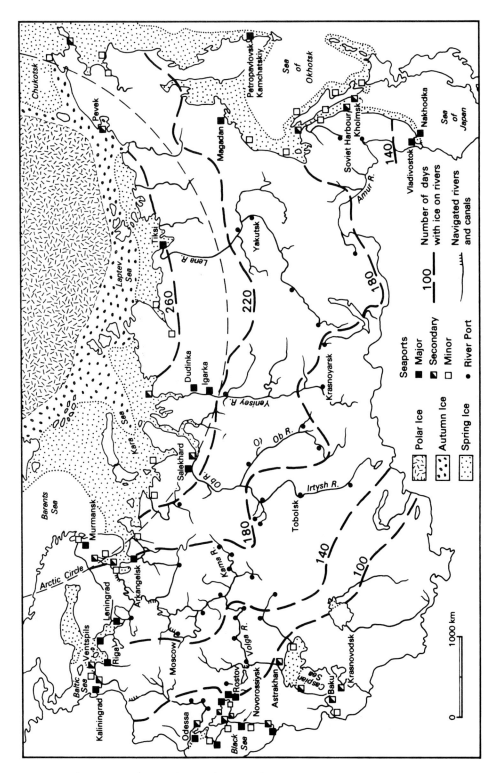

Fig. 11.2 Waterways, ports and icing of seas and rivers

Despite the moistness of the northern forests there is a fire danger in summer, and wide avenues alongside railway tracks are cleared as spark arresters and to prevent trees blown down in winter blizzards from fouling the track.

Southwards and eastwards, the plains become drier, turning to steppe and even desert. Low precipitation is exceeded by high summer evaporation, and sudden thundery downpours fall on parched ground unable to absorb the moisture, causing much loss by run-off, so that gullying easily occurs, especially where the vegetation cover has been broken. In winter, when the thin snow cover is quickly blown away, or even in a dry spring or summer, the parched, friable ground can be easily eroded by high winds sweeping unimpeded across these open plains. Roads and railways need to be protected against drifting dust and sand by long shelter belts, a notable feature of the landscape in the steppe and even in parts of the farmed forest belt. The first were laid out along the Moscow–Gorkiy railway in 1861 against snow drift; by 1900 some 3982 km of track were protected by shelter belts and by 1950 the length exceeded 37 450 km. (23 371 miles). In the desert, protection against sand drift often requires belts extending up to 120 m from the track.

In the northern forest, there is a rough mantle of glacial debris where drier morainic ridges provide routes between the marshy hollows. The southern plains beyond the limits of glaciation have a cover of loessic materials, sands and clays into which gullying to varying degrees has taken place. Where this process has extensively developed, the steep-sided troughs are obstacles to the ready construction of roads and railways and require numerous bridges or culverts, while routes tend to become sinuous to avoid gullies or to follow them along their marshy floors. With a deep mantle of loosely consolidated material and few solid rock exposures, a problem is to find suitable aggregate for track ballast or road surfacing. Stone work is replaced by bricks (often air-dried in the south), but concrete is nowadays commonly used.

In the mountains and uplands, in addition to several of these problems there are other difficulties, and the choice of routes is generally more clearly defined by major relief features. The dissected plateau of central Siberia and mountainous north-eastern Siberia, immense undeveloped areas, are largely without roads or railways and

rivers remain the main routeways linked together by convenient portages. The few roads are mainly for heavy lorries and there are primitive tracks and recognised 'winter roads'. Railway building can only be justified when traffic becomes large enough to warrant the high capital cost of construction. Numerous grandiose railway plans have been proposed for northern Siberia, but it is unlikely that many will warrant realisation. Essential communications are maintained by aircraft, and airfields are cheaper to build and to operate than land transport in these remote areas. Southern Siberia is a complex mass of old mountain structures and great tilted blocks of country, with particularly rough terrain east of Lake Baykal. The Trans-Siberian Railway circuitously traverses these Baykalian lands following river valleys and suitable low passes, while the new Baykal–Amur trunkline is being built through similar rough terrain. Here are found some of the steepest gradients on Soviet railways, while the Trans-Siberian, the Kuzbas–Tayshet, and now the Baykal–Amur lines account for a substantial share of the total length of railway tunnels in the Soviet Union. There are also problems caused by patches of permafrost, especially in the wet lands of the Amur valley.

The north-south parallel ranges of the Ural, traditional divide between Europe and Asia, present no major obstacle, because their central section is low and open and there are a number of useful if restricted routes following well-defined valleys, such as that along the Chusovaya river, while the main north-south rail artery runs along the more open country on the piedmont edge of the eastern Trans-Uralian peneplain. In southwest European Russia, the Carpathian mountains have convenient passes of great historical and strategic significance allowing easy access to the Hungarian plains. Modern arterial roads and railways follow the central Veretskiy Pass, but the Uzhokskiy and Jablonka Passes also remain important.

The Caucasian isthmus is dominated by the alpine ranges of the Great Caucasus separated from the rugged Little Caucasus by a broad trough blocked at its western end by the granitic Suram range. Railways from the north follow the eastern and western coastal shelves into Transcaucasia, and the main east-west line from Batumi to Baku via Tbilisi takes a narrow and steeply graded route through the **Suram range**.

Steep gradients are also found on the Tbilisi–Leninakan line, while on some branch lines reaching into the Great Caucasus, mainly to serve mountain health resorts, gradients are so steep that it is necessary to have rack working, adhesion being gained by gearing a cogwheel driven on an axle of the locomotive to a toothed rack set between the rails. The mountain tribes of the Great Caucasus were not finally subjugated by the Tsars until the latter part of the nineteenth century, and their military campaigns led to the building of some excitingly engineered roads across the mountains, notably the magnificent Georgian, Osetian and Sukhumi Military Highways.

In the mountains of Central Asia ancient routes linking China to Europe through the oases of the Silk Roads still exist, though some are little better than camel and mule tracks and in places follow gorge-like valleys along narrow ledges or even wooden catwalks, crossing torrents on simple but efficient wooden suspension bridges. The few railways found in these mountains usually follow well-defined valleys to reach large and fertile intra-montane basins like the Fergana valley, but some modern roads have been engineered into the high and rugged Pamir territory of Gorno–Badakhshan. One ancient route whose strategic importance remains is through the Dzungarian Gate, leading from Kazakhstan into Chinese Turkestan, followed by a road and by a part-completed project for a railway from Aktogay in the USSR to Lanchow in China.

In Siberia, the great streams which rise in the southern mountains reach gigantic size in their lower courses along the shores of the Arctic Ocean. Large ocean-going vessels can sail 720 km up the Yenisey to Igarka, where the river is still over 5 km wide, but the Lena delta has many shallow and winding arms, and on the Pacific coast the Amur estuary is closed by a shallow bar. The rivers of the central Siberian uplands cannot be easily navigated since they have ungraded profiles, with strongly flowing rapids and narrow gorges in places. The great rivers present serious obstacles across the routes of roads and railways and long bridges are common, extending not only across the river but also across the flood plain regularly inundated by spring floods. Bridges over 60 m long form two per cent of the total number but 22% of the total length: for example, the **Trans-Siberian Railway** crosses the Yenisey on a bridge 854 m long, while the bridge across the Amur at Komsomolsk is over 1.5 km long.

Vehicle ferries for cars and lorries have long been common on most rivers where traffic has not warranted building a bridge and many settlements grew up round such a ferry. In winter, a crude roadway is laid across the ice. There has also been the use of ferries for railway traffic—the classic example was the short-lived ferry across Lake Baykal until completion of the Trans-Siberian Railway round the south shore. From the early 1950s a train ferry across the Strait of Kerch has provided a short cut from the Ukraine to north Caucasia, and from the late 1950s a train ferry has operated across the Caspian Sea from Baku to Krasnovodsk. Elsewhere reference is made to replacement of the train ferry across the Amur at Komsomolsk by a bridge and the introduction of a train ferry from Vanino to Sakhalin; a later addition has been the Ilyichevsk–Varna train ferry linking the Soviet Union to Bulgaria. Change-of-gauge wagons can also enter the Soviet Union using the train ferry from Travemünde in West Germany to Hangö in Finland.

TRAFFIC AND OPERATION OF TRANSPORT

It is evident from Tables 11.1 and 11.2 that in the percentage share of total national traffic—i.e. the *effort* of carrying goods or people—the railways predominate even though their overall share has tended to fall.* As a goods haulier, the railway fits Russian conditions well, because it is able to handle effectively and cheaply bulk freights (coal, ore, etc.) with reliability over the long distances necessary in the Soviet Union, reflected in an average haul of coal of 676 km in the Soviet Union compared to 80 km in the United Kingdom. The largest proportion of railway passengers is in commuting traffic, but this has been increasingly eroded by expanding bus services, while long distance passenger traffic has likewise

* Traffic is the effort of carrying a given volume of goods or passengers over a stated distance, e.g. one tonne-kilometre represents the effort of moving one tonne of goods over one kilometre distance. Originating tonnage or passengers represent the volume of goods or people to be moved irrespective of the distance they have to be carried.

TABLE 11.1 PERCENTAGE SHARE OF THE PRIME HAULIERS IN TOTAL NATIONAL TRANSPORT, 1913–1978

	Percentage share of total for the country				
	1913*	1940	1950	1960	1978
Goods Transport					
Railways					
Traffic	60.4	85.2	84.4	79.7	57.7
Tonnage Originating	72.3	38.0	29.5	17.5	13.4
Roads					
Traffic	0.1	1.8	2.7	5.2	6.6
Tonnage originating	4.6	55.0	65.6	78.7	81.8
Shipping					
Traffic	16.3	4.9	5.5	6.9	13.9
Tonnage originating	6.8	1.9	1.1	0.7	0.8
Waterways					
Traffic	22.9	7.4	6.5	5.3	4.1
Tonnage originating	16.1	4.6	3.3	1.9	1.9
Airways					
Traffic	—	. . .†	0.02	0.02	0.05
Tonnage originating	—	. . .†	. . .†	. . .†	. . .†
Pipelines					
Traffic	0.1	0.7	0.7	2.7	17.6
Tonnage originating	0.2	0.5	0.5	1.2	2.1
Passenger Transport					
Railways					
Traffic	93.2	92.4	89.5	68.5	39.5
Passengers originating	94.4	66.5	51.1	14.5	5.9
Roads					
Traffic	. . .†	3.2	5.4	24.6	42.9
Passengers originating	. . .†	29.5	46.4	84.3	93.7
Shipping					
Traffic	3.0	0.8	1.2	0.5	0.3
Passengers originating	1.4	0.4	0.3	0.1	0.1
Waterways					
Traffic	4.0	3.5	2.7	1.7	0.7
Passengers originating	4.2	3.6	2.2	0.8	0.2
Airways					
Traffic	—	0.1	1.2	4.7	16.6
Passengers originating	—	. . .†	. . .†	0.1	0.2

* In contemporary boundaries   † Share too small to allocate.
Source: Narodnoye khozyaystvo SSSR v 1978.

fallen victim to air competition. There is reason to believe that the importance of railways has been artificially maintained by a strong railway lobby in Soviet government circles.

Compared to the Western world, road transport long made a poor showing in the Soviet Union, but growth in the 1970s was dramatic as big investment programmes in production facilities for vehicles and in roads began to pay off. Nevertheless, despite construction of such large plants as that for light vehicles at Tolyatti on the Volga and the Kama lorry plant at Naberezhnyye

TABLE 11.2 SOVIET FREIGHT TRANSPORT PERFORMANCE
(milliard tonne-kilometres)

	1928	1940	1960	1970	1980	1986
Rail	93.4	420.7	1504.3	2494.7	3439.9	3834.5
Sea	9.3	24.9	131.5	656.1	848.2	969.7
Inland waterway	15.9	36.1	99.6	174.0	244.9	255.6
Pipeline-oil	0.7	3.8	51.2	281.7	1216.0	1401.3 (O)
Pipeline-gas	—	—	12.6	131.4	596.9	1240.0 (G)
Road	0.2	8.9	98.5	220.8	432.3	488.5
Air	—	0.02	0.6	1.9	3.1	3.4
TOTAL	119.5	494.4	1898.3	3960.6	6781.1	8193.0

Source: Narodnoye khozyaystvo SSSR, various years.

Chelny, the USSR still produces less than twenty per cent of the world's motor vehicles. Private car ownership, though growing vigorously, also lags well behind the industrial countries of the West, but the length of hard-surfaced road in the Soviet Union still remains at less than a tenth that in the United States which has a smaller area. Long distance road transport is little developed because of the poor, inadequate road network and because it cannot compete with railways over the distances required, besides being more likely to suffer seasonal disruption. Nevertheless, the transfer of all hauls of less than 50 km to roads from the railways is being undertaken. Table 11.1 reveals, however, that the greatest proportion of the total *originating* tonnage and passengers is handled by road transport (including trams and trolley buses), reflecting the predominantly short hauls involved, especially in towns or as feeder movements to railways in the countryside.

A trolleybus line: Simferopol to Yalta, Crimea

The most striking change has been in traffic handled by waterways, the traditional carrier, whose proportion of total freight traffic has fallen from nearly a quarter in 1913 to a little under 5% currently, despite an increase in the originating tonnage and in traffic. An important factor has been the predominantly north-south alignment of the rivers while the present day major traffic flows are mostly on an east-west axis. The river is, of course, a cheap bulk carrier and should compete readily with railways, but unfortunately the long winter and spring period of disrupted river traffic and the summer low water period compare unfavourably with the reliability of the railway for 'moving belt' delivery and a much greater choice of route mobility. River systems are usually self-

contained and canal construction to join systems together offers relatively limited possibilities and could be done only at great cost. Regionally, however, rivers do remain important, as, for example, the Volga, the Lena and the Yenisey, the latter two serving as the prime arteries of movement for large areas of Siberia.

Apart from some restricted waters which suffer little or no ice, sea transport is also markedly seasonal, with the interruption varying from a few days to many months. Because Soviet seas are separated from each other by long coastlines under foreign control, coasting traffic between them is negligible, but even within the individual seas it is poorly developed, except the Black Sea

TABLE 11.3 TONNAGE OF MAIN COMMODITIES MOVED BY PRINCIPAL MEDIA (million tonnes)

	Rail	Sea	1970 I.W.	Road	Rail	Sea	1986 I.W.	Road
TOTAL	2,896.0	161.9	357.8	14,622.8[a]	4,077.6	249.5	648.7	26,984.8
				3,810.0[b]				6,653.0
Coal	647.2	9.3	17.6	77.0	781.3	9.5	20.5	
Petroleum	302.8	75.1[1]	33.5	31.2	419.7	102.0[1]	39.5	
Metals	141.6	6.7	2.0	109.6	210.0	16.6	4.6	
Timber	178.8	11.0[2]	91.2[4]	55.0	163.6	11.2[2]	50.8[4]	
Ores	245.6	13.6	—	13.0	334.7	17.1	—	—
Building Materials								
(mineral)	691.0	15.3	180.9	1,322.9	1,113.8	21.3	455.0	
Fertilisers	70.9	5.3[3]	—	5.9[3]	154.2	11.2[3]	—	—
Grain	106.1	6.5	6.8	116.3	145.0	20.2	6.8	

a All branches incl. farm transport   I. W. Inland waterways
[1] incl. other liquid cargoes
[2] incl. 0.4 m. tonnes by raft

b Public Service transport [p343]
[3] incl. other chemicals
[4] incl. 20–25 per cent floating (raft etc.).
Sources: Narodnoye khozyaystvo SSSR, various years: Transport i suyaź SSSR (1972).

and the Sea of Azov. The growing trading relations with the world at large are increasing the importance of sea transport and account primarily for the rise in its share of total traffic. The Soviet merchant fleet has grown rapidly since 1945 and comprises mostly excellent modern vessels, many built in other countries. To conserve foreign currency, the Soviet Union has tried to carry as many of its cargoes as possible in its own vessels or those of East European socialist countries like Poland and the German Democratic Republic. Soviet *bloc* shipping has also taken an aggressive policy to get into Western markets for shipping services.

The growth of air transport has been particularly striking and the Soviet Union, with its great physical difficulties and vast distances, has become an unusually air-minded country. Not only is air travel over such great distances immensely quicker than land travel, especially in territories not served by railways, but it is also in many instances cheaper than first class rail travel. Aircraft are ideally suited to journeys across immense Siberian lands lacking roads and railways, so that air transport has been recognised as a key to the economic development of these remoter areas, where the helicopter has begun to play a major role. If the Soviet Union were to offer positive encouragement for foreign airlines to operate across its territory which lies astride many excellent great circle routes from Europe to the Far East and Australasia, or from western North America to India and Africa, the role of air transport could increase very greatly. International connections are operated principally via Moscow by Aeroflot and foreign airlines, as for example, between Europe and Pakistan, India and the Far East across Central Asia or Siberia, while connections exist at Moscow between these services and North America and Africa.

Pipelines, a cheap means of moving gaseous and liquid freights, also reduce the burden on other forms of transport. It is estimated that the Friendship Pipeline from the Ural–Volga oilfields to the East European socialist countries has released some 10 000 20-tonne tank cars on the railways, while pipeline transport costs one third of railway traffic. The development of both the large Ural–Volga oilfields and the immense oil and gas deposits of western Siberia has resulted in the laying of long pipelines to refinery and consuming centres, just as long gas lines have been laid from Central Asia to the Ural industrial towns and from northern Caucasia to the

Moscow region and to Leningrad, as described in Chapter 8. Even in 1913 there were 1100 km of pipeline in Russia; by 1946, the length reached 4400 km and by 1966, 29 400 km to which could also be added 47 000 km of gaslines. By 1986 there were 81 500 km of oil pipelines and 185 000 km of gaslines.

In the Soviet Union, the share of the different means of transport in the total traffic is conditioned not only by their economic and technological suitability to handle particular traffics but also by the allocation of tasks and of resources determined by central planning policy, with special attention given to the eradication of wasteful crossflows or duplicated movements of the same or interchangeable goods. The planners have also aimed to minimise the demand on the limited transport capacity available by reducing as far as possible the contrasts in levels of development between planning regions. Wide differences still remain, however, and will continue to do so because of the widely variant potential of the different regions. Density of the transport network in any region is, therefore, a crude index of the degree of current development. It is not surprising that the densest transport network is in European Russia, south of Leningrad and west of the Volga, where on one sixth of the country's area live two-thirds of the Soviet population, the core of the so-called 'settled triangle' whose base lies between the Baltic and the Black Sea and whose apex rests in western Siberia.

In recent years Soviet literature on transport has been concerned to emphasise that within the Soviet Union there is a 'unified system of transport', where the role of each mode of transport is fully co-ordinated to eliminate wasteful effort. The emphasis is placed on transport as an integral constituent of the economic infrastructure, a vital link in the chain of production though itself not regarded as 'productive'. It is claimed that the systematic and proportional development of the Soviet economy conditions the rational development of transport, both in terms of inter-media relations and in the regional provision of a transport system. This system is regarded as 'unified' because each mode of transport is combined into a structure where it performs the tasks it can best undertake and it is not in unnecessary competition with other media. As noted, the effort is directed at achieving the requirements for movement of goods and people by the most economical

investment in the transport infrastructure: it is essentially a 'minimum input–maximum output' equation. Each movement is regarded as a specific flow from origin to destination for which the most economical 'media mix' is set up in relation to parameters such as distance, volume, frequency and orientation. All movement is classified on a scale ranging from 'local' to 'interregional', 'national or trunk' and 'international'.

We must consequently look at the overall transport map in this light. The basic framework is composed of railways, though in the remoter regions of Siberia and Central Asia, lorry roads and rivers form a continuation to these routes, and in places, they are also continued by shipping on the peripheral seas, as in the Black Sea, the Baltic, the Caspian, the Arctic basin and the Pacific coast. In the more thickly settled areas, particularly within the 'settled triangle' already described, this basic framework is augmented by an interstitial transport system of feeder services, mostly provided by road transport but including some branch railways and even rivers. Airways may be seen as superimposed on this framework of surface transport for special tasks, though in particularly remote regions (for example, in the Arctic and northern Siberia), air transport essentially provides a continuation of the main framework.

The railways
Though the first railway ran from Pushkin to Leningrad in 1837, it was not until completion of the Leningrad–Moscow mainline in 1851 and the Leningrad–Warsaw railway in 1861 (later extended to Vienna) that development began in earnest. These were quickly followed by lines built to replace old portages linking river systems and, during the latter 1860s, by lines from the grainlands of the Ukraine to carry grain to Baltic and Black Sea ports or to the Moscow region, where several short branches also served the central black-earth lands. By 1872 the Volga had been reached, after which railways began to appear in the Ural in the latter 1870s. In 1892, construction of the Trans-Siberian Railway from Chelyabinsk started: it had reached Krasnoyarsk five years later and Irkutsk in 1898. In 1905, the train ferry across Lake Baykal was replaced by a through railway route along the precipitous southern shore, but railway communication entirely across Russian territory to Vladivostok was not opened

until 1916 with completion of the Kuenga–Khabarovsk section, following the loss of control over the Chinese Eastern Railway across Manchuria via Mukden.

The last 20 years of the nineteenth century saw a great burst of railway building: there was rapid growth of the railway system in the industrial south, focused on the Donbas where in 1884 a vital line from the Donbas coalmines to the iron ore mines of Krivoy Rog had been completed. Between 1881 and 1899, the Trans-Caspian Railway from Krasnovodsk to Tashkent had been built in Central Asia, primarily for military reasons in the conquest of this territory. A link from this area to the Volga was established in 1905 by the opening of the Trans–Aral Railway, able to carry Central Asian cotton to the growing textile industry of the Moscow region. This was also a period of railway building in Transcaucasia, but these railways were not linked to the rest of the system in Europe until 1900. In 1913, 80% of the 70 000 km of route was in European Russia.

Unlike the rest of Europe and North America, where building trunk routes had virtually ended by 1914, trunk line construction continues in the Soviet Union and since the Revolution the Turksib (Turkestan–Siberian), Yuzhsib (South Siberian) and the long Sredsib (Central Siberian) railways have been completed and steady progress has been made on the greatest project of all, the new Baykal–Amur trunkline. In the 50 years

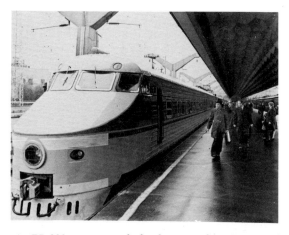

An ER 200 express ready for departure from Leningrad to Moscow on the USSR's busiest line.

of Soviet power, the length of railways has almost doubled, with the bulk of new building in the eastern regions, especially in western Siberia and Kazakhstan.

THE BAYKAL–AMUR TRUNKLINE

Indicative of the major part still played by railways in the Soviet transport scene is the fact that the central task begun in the tenth Five Year Plan has been to build a trunk railway some 3200 km long between the Lena river and the Amur at Komsomolsk on a trajectory some 150–300 km north of the existing Trans-Siberian Railway by 1983. The concept of this line dates back to the years when Count Witte was planning the original Trans-Siberian Railway, because at this period American interests were trying to wrest a concession from the Tsar for a railway along roughly the same northern alignment as the present project. This so-called Siberian–Alaskan Railway would have, however, branched in the Bureya basin, one route going south to Khabarovsk and the other north to the Amur delta at Nikolayevsk and then along the coast to the Chukotsk Peninsula.

The idea was again revived in the 1920s and an actual survey made and work began on branch lines north from the Trans-Siberian Railway to give access to points along the proposed trajectory from which constructional work would begin. By the mid-1930s the Never–Tynda, Izvestkovaya–Urgal and Volochayevka–Komsomolsk lines were being built, but work halted in 1941 and the Never–Tynda line was dismantled and the rails sent to build strategic railways near Stalingrad. The whole project had been revealed in the third Five Year Plan of 1938, where the details of the so-called Baykal–Amur Trunkline to duplicate the Trans-Siberian Railway were set out. Priority was given to the western section from Tayshet on the Trans-Siberian to the River Lena at Ust-Kut, for the railway was needed to carry constructional materials for the great hydro-electric barrage at Bratsk on which work was to begin, but the threat which the Japanese posed to the Amur section of the Trans-Siberian from their hold on Manchuria gave an added urgency to this new trunkline.

Little subsequent information about the project was released by the Russians and Western observers assumed work was continuing in a low key, using prisoner-of-war or forced labour. This was partly true: in 1945, news came of completion of a railway from Pivan opposite Komsomolsk across the Sikhote Alin to a new port, Sovetskaya Gavan, on the Tatar Strait. In 1947, it was announced that the line from Tayshet to Bratsk was complete and by 1950 the Lena river had been reached, while by the mid-1950s the line from Izvestkovaya to the Bureya coalfield near Urgal was ready.

The silence about work on the rest of the railway now seems to have been because of major revision of its trajectory, especially between the Lena and Lake Baykal and on the Vitim—Laba section, where increased knowledge of the topography, physical geography and resources wealth called for new surveys. The resurvey seems to have been completed about 1961 and resulted in a southward shift of the alignment compared to the pre-war original. Some of the change appears to arise from more advanced technology and better constructional methods now available that make tunnel-building more acceptable, so allowing distances to be shortened by avoiding circuitous valley routes and by cutting directly through mountains instead of finding trajectories with acceptable gradients across them. The basic plan remained, however, for a single track route, though this was now to be built for eventual double track electrification. Use of diesel instead of the original steam traction also allowed changes in the original plans for operating sections and for maximum gradients.

In 1973, the main project was reactivated with reconstruction of the line to Tynda on an improved trajectory from the Trans-Siberian Railway, while the train ferry across the Amur from Komsomolsk to Pivan was replaced by a 1.5 km-long bridge. As lines like that from Abakan to Tayshet and from Khrebtovaya to Ust–Ilimsk were completed, the work teams were transferred to the Baykal–Amur trunk route. Work on the easternmost sector from Urgal to Komsomolsk is already well advanced and the section from Ust–Kut to Tayura (site of a bridge across the Lena) and some way beyond is complete, though the route from the Lena to Nizhne-angarsk is one of the most difficult sections, requiring a bridge across the Lena and a trajectory through the Baykal mountains. Since 1971 work has been under way on several major civil engineering structures notably a 15 km-long tunnel under the North Muy range and a rather

shorter one through the Baykal mountains, though it now appears that preliminary work on these had begun in the late 1950s. To cope with rising traffic supplying the construction gangs at the western end, the Tayshet–Ust-Kut line has been double tracked and electrified.

The task of building this railway is immense, for it runs through virgin tayga and across large swamps, besides having to cross seven mountain ranges with some exceeding 3000 m in elevation. About 40% of the route is across ground affected by permafrost, while between Lake Baykal and the Olekma river seismic problems have to be overcome. Constructional and operational difficulties arise from the climate, with temperatures below −50°C in winter and over 35°C in summer. In such empty country, all constructional workers have to be brought in and housed in special camps, so that provision of the necessary infrastructure (e.g. living accommodation, medical, welfare and educational facilities as well as roads) has to be completed before major work on the railway itself can begin. Some survey and work teams in the remotest parts are entirely dependent on air support (mostly by helicopter). Work has been pressed into the tayga from the railheads at Ust-Kut, Tynda, Urgal and Komsomolsk, but the underdevelopment of the territory through which the line is to pass is such that a main settlement is planned approximately every 50 km, with stations and passing places at 20 km intervals. The double track passing places will be between 3 km and 12 km in length, to make passing and overtaking at speed possible. All traffic will be centrally controlled, with radio communication used on trains and at stations. Because of the extreme climatic conditions, special diesel and electric locomotives will be needed.

Even now the extension of this line westwards from Bratsk to the foot of the Ural mountains is being discussed, while the bridge across the Amur at Komsomolsk and a train ferry from Vanino near Sovetskaya Gavan to Kholmsk on Sakhalin Island (whose railway system has been expanded by completion of a north-south trunk route) suggest a new interest in the Far East.

RAILWAY OPERATIONS

Operating conditions on Soviet railways are eased by the generally low gradients prevailing, so that three-quarters of the route length has gradients gentler than 1:166 while less than one-fiftieth of the route length has gradients steeper than 1:60; some three-quarters of the route is straight track. Over large areas of the plains, the fact that Russian railways have been constructed as cheaply as possible has had little adverse effect on gradients or the radius of curves, but in rougher terrain such cheapness of original construction has made it impossible to avoid severe gradients and sharp sinuous curves. This makes long-term operating costs higher than if more had been spent initially on civil engineering works and keeps down permissible speeds and axle loads and consequently the density of traffic.

Rapid growth in traffic in the 1950s and 1960s had demanded heavier, faster and more frequent trains without the expensive provision of alternative routes or doubling of track, except in special cases, so that faster and more powerful motive power was seen as an important clue to the solution of the problem. The steam locomotive at this time reigned supreme (Table 11.4) except for a few routes where special circumstances prevailed, because its operating reliability, simple maintenance and its initial cheapness to build, besides the ready availability of fuel, more than offset its low thermal efficiency and water supply problems in arid or permafrost areas or even its limited operating radius. One solution was obviously more powerful steam locomotives, but if these had rigid frames it meant much heavier axle-loads and greater radius curves (including points and station trackwork) which demanded strengthening the track and civil engineering works, using better ballast and heavier rails. Time and cost and the added demand for steel to carry

out such work eliminated such a solution, because even the need to lengthen passing places and sidings for longer trains and to provide additional passing places for faster and more frequent services imposed sufficient strain on constructional work and rail production. Existing track standards could have been maintained by using articulated steam locomotives, but these were disproportionately more costly to build and more complicated to maintain, while their power/weight ratio was not compensatingly better.

A strikingly better power/weight ratio could only be obtained using electric or diesel traction. A rapid improvement in the supply of petroleum, as output in the Ural–Volga oilfields expanded, encouraged the use of diesel locomotives, either as an interim measure on routes marked for electrification or as a replacement for steam on lightly loaded routes where electrification was not considered economic. Diesel locomotives had been used in the early 1930s in the oil-producing areas of Trans–Caspia, but they found little favour until after 1945 when Russian-built locomotives were developed from imported American proto-types. These locomotives have demonstrated a better thermal efficiency in cold weather and make possible a 90% water saving (important in water shortage areas) compared to steam locomotives.

Blessed by Lenin, electrification has always had substantial support in the Soviet Union, so that electric traction was planned for main trunk routes where the high cost of installation could be quickly recovered. It has, however, been dependent on current generating capacity available along the selected routes, and the several large power stations under construction in southern Siberia favoured the choice of the Trans-Siberian route as the largest single project envisaged, with the optimal conditions of very heavy traffic and available current supply. Similar factors have underlain electrification of arterial routes in European Russia, and this has also been applied on commuting lines around some of the largest towns. The Soviet authorities have favoured electrification at 25 KV, because this system allows industrial current to be used without costly substations and is also economical in copper for the overhead wires (copper has been in relatively short supply in the Soviet bloc). Electrified sections still account, however, for only 30% of the

TABLE 11.4 PROPORTION OF FREIGHT TRAFFIC HANDLED BY DIFFERENT TYPES OF RAILWAY TRACTION

Year	All forms of traction	Electric	Diesel	Steam
1940	100	2.0	0.2	97.8
1950	100	3.2	2.2	94.6
1960	100	21.8	21.4	56.8
1966	100	42.0	46.8	11.2
1975	100	51.2	48.2	0.6

Source: Nikolskiy, I. V. *Geografiya transporta,* Moscow 1979.

total route, but they handle, nevertheless, 52% of all freight traffic.

Soviet railway operation is a mixture of American and European practice, with a rising frequency of heavier and faster trains, but as is characteristic of most railway systems, traffic density is very unevenly spread over the network. Some 86% of all freight traffic is carried by 46% of the total route length, while the 28% of route length which is double track handles no less than 67% of all freight traffic. Loadings of some sections are particularly heavy and the average freight traffic density on Soviet railways is among the world's highest.

Tables 11.3 and 11.5 reflect the predominance of heavy bulk goods on the railways, but, as might be expected, the share of coal in the traffic has fallen and its place has been taken by petroleum, while compared with pre-Soviet times, grain has also declined, though there has been an increase in mineral building materials and chemical fertilisers. Despite attempts to keep the length of haul down, there has been a slow upward trend, with some goods moving very great distances (e.g. cotton moving over 3000 km) but all hauls are exceptionally long by western European standards. Careful planning of freight movements on such hauls is important because of the great amount of empty running which may result if return freights are not available, though some wagons (e.g. tank cars) necessarily have to be allowed to run empty on return. Originating tonnages show some remarkable concentrations at a

few goods yards or railway directorates on the system: reputedly half the coal loadings are by the railway directorates of the Donets and Western Siberia (Kuzbas), while 20% of all ferrous metal semi-finished goods are shipped from three yards (Magnitogorsk, Sartan in the Donbas and Zaporozhye), and 55% of all iron ore shipments come from the Dnepr Directorate.

Passenger traffic, (Tables 11.6 and 11.7) regarded as less important, has risen by little over ten times since 1913 compared to a 50 fold increase in freight traffic. A third of the passenger traffic is in commuting and over 90% of the originating passengers are commuters, with an average journey of 20 km. The average non-commuting journey is 598 km. The Moscow commuting sphere, served by 1200 pairs of trains a day, extends up to 167 km (e.g. to Kalinin) and handles well over 750 000 travellers in each direction daily. Passenger service frequencies otherwise are generally poor. On secondary lines there is usually one train a day (often a mixed train) in each direction. On trunk routes there may be between seven and twenty trains in each direction daily, though only on selected sections do the best trains exceed 80 km/hr. Moscow, with nine terminal stations, dispatches 200 long distance trains daily with about 150 000 passengers in the summer season, a much busier season for travellers than winter. Short distance railway journeys are being eroded by buses, while very long distance trips are being taken over by air competition. A problem of long distance trains is that one train in

TABLE 11.5 GOODS TRAFFIC ON RAILWAYS BY TYPES OF FREIGHT

	Total traffic (milliard tonne-km.)	Coal and coke %	Petroleum goods %	Ferrous metals %	Wood %	Grain %	Ores %	Mineral building materials %	Mineral fertiliser %
1913	65.7	19	5	—	8	15	—	—	—
1928	93.4	20	7	5	12	16	3	—	—
1940	415.0	26	9	6	11	8	5	7	1
1950	602.3	29	9	8	12	5	5	8	1
1960	1504.3	22	14	7	14	6	5	10	2
1970	2494.7	17	14	8	12	4	7	12	3
1980	3439.9	18	13	8	7	4	7	13	4
1986	3834.5	17	12	8	7	4	6	15	4
1987	3824.7	18	12	8	7	4	6	15	4

Source: Narodnoye khozyaystvo SSSR, various years.

TABLE 11.6 SOVIET PASSENGER
TRANSPORT PERFORMANCE
(milliard passenger-kilometres)

	1928	1940	1960	1970	1980	1987
Rail	24.5	100.4	176.0	273.5	342.2	402.2
Sea	0.3	0.9	1.3	1.6	2.5	2.3
Inland Waterway	2.1	3.8	4.3	5.4	6.1	5.7
Bus, etc.	0.2	3.4	61.0	202.5	389.8	470.61
Air	0.0	0.2	12.1	78.2	160.6	204.2
TOTAL	27.1	108.7	254.7	561.2	901.2	1085.0

TABLE 11.7 NUMBER OF PASSENGER
JOURNEYS (millions)

	Rail	Inland waterway	Bus, etc.	Sea
1940	1377	73.4	590.0	9.7
1960	2231	119	11 315.6	26.7
1970	3354	145.2	27 343.8	38.5
1980	4072	138.0	42 175.0	51.7
1987	4360	128.0	49 983.0	50.0

Source: Narodnoye khozyaystvo SSSR various dates.

each direction daily may need as many as 20 or more sets of carriages to operate the service. The through carriages from Adler in the Caucasus to Vladivostok take 204 hours for the journey!

A dominant T-shaped distribution is seen in the principal flows of both goods and passenger traffic (Figs. 11.3 and 11.4). There is a main north-south artery, with interchange between the industrial South (chiefly the Donbas) and the North-West (mostly from Leningrad and environs), passing through the Central Industrial region at the western end (chiefly around Moscow and its satellites) of the major east-west artery. The main east-west artery from the Moscow region joins the Volga lands to the Ural and Siberia. The bulk of the remainder of the system acts as a feeder to this T-shaped arterial core. Railway traffic between the Soviet Union and its neighbours, with the exception of Finland, has been hampered by a difference in gauge, with 1524 mm in Russia* and elsewhere 1435 mm, European

* Since January 1972, this 'traditional' guage has been modified to 1520 mm, a simpler metric measure.

standard gauge; but since 1945 efficient methods of changing the gauge on wagons and carriages have been developed. There are now through passenger workings from Moscow to several European capitals and these are also possible to Peking and Pyongyang. Increasing shipments of coal and ores to Poland, Hungary, Czechoslovakia and the German Democratic Republic have been handled by large Soviet hopper wagons run on to overhead gantries at frontier points (e.g. Chop, Przemysl and Terespol) to empty into European standard gauge wagons, but the large demand of the new East Slovakian Ironworks at Košice for ore and coal has been solved by building an 80-km long broad-gauge railway across the Czechoslovak border to the plant. Soviet railways have been active in promoting use of containers and a regular service of container trains across the Trans-Siberian Railway now handles traffic between Japan and Europe. Highly competitive in cost and delivery time with sea transport, the main constraint is the capacity of terminal facilities and trains to cope with rising demand. To speed railway freight in transit from the Soviet Union, direct train-ferry routes have been opened across the Black Sea from Ilyichevsk in the Ukraine to Varna in Bulgaria, and across the Baltic from Lithuanian Klaypeda to the Island of Rügen in East Germany. Some Soviet traffic also uses a train-ferry between Finland and West Germany. In the Far East, a train-ferry also joins the mainland to the railways of Sakhalin across the Strait of Tartary.

The roads

The earliest 'roads' in Russia were vaguely defined tracks (trakty) that shifted across broad avenues of country as sections became impassable because of mud, dust or rutting. The first ballasted road did not appear until 1817 while no true road linked Moscow and Leningrad until 1834. Traditionally, roads served only as feeders to the rivers or across portages. Even today, motorable roads do not penetrate into every corner and do not even link the country right across from west to east. Road transport is most important in and around towns and in the less developed parts of the country like Siberia or Central Asia, where railways have not yet been built and where roads are frequently described as 'routes without rails'. There are still only 31 km of hard surfaced road per 1000 km^2. This varies between republics,

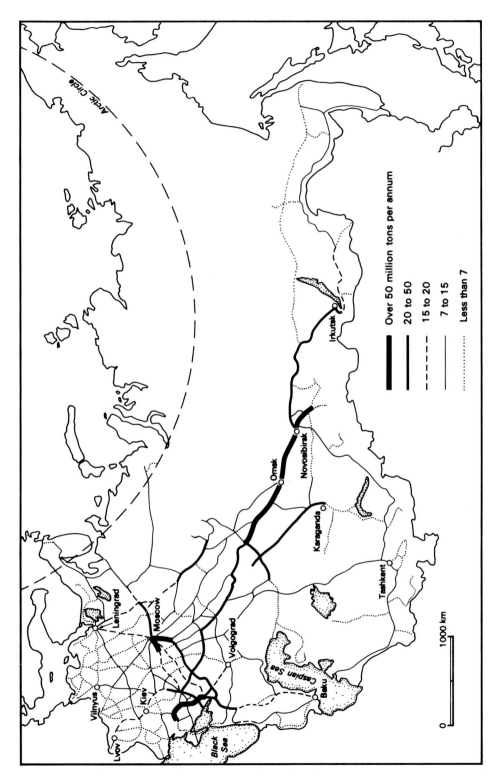

Fig. 11.3 Dominant flows of railway freight traffic, annual figures based on Soviet sources relating to period 1960–65, no later figures having been traced

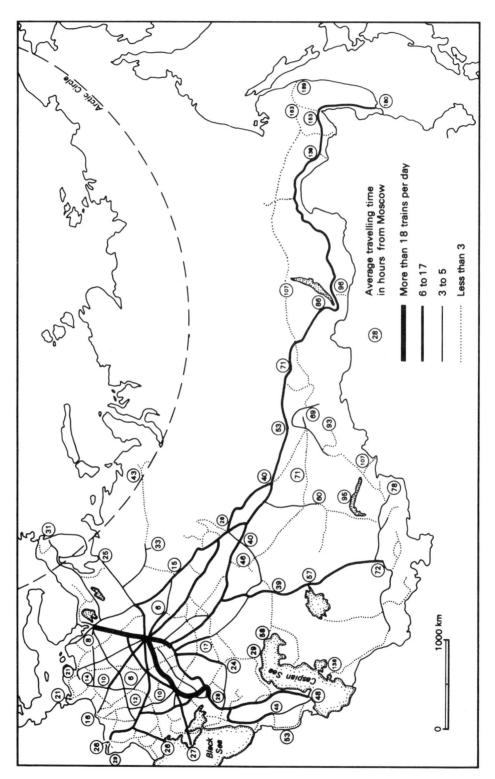

Fig. 11.4 Railway passenger services and isochrones centred on Moscow

A busy section of the Baykal-Amur mainline near Zheleznogorsky-Ilimsky in winter conditions typical of the Siberian mountains.

A public service bus in the mountains, Tadzhikistan

the swamps 'winter roads' are roughly staked out across what in summer is impassable morass. Unembanked grit or earth roads turn into quagmires in the *rasputitsa,* as the German motorised columns found to their cost in the campaigns of 1941–3. Snow and ice add to winter hazards, while in arid areas there are hazards from gullying, washouts and drifting sand, and roads are often lined for great distances by shelter belts.

The country west of the Volga contains a widely meshed but expanding system of roads designed for comparatively fast traffic over long distances and used by inter-city lorries and buses, but once away from these roads, most country roads are merely tracks. East of the Volga, the mesh of roads is much wider than west of the river and few high quality routes exist. In Siberia, a reasonable system exists in the south, notably along the railway zone, but even here some local systems are interlinked only by a single road while in eastern Transbaykalia and Amuria, the local systems are not even interlinked and movement between them depends on the railway. A better developed system exists in the Maritime Kray of the Far East, radiating from Vladivostok, and the Ussuri Highway runs north to Khabarovsk. In

however, from 14 km per 1000 km² in the Turkmen SSR to 534 km per 1000 km² in the Estonian SSR. Choice of surface is, however, difficult since intense winter frost and bad drainage over large parts of the country are peculiarly damaging: preference remains for grit rather than for tar or cement, though dust is a problem in summer. In the northern forest, 'corduroy' roads made of tree trunks covered by grit are laid and in north-eastern Siberia, a road runs from Magadan to the mines of the upper Kolyma and to Yakutsk, from which a good road (the Aldan Highway) runs to the Trans-Siberian Railway at Never. Regular if infrequent bus services operate on these roads in Siberia. In Central Asia and in Caucasia are numerous interlinked local systems carrying bus and lorry traffic and reference has already been made to the well-engineered mountain highways in these areas.

In both town and country, traffic is dominated by buses and lorries, and light motor cars are mostly taxis, but trams and trolley buses operate in many towns. One of the greatest social pressures on government is to ease ownership of private cars, which still lags behind the West despite increased manufacture. The local nature of road haulage is reflected in an average goods haul of 14 km and an average bus journey of under 7 km. Of the goods originating, building materials and the collection and distribution of goods to and from railway stations and river quays are the main items. Horse carts are still seen in provincial towns and the countryside, while camels are used in Central Asia, and in the tundra

and forest lands of Siberia, reindeer and dog teams remain a common means of native transport.

The waterways

From the earliest times rivers were used in the Russian lands: Scandinavian contact with Byzantium was first established via the Volkhov, Lovat and Dnepr rivers—in Nestor's words, the 'route from the Varangians to the Greeks'—while the later conquest of Siberia similarly moved along one river system and across a convenient portage to the next. Several of the great historical towns of Russia (e.g. Smolensk) owe their importance to being astride portages. Until the coming of the railways, rivers like the Dnepr and the Volga carried grain from the lands in the south producing surpluses to the food-deficit areas of the north: the Volga boatmen's song records the harshness of getting the heavily laden boats upstream. To link the systems together, a few simple canals were built in the late eighteenth and early nineteenth centuries, but these fell into disuse in the railway age. Some have been replaced by modern canals, where traffic has justified the great

The harbour at Yalta

cost, though these have also commonly been associated with electricity generation or irrigation as well as navigation. Examples are the Volga-Don Canal (1954), making possible voyages from the Volga and Caspian to the Sea of Azov and the Black Sea; and the Volga-Baltic canal (1964), which replaced the old Mari canal, opening up the possibility of a voyage from the Caspian via the Volga to the Baltic. In the interwar years, a series of short canals were built to join together the lakes of the Karelian isthmus and form a waterway from the Baltic to the White Sea, while a current plan among other schemes projects a major waterway link from the Dnepr via the Pripyat and Bug to the Vistula.

Improvements in navigation and large hydro-technical schemes sometimes create undesirable side-effects. Even modest dredging or straightening out meanders can affect the local water table, but large schemes usually include barrages that inundate extensive areas which have to be cleared of settlement or even forest. By slowing the flow of water, such schemes can mean longer periods with ice, or, by altering the thermal regime of the water, affect fauna and flora. It has been claimed that the vast new Volga barrages have held back so much water from the Caspian that they have helped to accelerate its naturally induced fall in water level, so that former anchorages like Prorva are now dry and a special channel has to be dredged to the Volga delta port of Astrakhan. The reduced inflow has meant less nutrients for fish, while some species spawning in the Volga have had their breeding habits upset by the change in the thermal gradient.

Two-thirds of all river-borne freight traffic is on the Volga–Kama waterways. The two principal freights are timber (including firewood) and mineral building materials (including cement), but on the Volga–Kama system, petroleum products are an important element: average length of haul is about 400 km. Despite the cheapness of river transport and its ability to handle very large quantities, official encouragement to use rivers and suitable tariff-rigging have not been strikingly successful and river transport has shown the lowest growth rate of all forms of transport. Unfortunately the long winter freeze and other delays interrupt traffic and users have to hold large stocks to tide them over or rely on available surplus railway capacity at such times, both difficult to achieve in the Soviet economy. River

passenger traffic is also under competition from faster hauliers, unless these are absent (as in parts of Siberia), but average journey distance is surprisingly short at about 44 km.

Sea transport

Sea transport plays relatively little part in the internal transport system of the Soviet Union, but a substantial increase in goods and passengers moved reflects the growing contact with the outside world. Despite the great length of the Soviet coastline, some 42 000 km, giving a frontage on several seas, political and economic conditions, quite apart from the problems of physical background, have made sea transport relatively unattractive. Traffic from one Soviet sea to another (e.g. from the Baltic to the Black Sea) has been insignificant because of the long and circuitous distances between them. The Northern Sea Route along the Arctic coast, with serious inherent physical difficulties confronting navigation, has not provided a really usable 'Soviet Canal' from western to eastern waters. Transport on both a local and an international scale has therefore been organised generally on the ports of each sea individually.

In the Black Sea—Sea of Azov, 55% of the traffic is to non-Soviet ports and 45% to Soviet ports on the same or other seas. Some 80% of the traffic is petroleum, ores or coal: Tuapse, Batumi and Novorossiysk send petroleum to Odessa, Kherson and other Ukrainian ports, while coal moves from Zhdanov to Odessa and Poti, with manganese ore shipped from Poti to Ukrainian and Danubian ports; but there are also shipments of cement and grain from northern Caucasian ports, and Transcaucasian ports send semi-tropical fruits and tea in return for grain, salt and timber from Ukrainian ports and the Don estuary. Odessa is a major port for overseas passenger journeys.

The Baltic is particularly important for its traffic to North Sea and Atlantic ports. About 40% of all coal, grain and metals exported from the Soviet Union pass through Baltic ports, as well as 30% of all timber and manufactured goods, while the Baltic handles Ural–Volga petroleum brought by pipeline to a terminal at Ventspils. Leningrad is one of the largest Soviet ports. From the Barents and White Seas are shipped large amounts of northern timber and minerals such as apatite from the Kola peninsula, but 70% of the traffic is coastwise.

The seas of the Pacific coast supply outposts of settlement in north-east Siberia and also maintain services with Sakhalin, so that 80% of the turn-over is local coastwise traffic. Timber, petroleum, fish products, salt and general goods are the main cargoes. Vladivostok, the main port, is nowadays closed to foreign vessels, which use Nakhodka, fortunately ice-free longer, and another new port has been built nearby at Vostochnyy. An oil terminal at Moskalvo on Sakhalin ships petroleum to Japan and south-east Asian buyers.

The Northern Sea Route, operated by scheduled services since 1935, is open for 70–120 days each year, though the shallow Vilkitskiy Strait may be closed throughout some summers by grounded icebergs and the Laptev Sea has very bad ice and fog conditions most summers. The route is used by about 100 ships each season, even though comparatively few pass through its whole length. Igarka and Dudinka on the Yenisey ship wood and minerals, while the eastern parts also ship minerals but most services are for victualling Arctic settlements. Attempts to get more use of the costly investment in the Northern Sea Route by opening it to foreign vessels have not been successful, because the charges for pilotage, etc., and the conditions of passage have been unattractive to foreign shippers between the Atlantic and Pacific. With the Suez Canal back in use, and quick services for containers across the Trans-Siberian Railway, this arctic route will be even less of interest. Of the traffic on the inland Caspian Sea, 85% is petroleum, but there are also movements of wood, minerals, fish and salt, and there is also some passenger traffic.

Air transport

In time and monetary saving, great distance favours air transport, so that an average haul of 970 km for passengers and 1310 km for goods reflects this advantage, though the number of passengers and the tonnage originating are both small compared to the railways or roads. Physical conditions in the Soviet Union favour flying: large flat spaces are available for airfield construction; there is a lack of serious relief obstacles; and there are long periods of anticyclonic conditions with still air and good visibility in winter, though greater turbulence and poorer visibility characterise summer.

Since the first scheduled civil airline opened in 1923 from Moscow to Gorkiy, state policy has encouraged air-mindedness. The state airline, *Aeroflot,* one of the world's largest operators, has a virtual monopoly of both internal services and Soviet international traffic, but there are a few specialised operators, like the Chief Administration of the Northern Sea Route. In 1986 1 156 000 km of route were flown (972 000 km within Soviet borders), moving over 116 million passengers and 3.4 million tonnes of goods and mail. Besides the main trunk routes flown by jet airliners, there is an elaborate system of secondary routes served by more modest machines, and aircraft are widely used in maintaining contact with Arctic stations, winter ice observations at sea, forest fire surveillance, agricultural work (as noted in Chapter 7), helping reindeer herders to find pasture and muster their herds, while aircraft are reputedly even used in wolf-hunting! The principal international traffic centre is Moscow, which has four airports, but other important internal centres are the Union Republic capitals and such cities as Novosibirsk, Sverdlovsk, Irkutsk, Khabarovsk, Kharkov, Kuybyshev and Volgograd. Equipment is supplied by a well established Soviet aircraft industry as discussed on page 172.

A study of transport is a useful way of appreciating the problems inherent in the vastness of the USSR, for distance remains one of the most difficult problems to surmount in welding the huge and differing major economic regions of the country into an effective economic and political unit spread across continental dimensions.

In making comparisons between the Soviet Union and other parts of the world, care must be taken to make allowance for these dimensional contrasts and the harshness of the physical environment. It would be false to think, for instance, that continued predominance of the railway indicates technical backwardness—it has been shown that for physical and technical reasons neither road nor river transport can serve the country's needs as effectively as the railway raised to a high technical level.

Gorbachev's reforms are certain to shift the balance in the Soviet transport system. The long-present pressure for greater personal mobility will tax a system where freight transport has always enjoyed priority over passenger movement. The persistent pressure, likely to intensify, for greater

The Antonov 124, the world's largest transport aircraft, is the latest in a long line of giant aeroplanes for both civil and military use over the vast distances of the Soviet Union.

personal ownership of motor vehicles, one of the Soviet citizen's most cherished hopes, will be hard to deny. It will doubtless see a fierce rearguard action by the 'railway lobby', one of the most powerful inner political pressure groups since pre-war times. The most striking change could come if Soviet citizens were to be allowed to indulge in one of their dearest dreams, the ability to travel as freely as westerners beyond the boundaries of their own country: to cater for such a traffic would require massive investment in long-neglected facilities. The shortcomings of the present Soviet transport system have been surprisingly revealed by the new openness of *glasnost'* through reports of major aviation and railway disasters, events seldom if ever previously reported.

BIBLIOGRAPHY

Ambler, J., Shaw, D. J. B. and Symons, L. (eds.) (1985), *Soviet and East European transport problems*, Croom Helm, London.

Armstrong, T. (1975), 'The northern sea route,' in Symons and White (eds.), 1975, pp. 127–141.

Biryukov, V. (1980), 'The Baykal-Amur mainline—a major national construction project,' *Soviet Geography*, **21**, pp. 225–270.

Briliant, L. A. (1975), *Geografiya morskogo sudokhodstva*, Transport, Moscow.

Crouch, M. (1979), Problems of Soviet urban transport, *Soviet Studies*, **31**, pp. 231–256.

Crouch, M. (1985), 'Road transport and the Soviet economy in Ambler,' J., Shaw, D. J. B. and Symons, L. (eds.), pp. 165–188.

Danilov, S. K. (1977), *Ekonomicheskaya geografiya transporta SSSR*, Transpechat, Moscow.

Dubrowsky, H. J. (1975), *Die Zusammenarbeit der RGW—Länder auf dem Gebiet des Transportwesens*, Berlin.

Galitskiy, M. I. *et al.* (1965), *Ekonomicheskaya geografiya transporta*, Moscow.

Garbutt, R. (1950), *Russian railways*, Sampson Low, London.

Greenwood, R. H. (1975), 'The Soviet merchant marine,' in Symons and White (eds.), pp. 106–126.

Hunter, H. (1987), 'Tracing the effects of sectoral transport demands on the Soviet transport system,' in Tismer, J. F., Ambler, J. and Symons. L. (eds.), pp. 1–31.

Kalinin, U. K. *et al.* (1970), *Obshchiy kurs zheleznikh dorog*, Moscow.

Kazanskiy, N. N. *et al.* (1969), *Geografiya putey soobshcheniya*, Moscow.

MacDonald, H. (1975), *Aeroflot, Soviet air transport since 1923*, Putnam, London.

Malashenko, V. (ed.) (1977), *The great Baykal–Amur railway*, Moscow.

Medvedkova, E. A. and Misevic, K. N. (1978), 'Die Erschliessung der BAM-Zone unter ökonomisch–geographischen Aspekten,' *Petermanns Geographische Mitteilungen*, **122**, pp. 37–43.

Mellor, R. E. H. (1964), 'Some influences of physical environment on transport problems in the Soviet Union,' *Advancement of Science*, **20**.

Mellor, R. E. H. (1976), *Die Sowjetunion. VI–das Transportwesen.* Harms Erdkunde, List Verlag, Munich.

Mellor, R. E. H. (1975), 'The Soviet concept of a unified transport system and the contemporary role of the railways,' in Symons and White (eds.), 1975, pp. 75–105.

Miller, E. B. (1978), 'The Trans-Siberian landbridge—a new trade route between Japan and Europe,' *Soviet Geography*, **19**, pp. 222–243.

Nikolskiy, I. V. (1960, 1978), *Geografiya transporta SSSR*, Moscow, 1960, 2nd edition, 1978.

North, R. N. (1979), *Western Siberia transport in Tsarist and Soviet development*, Univ. of British Columbia Press, Vancouver.

Parker, W. H. (1979), *Motor transport in the Soviet Union*, Research Paper 23, School of Geography, Oxford.

Shafirkin, B. I. *et al.* (1971), *Ekonomicheskiy spravochnik zheleznodorozhnika*, Moscow.

Spring, D. (1975), 'Railways and economic development in Turkestan before 1917,' in Symons and White (eds.), 1975, pp. 46–74.

Symons, L. (1985), Soviet air transport, in Ambler, Shaw and Symons, (eds.), pp. 144–164.

Symons, L. (1975), 'Soviet civil aviation—objectives and aircraft,' in Fallenbuchl, A. M., *Economic development in the Soviet Union and Eastern Europe, I, Reforms, technology and income distribution*, Praeger, New York, 1975, pp. 221–237.

Symons, L. and White, C. (eds.) (1975), *Russian transport: an historical and geographical survey*, Bell, London.

Thomas, B. (1988), *Trans-Siberian handbook*, Roger Lasceues, Brentford.

Tismer, J. F., Ambler, J. and Symons, L. (eds.), (1987), *Transport and economic development—Soviet Union and Eastern Europe*, Osteuropa Inst. and Verlag Duncker and Humblot, Berlin.

Tonyaev, V. I. (1977), *Geografiya vnutrennykh vodnykh putey*, Transport, Moscow.

Tupper, H. (1965), *To the Great Ocean*, Secker and Warburg, London.

Westwood, J. N. (1964), *History of Russian railways*, Allen and Unwin, London.

Westwood, J. N. (1987), 'Soviet inland waterways,' prospects for offering relief to the railways' in Tismer, J. F., Ambler, J. and Symons, L. (eds.)

12 The Regions

The larger a country's territory, the more difficult it is to organise and develop that territory in a planned, integrated and comprehensive manner, a fact which applies whatever form of economic and political system may be adopted. In the case of the Soviet Union, which is the world's largest state and where a planned and highly centralised economy is one of the most noteworthy features, these difficulties are particularly significant. It is not surprising therefore that, throughout the period since the Revolution, special attention has been paid in the USSR to the matter of regionalisation and numerous systems have been devised for dividing the country into regions of various kinds as a framework for administration, planning and economic development.

ADMINISTRATIVE DIVISIONS

Administrative divisions of the USSR fall into two categories: those which, in addition to their administrative function, also have political status and are directly represented in the organs of central government, and those which are solely administrative and are not thus represented. In the first category we have the *Soviet Socialist Republic*, the *Autonomous Soviet Socialist Republic*, the *Autonomous Oblast* and the *Autonomous Okrug*, while the second category involves two units, the *Oblast* and the *Kray*. The two types together form a complex, multitiered hierarchy, which is shown diagramatically in Fig. B (p. 4). Within this hierarchy, each level is subordinate to the one above it and the degree of administrative

autonomy which an area enjoys depends on its position in the hierarchy.

Political-administrative divisions
Table 12.1 gives a full list of the 53 political-administrative divisions within the Soviet Union, which are mapped in Fig. A (p. 2). At the summit of the hierarchy, of course, stands the Soviet Union itself or, to give it its full title, the Union of Soviet Socialist Republics (USSR). The highest legislative and administrative body of the USSR is the Supreme Soviet, which comprises two chambers: the Soviet of the Union and the Soviet of Nationalities. The Soviet of the Union is elected by universal adult suffrage, and for this purpose the whole country is divided into electoral districts, each with a population of about 300 000. Membership of the Soviet of Nationalities, however, is based on the division of the country into the various political-administrative units discussed below, which are themselves based primarily on the nationality of the local population. Each Union Republic sends 32 deputies to the Soviet of Nationalities, each Autonomous Republic sends 11, Autonomous Oblasts five and Autonomous Okrugs one member each. The aim of this system is to ensure that, since all measures must be passed by both houses, the interests of the smaller nationalities are not overwhelmed by those of the larger groups. Thus, for example, both the Estonian republic, with a population of about 1.5 million, and the Russian republic, with 144 million, are equally represented in the Soviet of Nationalities, to which they send 32 members each. In the Soviet of the Union, on the other hand, the

TABLE 12.1: POLITICAL-ADMINISTRATIVE DIVISIONS OF THE USSR, 1 JANUARY 1986

Key No.*	Division	Area 000 km^2	Population (000's)	Capital	
	USSR	22 402.2	278 784	Moscow	(8 714)
Soviet Socialist Republics					
I	Russian SFSR	17 075.4	144 080	Moscow	(8 714)
II	Ukrainian SSR	603.7	50 994	Kiev	(2 495)
III	Belorussian SSR	207.6	10 008	Minsk	(1 510)
IV	Lithuanian SSR	65.2	3 603	Vilnius	(555)
V	Latvian SSR	63.7	2 622	Riga	(890)
VI	Estonian SSR	45.1	1 542	Tallin	(472)
VII	Moldavian SSR	33.7	4 147	Kishinev	(643)
VIII	Georgian SSR	69.7	5 234	Tbilisi	(1 174)
IX	Azerbaydzhanian SSR	86.6	6 708	Baku	(1 722)
X	Armenian SSR	29.8	3 362	Yerevan	(1 148)
XI	Kazakh SSR	2 717.3	16 028	Alma-Ata	(1 088)
XII	Kirgiz SSR	198.5	4 051	Frunze	(617)
XIII	Tadzhik SSR	143.1	4 648	Dushanbe	(567)
XIV	Uzbek SSR	447.4	18 487	Tashkent	(2 077)
XV	Turkmen SSR	488.1	3 270	Ashkhabad	(366)
Autonomous Soviet Socialist Republics					
16	Bashkir ASSR	143.6	3 870	Ufa	(1 077)
17	Buryat ASSR	351.3	1 014	Ulan-Ude	(342)
18	Dagestan ASSR	50.3	1 753	Makhachkala	(311)
19	Kabardino–Balkar ASSR	12.5	724	Nalchik	(231)
20	Kalmyk ASSR	75.9	325	Elista	(83)
21	Karelian ASSR	172.4	787	Petrozavodsk	(259)
22	Komi ASSR	415.9	1 228	Syktyvkar	(218)
23	Mariy ASSR	23.2	731	Yoshkar–Ola	(236)
24	Mordov ASSR	26.2	964	Saransk	(315)
25	North Osetian ASSR	8.0	616	Ordzhonikidze	(308)
26	Tatar ASSR	68.0	3 537	Kazan	(1 057)
27	Tuvinian ASSR	170.5	284	Kyzyl	(77)
28	Udmurt ASSR	42.1	1 571	Ustinov	(620)
29	Checheno–Ingush ASSR	19.3	1 225	Groznyy	(399)
30	Chuvash ASSR	18.3	1 320	Cheboksary	(402)
31	Yakut ASSR	3 103.2	1 009	Yakutsk	(184)
32	Kara–Kalpak ASSR	164.9	1 108	Nukus	(146)
33	Abkhaz ASSR	8.6	530	Sukhumi	(128)
34	Adzhar ASSR	3.0	382	Batumi	(133)
35	Nakhichevan ASSR	5.5	272	Nakhichevan	
Autonomous Oblasts					
36	Adygey A Ob	7.6	423	Maykop	(142)
37	Gorno–Altay A Ob	92.6	179	Gorno–Altaysk	
38	Yevreysk (Jewish) A Ob	36.0	211	Birobidzhan	(80)
39	Karachayevo–Cherkess A Ob	14.1	396	Cherkessk	(105)
40	Khakass A Ob	61.9	547	Abakan	(148)
41	South Osetian A Ob	3.9	99	Tshinvali	
42	Nagorno–Karabakh A Ob	4.4	177	Stepanakert	
43	Gorno–Badakhshan A Ob	63.7	149	Khorog	

TABLE 12.1: CONTINUED

Key No.*	Division	Area 000 km²	Population (000's)	Capital
*Autonomous Okrugs**				
44	Aga–Buryat A Ok	19.0	77	Aginskoye
45	Komi–Permyak A Ok	32.9	162	Kudymkar
46	Koryak A Ok	301.5	39	Palana
47	Nenets A Ok	176.7	53	Naryan–Mar
48	Taymyr (Dolgano–Nenets) A Ok	862.1	54	Dudinka
49	Ust-Orda Buryat A Ok	22.4	129	Ust Ordinskiy
50	Khanty–Mansiy A Ok	523.1	1 047	Khanty–Mansiysk
51	Chukot A Ok	737.7	155	Anadyr
52	Evenki A Ok	767.6	21	Tura
53	Yamalo–Nenets A Ok	750.3	383	Salekhard

* As shown in Fig. A, p. 2.
† National Okrugs until 1977
Towns whose populations are not given have fewer than 50 000 inhabitants
Source: Narodnoye Khozyaystvo SSSR v 1985 godu, Moscow 1986.

Russian republic will have over 400 members and the Estonian republic five or six.

The USSR is a federation of 'Union Republics', at present numbering 15. The largest of these, the Russian republic, has been referred to as 'a Union within a Union' on account of the large number of subordinate units which it contains, a situation reflected in its cumbersome title of the Russian Soviet Federated Socialist Republic (RSFSR). This covers about three-quarters of the country, including the Russian homeland in the east European plain together with the whole of Siberia and the Far East, regions of Russian colonisation where people of Slav origin are in an overall majority. The RSFSR has large thinly populated areas and its inhabitants comprise some 52% of the total Soviet population. 82% of those living in the RSFSR are Russians; the remaining 18% include representatives of virtually every other Soviet nationality, some numbering several millions.

The remaining 14 Union Republics, officially Soviet Socialist Republics (SSRs), fall into a number of regional groupings, the arrangement of which reflects the distribution of the larger and more advanced non-Russian peoples of the USSR as discussed elsewhere in this volume (Chapter 6). The 15 Union Republics purport to be partners in a voluntary political association and, in theory at

least, each has the right to secede from the Union. This theoretical right has a significant effect on the political-administrative structure of the USSR for it means that no SSR can be established which does not have a boundary with the outside world. Political divisions in the interior of the country can never be raised to SSR status since they would then have the right to secede, thus creating the possibility of the 'political anomaly' of an independent state completely surrounded by the remaining territory of the USSR, a situation unacceptable to Soviet theorists.

Subordinate to the Union Republics within which they lie are the Autonomous Soviet Socialist Republics (ASSRs). At present these number 20, of which 16 are in the RSFSR, two in Georgia and one each in Azerbaydzhan and Uzbekistan. These units, too, tend to occur in clusters, indicating the distribution of important minority groups. There are six ASSRs in the zone between the middle Volga and the Urals, a region inhabited by various Finno-Ugrian and Turkic peoples, another eight in the Caucasus and Transcaucasia, two in the north European section of the RSFSR and three in Siberia. Finally there are eight Autonomous Oblasts (AOb) and 10 Autonomous Okrugs (AOk.; until 1977 known as National Okrugs, NO), representing smaller and less advanced minority groups. The majority of these are in Siberia and there are several in the Caucasus. Autonomous Okrugs occur only in the RSFSR.

Although these political-administrative divisions are named after particular national groups, it should not be imagined that each is inhabited solely by one group or that it contains all the members of one group. Long-continued migration movements within the Soviet Union have ensured that every areal unit has an ethnically mixed population. In particular, since the dominant migration movements have been from the European to the Asiatic parts of the country, the Slavs, especially the Russians, are found in sizeable numbers in most areas. In the case of the SSRs, the titular group (i.e. the group after whom the republic is named) is in a majority in all but two cases, the exceptions being Kazakhstan, where Kazakhs are outnumbered by Russians, and the Kirgiz SSR where the Kirgiz, though still the largest group, form less than half that republic's population. In the great majority of the lower-grade units, the titular group is very much in a minority. An extreme example is provided by the Yevreysk (Jewish) AOb, which had a 1979 population of 188 710 of whom only 10 166 were in fact Jews (0.6% of all Soviet Jews) and 158 765 were Russians. A further point is that the system of political-administrative divisions is flexible and an area may be raised to a higher status as its population grows and its economy develops. Demotion is also possible, and a number of ASSRs were abolished in the late 1940s on the grounds that their populations had collaborated with the enemy. However, the present pattern of political-administrative divisions has remained unchanged since the early 1960s.

Administrative divisions

In addition to the four types of political-administrative division discussed so far, there are two other units which are purely administrative and economic in function and are not represented in the Soviet of Nationalities, though they often form electoral districts sending deputies to the Soviet of the Union. These are the Oblast and the Kray, of which there were 123 and 6 respectively in 1986. Those parts of the RSFSR which do not enjoy ASSR status, together with the majority of SSRs (the exceptions are the smaller republics— Lithuania, Latvia, Estonia, Moldavia, Georgia, Azerbaydzhan and Armenia) are divided into

oblasts, which are the basic units of local government. The boundaries of these units are shown in Fig. A. The organising centre of an oblast is usually an industrial centre of some importance from which the oblast takes its name (Moscow oblast, Kiev oblast, Karaganda oblast, etc.). The half dozen krays, which occur only in the RSFSR, now have functions similar to those of the oblast. The lowliest political units, the AOb and the AOk, are in fact subordinate, as far as local government is concerned, to the ASSR, oblast or kray in which they are situated.

Below the level of the ASSR, oblast and kray are a variety of smaller units. The country is divided into 3224 rayons which contain 42 312 village soviets, 2170 towns (the larger of which are further subdivided into urban rayons) and 3961 'settlements of urban type'.

ECONOMIC DIVISIONS

In addition to the system of political and administrative regionalisation outlined above, there is a second, parallel system by which political and administrative divisions are grouped to form various kinds of economic regions. The details of this second type of regionalisation have undergone several major changes during the Soviet period, reflecting changes in the attitudes of Soviet planners towards the principles on which economic regions are based. Throughout most of the period since the Revolution there have been at least two levels of economic regions, of which the more stable has been the division of the country into a relatively small number of large units known as Major Economic Regions. Their actual number has varied between 13 (Fig. 12.1) and 23; the present division is into the 20 units shown in Fig. 12.2, which have changed very little since the mid-1960s. These are essentially economic planning regions and are also the basis for the tabulation of economic and other data in various official statistical publications. In addition, Soviet geographers frequently use the Major Economic Regions as the main regional divisions in their textbooks.

Economic-administrative areas

The lower tier of economic regions, because they have usually consisted of small groups of administrative units, are often referred to as economic-

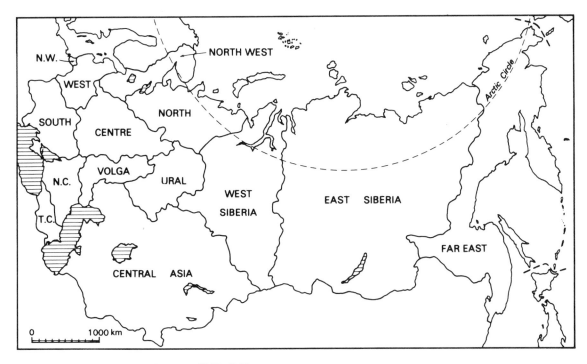

Fig. 12.1 Major Economic Regions, 1940–1960

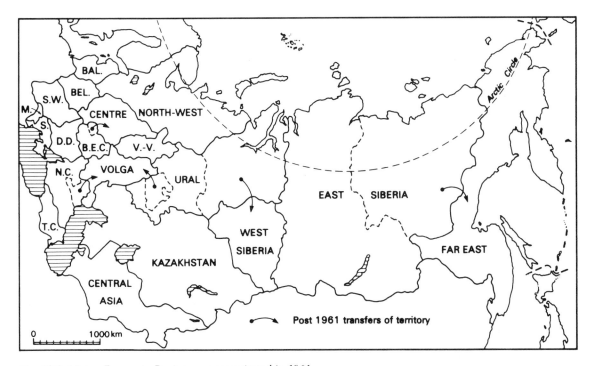

Fig. 12.2 Major Economic Regions as reconstituted in 1961

administrative areas, and it is these which have been subject to the most frequent changes, particularly in the 1950s and 1960s.

Until 1957, the economic-administrative areas were simply the smaller political and administrative units already described, namely the oblasts, krays, ASSRs and small SSRs. In practice, under the highly centralised forms of political and economic organisation characteristic of the Stalin period, these units had very little economic autonomy. They were responsible only for the organisation of agriculture and of 'industries of local importance', the greater part of their economic activities being under the direct control of the various government ministries in Moscow.

In 1957, however, as a result of Khrushchev's desire to decentralise control of the economy, the economic powers of economic-administrative areas were much enhanced by the establishment of the *Sovnarkhoz* system, under which the country was divided into 105 Sovnarkhoz regions (Fig. 12.3). The majority of these comprised a single oblast, kray, ASSR or SSR, but in a few cases several oblasts were amalgamated to form one of these units. Each Sovnarkhoz region was given

planning, managerial and budgetary responsibility for practically all forms of economic activity within its boundaries, and only a few items, such as the construction of major hydro-electric plants and the defence industries, remained under centralised control. At the same time, the various production ministries, each responsible for a particular industry throughout the country, were abolished.

In the early 1960s, it was decided that the decentralisation of economic control had gone too far, and that the Sovnarkhoz regions were too small for efficient management of the economy. Consequently they were re-grouped into 47 larger units known as Industrial Management Regions (Fig. 12.4) and for a while it appeared likely that these would provide a long-term solution to the regionalisation problem and would perform both planning and managerial functions. In 1965, however, there were further moves back towards centralisation. The regional economic councils which had been in charge of the 47 regions were abolished and the central industrial ministries re-established. Since the mid-1960s, the economic powers and functions of the administrative

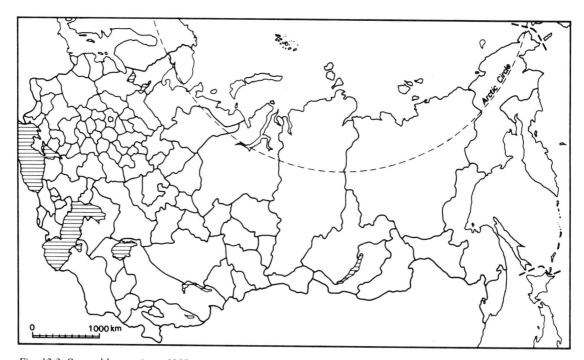

Fig. 12.3 Sovnarkhoz regions, 1957

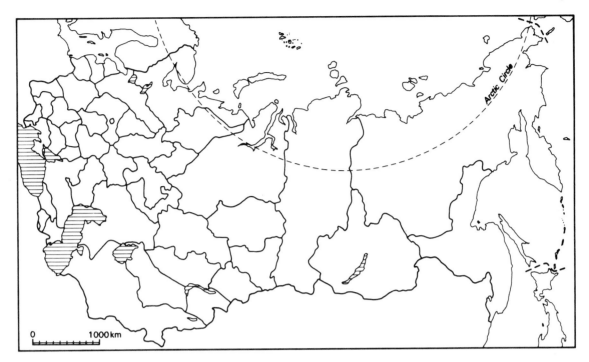

Fig. 12.4 Industrial management regions, 1963

divisions have again been very restricted and the control of industry has remained in the hands of some two dozen ministries, each responsible for a particular industry over the USSR as a whole. To avoid the excessive centralisation of the Stalin period, however, some of the powers of these All-Union Ministries have been devolved to Union-Republican Ministries responsible for a particular industry within a single republic, and much greater responsibility has been given to the managements of individual farms, factories and other economic enterprises.

Major Economic Regions

Throughout these changes, the Major Economic Regions have remained in being, though their boundaries have been altered on several occasions (Figs. 12.1, 12.2 and 12.5), and are still used for economic planning (as distinct from management) purposes and as 'standard regions' for data tabulation. Before proceeding to a discussion of each of these regions, a brief consideration of the general pattern (Fig. 12.5) is necessary. It will be seen that, in many cases, the boundaries of the

Major Economic Regions coincide with those of the Union Republics, and it is in fact a general rule that political-administrative boundaries must be preserved intact in any system of economic regionalisation. Thus, while an economic region may be a sub-division of an SSR, or may unite several SSRs, no economic region may be established which takes part of one republic and joins it to another republic.*

These regions vary greatly in size and there is a marked contrast between the relatively small, compact units of the more densely settled European part of the country and the small number of very large units in the Asiatic sector. As Table 12.2 indicates, the variation in population size is less marked. While regions (excluding the special case of Moldavia, which is 'a republic outside the system of Major Economic Regions') range in area from 6.2 million km² (Far East) to 110 700 km² (South), their populations vary

* A single exception to this rule has been allowed in the case of the Kaliningrad oblast, a detached portion of the RSFSR annexed from Germany in 1945, which has been placed in the Baltic economic region.

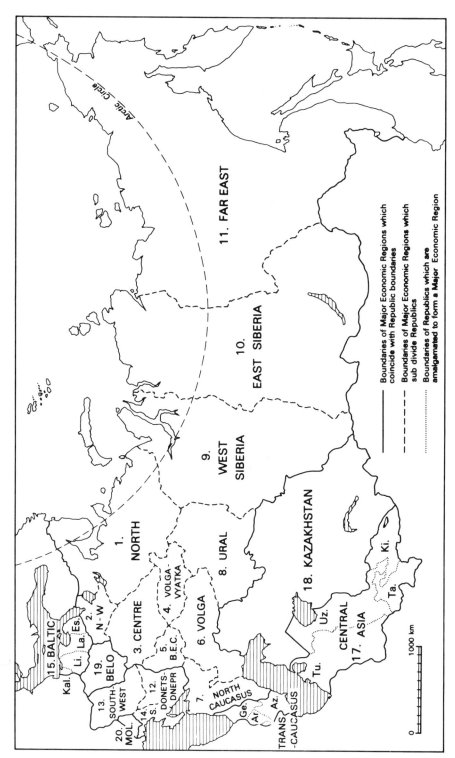

Fig. 12.5 Major Economic Regions (current)

TABLE 12.2: MAJOR ECONOMIC REGIONS, 1 JANUARY 1986

Region	Area 000 km²	%	Population 000	%	Density per km²	Urban %
1. North	1 465.3	6.6	6 003	2.2	4.1	76.6
2. North-west	196.5	0.9	8 134	2.9	41.4	86.3
3. Centre	485.2	2.2	29 795	10.7	61.4	82.0
4. Volga–Vyatka	263.2	1.2	8 362	3.0	31.8	68.1
5. Black Earth	167.7	0.8	7 652	2.7	45.6	58.8
6. Volga	536.4	2.4	16 081	5.8	30.0	72.3
7. North Caucasus	355.1	1.6	16 340	5.9	46.0	57.2
8. Ural	824.0	3.7	19 980	7.2	24.2	74.2
9. West Siberia	2 472.2	11.1	14 358	5.2	5.8	71.9
10. East Siberia	4 122.8	18.5	8 875	3.2	2.2	72.0
11. Far East	6 215.9	27.9	7 651	2.7	1.2	76.6
12. Donets–Dnepr	220.5	1.0	21 568	7.7	97.8	78.6
13. South-west	269.8	1.2	21 938	7.9	81.3	53.7
14. South	110.7	0.5	7 488	2.7	67.6	66.1
15. Baltic	189.1	0.8	8 616	3.1	45.6	69.8
16. Transcaucasia	186.1	0.8	15 304	5.5	82.2	57.1
17. Central Asia	1 279.3	5.7	30 456	10.9	23.8	40.9
18. Kazakhstan	2 715.1	12.2	16 028	5.7	5.9	57.5
19. Belorussin	207.7	0.9	10 008	3.6	48.2	63.1
20. Moldavia	33.7	0.2	4 147	1.5	123.0	45.7

between 6 million (North) and 30.5 million (Central Asia).

The attitude of Soviet economic planners towards the form and functions of the Major Economic Regions has varied. At different times they have extolled the virtues of both regional self-sufficiency, whereby each region attempts to produce the widest possible range of agricultural and industrial commodities thus lessening its dependence on other regions and reducing the volume of inter-regional freight traffic, and of regional specialisation with each region concentrating its efforts on producing, for the country as a whole, those commodities which it can most easily supply. In reality, neither complete regional self-sufficiency nor complete regional specialisation is possible, or even desirable. The resource base of each region is unique, and there are inevitable shortages of particular commodities in each region which can only be made good by 'imports' from other regions. At the same time it would be most illogical, in a country the size of the USSR, to concentrate production of a particular commodity in a single region if that commodity is required throughout the country. In practice, the 'choice' between regional self-sufficiency and

regional specialisation has never been a true choice but a matter of emphasis on one or other of the two extremes. As a broad generalisation we can say that, in the inter-war years, the emphasis favoured regional self-sufficiency and this was reflected, for example, in the establishment of small steel industries in the Far East and Central Asia. Since the mid-1950s, much more attention has been paid to relative costs of production (including transport costs) at different locations, resulting in a greater emphasis on regional specialisation. As evidence of this, we may note the division of the Centre region of the inter-war years into a smaller, predominantly industrial Central Region and a mainly agricultural Black Earth Centre, or the separation of the industrial Donets–Dnepr region of the Ukraine from the predominantly agricultural South-west and South.

Recent statements by Soviet authors bear witness to the fact that, while regional specialisation remains the dominant theme, each region aims at a diversified economy. Implicit in their arguments is the idea that no system of regions can be permanent, and that boundaries must change as the economy develops and new patterns of economic activity emerge. It remains to discuss the present regions individually, to indicate their relative importance to the Soviet economy and to assess the extent to which they fulfil the requirements of the theorists.

THE MAJOR ECONOMIC REGIONS

Since the early 1960s, although there have been several transfers of territory between Major Economic Regions (see Fig 12.1), there has been only one major change in their number and nomenclature: in 1982, the former north-western region was subdivided to form the northern and northwestern regions shown on Figure 12.5. The discussion which follows is organised on the basis of the 20 units (19 Major Economic Regions, plus Moldavia) existing in 1987.

(1) The North

This is the largest of the European regions and also the one with the lowest population density and the smallest proportion of cultivated land. Situated entirely to the north of the 60th parallel, it includes Karelia, the Kola peninsula and the Arctic drainage basins of the Northern Dvina and Pechora rivers. The region is remote and inhospitable, with a coastal tundra belt and large areas of coniferous forest and swamp. Consequently, agriculture is of little significance and the traditional activities are reindeer herding, fishing, hunting and lumbering. However, the region also contains valuable mineral resources which have been vigorously exploited over the past fifty years, including coal, oil and natural gas in the Pechora basin (Komi ASSR), bauxite in the valley of the Onega river (Arkhangelsk oblast) and nickel, iron ore and apatite in the Kola peninsula. There are numerous small hydro-electric stations in Karelia.

Thus the northern region is most important as a source of primary products, notably timber, fish, furs, coal, oil, natural gas and metallic minerals. In addition, it plays a significant role in the country's foreign trade through its ports of Murmansk and Arkhangelsk. The development of the region has involved the construction of long-distance transport links. The railways to Murmansk and Arkhangelsk carry imports and exports and tap the minerals of the Kola penin-

sula; the Vorkuta-Kotlas line, built during the Second World War and now paralleled by oil and gas pipelines, carries fuel and timber from the remote north-east to the Leningrad industrial complex, to the Baltic republics and to the Centre.

(2) The North-West

This title is now applied to the small region comprising the Leningrad, Novgorod and Pskov oblasts. Although wholly within the tayga belt, the region is quite densely settled and contains considerable areas of farmland. However, agricultural potential is limited by the harsh climate and the large stretches of poorly drained land so that much food has to be brought in. The Novgorod and Pskov oblasts remain predominantly rural and 80% of the region's population live in the Leningrad oblast. Leningrad itself, with a population approaching five million is the second city of the Soviet Union, contains 60% of the entire population of the north-western region and is a major industrial complex concerned with a wide variety of engineering activities, particularly those of the more highly skilled kind, and produces ships, machinery, electrical equipment, chemicals, textiles and consumer goods.

(3) The Centre

The Centre, which appears quite small on a map of the Soviet Union but is in fact nearly as large as France, is among the most densely populated and highly urbanised regions. Its pre-eminence in so many branches of the Soviet economy has been achieved despite a lack of natural resources. The Centre has no coal, oil or natural gas, though there are large deposits of lignite and peat, and its hydro-electric potential is small. There are only modest iron ore deposits and no other minerals apart from mineral salts. Lying for the most part in the mixed forest belt, the Centre has a greater agricultural potential than areas further north but climate and soil conditions are inferior to those of more southerly regions.

The Centre is a region which benefits more from its location than from its natural wealth. It owes its importance, as its name suggests, to its centrality with respect, not to the territory of the Soviet Union as a whole, but to the more densely settled and highly developed European section, and to its historical function as the organising centre of the Russian Empire and the USSR. It was this region which became the political focus of the Empire from the sixteenth century onwards, and its centrality was confirmed, after the Revolution, by the choice of Moscow as the capital and by the establishment there of the most important organs of the highly centralised Soviet state.

The centrality of the region is most clearly expressed in the pattern of communication, both old and new, all of which tends to focus on Moscow. At the time when movement across the European plain was largely by water, the region had easy access to the main arteries, particularly the Volga, and it contains some of the earliest canals built to link the various rivers. When, in the late nineteenth century, rivers and canals were largely (but never entirely) replaced by railways, the network took a radial form centred on Moscow, which thus gained direct links with all parts of the Empire. Similar developments have occurred in recent decades with respect to the modern road system—it was the roads radiating from Moscow which were the first to be improved. Again, the highly developed network of internal airlines focuses on Moscow and provides a means of rapid contact with all parts of the Union. Other forms of modern 'transport'— electric power lines, oil and gas pipelines—though they cannot perhaps be said to centre on Moscow, feed into the region the power supplies necessary to maintain its varied and growing industries. Thus Moscow sits at the centre of a spider's web of communications and the Centre as a region benefits enormously.

Until the opening up of the steppelands in the nineteenth century, the Centre was the most densely populated part of the Empire, and its large population provided the necessary labour for the development first of handicrafts and then of mechanised factory industry. This has given the region a tradition of skilled craftsmanship and technological innovation which has enabled it to take the lead in many of the most sophisticated branches of modern industry.

The growth of urban populations as a result of industrialisation stimulated agricultural development. Some 30% of the land area is under crops, a proportion exceeded only in the steppelands, and the land is devoted to what are, by Soviet standards, highly productive systems of mixed farming, producing large quantities of meat, milk and vegetables. There is scope for further expansion of agriculture, particularly by means of swamp reclamation, but it is highly unlikely that the Centre could ever approach agricultural self-sufficiency and large quantities of foodstuffs, particularly grain, must be brought into the region.

The heavy industrial base is relatively small and the region produces less than 10% of the country's steel. Although the iron ores of the Tula district support the production of some pig iron, there are large 'imports' of ore, pig iron, steel and coal, mainly from the Ukraine, though metal comes increasingly from the Black Earth Centre (see below). The Ukraine is also the main source of grain, so that the Centre's main inter-regional linkages are with its southern neighbours.

The Centre contributes about a quarter of all Soviet engineering products, among which the most important are machine tools, scientific instruments, transport equipment, machinery for the textile and other light industries, electrical equipment and consumer goods of all kinds. The region is also responsible for about 70% of Soviet production of textiles. Particularly noteworthy is its dominant position in the cotton industry, which persists despite the need to import all the raw material from Central Asia. Over the past two decades, with the building of oil and gas pipelines, the Centre has also become of major importance in the rapidly-expanding chemical industry.

The Centre thus has links with practically every other region of the USSR, drawing food, fuel and raw materials from a very wide area and supplying its manufactures to all parts of the country.

(4) The Volga–Vyatka Region

There would appear to be little logic in the separation of this region from its neighbours. Its western part (the Gorkiy oblast) straddles the Volga and is in many ways an eastward continuation of the Central industrial region. Gorkiy, with a population of 1.4 million, is the seventh largest city in the Soviet Union and is a major engineering centre and a large-scale producer of motor vehicles. Power is derived from large hydroelectric stations on the Volga and from thermal stations based on local peat and on oil from the Volga–Ural field.

The more easterly parts of the Volga–Vyatka region are relatively thinly settled and less developed economically. This applies particularly to the large Kirov oblast, which is, however, an important source of timber. Agriculture is fairly intensive along and to the south of the Volga, supplying meat, vegetables and dairy products to the towns of the region, but is much less productive to the north and east. Like the Centre, the Volga-Vyatka region as a whole relies heavily on outside sources of food and raw materials, supplying manufactured goods in return.

(5) The Black Earth Centre

One of the smallest of the 20 regions, the Black Earth Centre is among the more densely populated. At the same time it is one of the least urbanised—although the urban proportion has much increased in recent years—and has a higher proportion of its land under crops than any other region. As its name suggests, the Black Earth Centre lies mainly in the zone of chernozem soils and the wooded steppe and steppe vegetation belts. Thus conditions for agriculture are, by Soviet standards, particularly favourable and the region is characterised by highly productive mixed farming which supplies large quantities of wheat, sugar beet, sunflower, meat and dairy products and supports a dense rural population.

Although agriculture is still the dominant economic activity in this region, the situation is changing. Small-scale iron mining has long been carried on in the vicinity of Lipetsk and during the past decade this activity has expanded rapidly in several localities on the basis of the vast reserves of ferruginous quartzite contained in the 'Kursk Magnetic Anomaly'. Situated midway between the Donbas and the Centre, the Kursk Magnetic Anomaly already supplies iron ore to both these and is also supporting increased steel production in the Black Earth Centre itself. A major integrated plant has been built at Lipetsk and others are planned. There remain the problems of power supplies, which are a major weakness, but there is easy access to Donbas coal and oil and gas pipelines from the Volga and North Caucasus pass through the region. One of the earliest atomic power stations in the USSR was built near Voronezh.

The towns of the Black Earth Centre are growing rapidly and the large rural population has provided a labour force for industrialisation, leading to a high rate of rural depopulation. At present, engineering, chemicals and food processing are the main branches of manufacturing industry.

(6) The Volga Region

This large region comprises a collection of administrative divisions ranged along the Volga river from Kazan to the Caspian Sea.

The Volga river has long been a major transport artery linking the industrial regions of the Urals (via the Kama and Belaya tributaries) and Centre with the grainlands of the steppe and, via the Caspian sea routes, with the Caucasus and Central Asia. This link function was enhanced in the 1950s by the construction of the Volga-Don canal connecting the Volga region to the Donbas and the Black Sea. Traffic along the river thus included industrial raw materials, oil, foodstuffs and manufactured goods from many different regions. In addition, major rail routes cross the Volga, connecting the European regions with Siberia, Kazakhstan and Central Asia. Important urban centres, such as Kazan, Kuybyshev, Saratov and Volgograd, developed at these crossing points, where trans-shipment between rail and water took place.

Over the past 40 years, however, both the nature of the regional economy and its significance to the economy of the USSR as a whole have changed dramatically, mainly as a result of its emergence as a major source of power, producing great quantities of oil, natural gas and electricity. The development of the oil resources of the Volga region was greatly accelerated during the Second World War when the Caucasian fields appeared likely to fall to the enemy. During the 1950s and 1960s, as oil made up an increasing proportion of Soviet energy supplies and as the Caucasian fields proved unable to maintain their share of the rapidly-rising total output, the Volga–Ural field was developed at an even greater rate. By the early 1970s it accounted for at least 70% of Soviet oil production. Output peaked around 1975 and has since declined significantly in absolute as well as in relative terms. In the mid-1980s the region was producing about a quarter of all Soviet oil and the lead had passed to West Siberia (see region 9). In addition to the development of oil and natural gas, there has been a massive development of the region's hydro-electric potential. Half a dozen major stations are now in operation, with others planned. Each involves a large barrage across the Volga, which now resembles a string of narrow lakes, and this, incidentally, has improved navigation and permitted the irrigation of large areas of the dry steppe.

A village on river terraces in a Caucasus mountain valley

From a region once deficient in power, the Volga region has been transformed into a major power base, supplying oil, gas and electricity to the Urals, Centre, Ukraine and more distant regions by river, rail, pipeline and cable. The development of power resources has led to other forms of industrialisation, and the region has become important in a variety of industries, which fall into three main groups. The first of these is concerned with supplying equipment used in power production, such as hydro-electric turbines and oil drilling and pumping machinery, items which the region now produces not only for its own needs but also for those of other regions. Secondly there are industries which have been attracted by the availability of electric power, including various forms of non-ferrous metallurgy and engineering. Noteworthy among the latter is the large motor vehicle plant, the biggest in the Soviet Union, at Tolyatti near Kuybyshev. Finally, and most rapidly growing of all, there are

the chemical industries based on oil and natural gas.

The Volga region's industrial activities are found mainly in a series of clusters strung out along the river, of which by far the most important is that in and around Kuybyshev and its satellite towns.

(7) The North Caucasus Region

This is a region with several diverse elements and little internal unity. In the north, the Rostov oblast includes an extension of the Donbas coalfield industrial zone and the port cities of Rostov and Taganrog. This part of the region is highly urbanised and industrialised; its separation from the neighbouring Donets–Dnepr region of the Ukraine appears illogical and is imposed by the presence of the political boundary between the Ukraine and the RSFSR.

The remainder of the North Caucasus region is predominantly agricultural. Farming is most intensive in the west, particularly in the Kuban valley and along the foot of the main Caucasian range, producing wheat, maize, vines, rice, tobacco and fruit. The intensity diminishes eastwards towards the Caspian as conditions become drier. This part of the region is by no means devoid of industrial resources. Along the hillfoot zone is a string of small oilfields, which are still significant producers although their relative importance is much diminished, and the gasfields of the Krasnodar and Stavropol krays produce about 5% of Soviet natural gas. The mountains themselves supply non-ferrous metals, notably lead and zinc, together with hydro-electricity. No major industrial complex has developed (except in the Rostov oblast) but the several medium-sized towns of the region have oil refining, food processing, chemical and light engineering activities. There are several health and holiday resorts in the foothills and on the Black Sea coast.

(8) The Ural Region

The Ural economic region comprises the Kurgan, Orenburg, Perm, Sverdlovsk and Chelyabinsk oblasts, together with the Bashkir and Udmurt ASSRs, territories which cover the central and southern parts of the Ural ranges and extensive foothill and lowland zones on either side. As the figures in Table 12.2 indicate, the overall population density is low when compared with the European regions to the west, but there is a high level of urbanisation. The greater part of the region lies in the tayga zone, but there are extensive areas of steppe in the south, and it is here that most of the agricultural land is to be found. Although a major producer of grain and livestock products, the region is by no means self-sufficient in foodstuffs.

The Ural region plays a fundamental role in the industrial economy as the Soviet Union's 'second metallurgical base' (i.e. second after the Donets–Dnepr region) and produces nearly a third of the country's pig-iron and steel. Prior to the Revolution, despite a long tradition of iron production, the Urals were of minor importance and the development there of the second metallurgical base is the product of a planning decision made in the 1930s. Among the many ore bodies developed since then, the most outstanding have been the high grade magnetites of Magnitogorsk and Nizhniy Tagil which, until recently, supported the bulk of the region's iron and steel

industry. In addition, there are numerous valuable sources of other metals, notably copper, aluminium, nickel, chrome and platinum.

In contrast to its richness in metallic minerals, the Ural region is notably deficient in sources of energy. Although there are several sources of coal and lignite, these are quite inadequate to support the heavy industries of the region. There is a marked shortage of coking coal, which is brought in large quantities from the Kuzbas, some 1800 km away in West Siberia, and from Karaganda, 1100 km away in northern Kazakhstan. These linkages were established in the 1930s, and the movement of iron ore in the opposite direction permitted the establishment of iron and steel industries on the two coalfields. Although the Urals no longer send iron ore to the Kuzbas or Karaganda, the railway lines involved in the movement of coking coal to the Urals continue to be among the most heavily used in the USSR and indeed in the world.

By the 1960s it had become apparent that the Ural region, already deficient in coking coal, could no longer depend for its iron solely on its own ore deposits, some of which were nearing exhaustion. Alternative, though lower-grade sources have been found within the region, notably at Kachkanar, but there is a growing reliance on ore from the neighbouring Kazakh republic.

As might be expected, the manufacturing industries of the region are mainly in the heavy engineering group, machinery and equipment being supplied to many parts of the country. The lighter branches are less well represented, but timber-processing and chemicals are important, the latter based on local mineral salts and the oil and gas of the Volga–Ural field. Oil wells in the Perm and Orenburg oblasts and the Bashkir ASSR account for about 10% of Soviet output. Considerable quantities of oil and gas are brought into the region by pipelines from the Volga–Ural, Uzbek and West Siberian fields; a major new source of natural gas was discovered about ten years ago in the Orenburg oblast.

The Ural region provides a particularly good example of a region with a changing resource base. The presence of high grade iron ores, together with the strategic advantages of a location in the interior of the USSR, led to the decision to develop the region as a metallurgical base. Once this decision had been made, the fuel deficit was overcome by the establishment of links with neighbouring energy-rich regions. Today, the mineral resources on which industrialisation was based are no longer sufficient and new inter-regional links have been made to overcome this problem.

(9) The West Siberian Region

Siberia and the Far East, which together extend from the Urals to the Pacific, constitute well over half the territory of the USSR but contain little more than 10% of the population. About a fifth of this area and nearly half its inhabitants are in West Siberia. The latter falls clearly into two sections. The southern part, extending from the Trans-Siberian Railway to the Kazakh border and the Altay ranges, contains the bulk of the population, agricultural land and industrial capacity of West Siberia. Northwards to the Arctic Ocean, a distance of 1800 km, lies the vast, thinly settled west Siberian lowland. Here, the Tyumen oblast, one of the largest administrative divisions in the country, covers an area of 1.4 million km^2 and has a population of only 2.7 million; the large Khanty–Mansiy AOk and Yamalo–Nenets AOk within this oblast cover 1.2 million km^2 and have only 1 430 000 inhabitants. Until quite recently, the contribution of this area to the national economy was very small. Agriculture is virtually absent, lumbering is confined to the more accessible southern districts and the area's mineral resources were largely undeveloped. The significance of this zone has, however, been greatly increased over the past two decades by the exploitation of its huge resources of oil and natural gas, which are even greater than those of the Volga–Ural field. In 1985, the Tyumen oblast produced some 60% of all Soviet oil and 58% of the country's natural gas. Development of these resources in this remote area, with its extremely difficult physical environment, presented major technical problems, particularly in the case of natural gas, which is found mainly in the extreme north, and involved large-scale investment in pipelines to carry the gas and oil to other regions. Industrial development is not likely to proceed beyond the extractive stage and manufacturing industry· is unlikely to be established on the West Siberian oil and gas fields.

The southern section of the West Siberian region is by no means uniform, but is given a certain unity by the Trans-Siberian Railway which runs across it and links its various parts. As far east as the Ob, the predominant activity is agriculture, which expanded rapidly in the 1950s and 1960s as a result of the 'virgin lands' scheme. This zone is devoted to extensive grain growing and livestock rearing and sends grain and meat to the European regions and to eastern Siberia. The most important element here is the Kuzbas coalfield, already mentioned in respect of its links with the Urals. The Kuzbas is a particularly rich source of coking coal, which it supplies not only to the Urals but also to Central Asia, and has the largest iron and steel industry in the eastern regions. Major regional centres such as Novosibirsk, Omsk and Barnaul have varied manufacturing industries, based on Kuzbas coal and hydro-electric power from the Ob. In addition, the Altay ranges are a source of non-ferrous metals.

(10) The East Siberian Region

This region comprises the lowlands along the Yenisey, the western part of the central Siberian plateau and the mountains and basins of pre-and trans-Baykalia. As in western Siberia, there is a large thinly-settled northern zone and a smaller, more developed southern section along the Trans-Siberian Railway. The north, particularly the plateau section, is believed to contain valuable reserves of a variety of minerals and there are vast reserves of coal in the Tunguska and Taymyr basins, but all these remain virtually untouched. An exception is provided by the Norilsk district, near the mouth of the Yenisey, where a non-ferrous metal mining and smelting complex produces copper and nickel. This is linked to the rest of the country only by the Arctic sea route and by air.

The more developed areas of the south occur in a series of pockets along the Trans-Siberian Railway. This zone is most noteworthy for its large energy resources, which include the bituminous coal of the Cheremkhovo district (Irkutsk oblast), the lignites of the Kansk–Achinsk field (Krasnoyarsk kray) and the hydro-electric power of the Yenisey and Angara rivers. Prior to the 1950s, the most important industrial zone was along the upper Angara in the Irkutsk oblast, which had some significance in the engineering, chemical and timber industries. Over the past 25 years, a number of major developments have occurred in East Siberia, mainly in the field of energy production. The largest hydro-electric stations in the country have been built at Bratsk on

Tour buses at Lake Ritsa in the western part of the Caucasus mountains.

the Angara and Krasnoyarsk on the Yenisey and others are under construction. The Kansk–Achinsk lignites are used on a large scale in thermal-electricity generating stations. The net result has been to make East Siberia a major power base—its per capita electricity production is the highest in the country—and this has attracted power-hungry industries such as aluminium smelting and timber-based chemicals. Important sources of iron ore and non-ferrous metals have been discovered, but their development has been less rapid than was at one time predicted, considerable difficulty being experienced in maintaining the necessary labour supply.

The agricultural potential of the region is considerable, but development has been hindered by the small size of the rural population, which totals only 2.5 million, and by physical difficulties. Consequently there is a large import of basic foodstuffs, mainly from West Siberia and Kazakhstan.

(11) The Far Eastern Region

In comparison with eastern Siberia, the Far East is yet more remote and thinly settled; covering more than a quarter of the Soviet Union, it contains less than 3% of the country's population—7.6 million people in an area two-thirds the size of the United States. In parts the Far East has a rather more favourable physical environment than the rest of Siberia and it has a rich but generally undeveloped resource base. Coal is present in large quantities in the Lena basin but only in relatively small amounts in the south. There is a large hydro-electric potential, little of which has so far been developed, on the Amur and its tributaries. Varied mineral deposits exist in the north, but exploitation has so far been confined to the more valuable items such as the diamonds of Mirnyy in the Vilyuy valley and the gold of the Lena valley. Oil is produced in Sakhalin, but reserves are small, while the extent of the oil and gas resources of the Lena basin has not yet been assessed.

As in the rest of Siberia, most of the settlement and economic activity occur in the southernmost districts; the Yakut ASSR together with the Magadan and Kamchatka oblasts cover 4.8 million km² but have a combined population of only two million.

A small steel plant was built at Komsomolsk in the 1930s, but this remains a very minor producer. Most manufactures, including basic items of industrial equipment as well as consumer goods, have to be brought in from the west and, despite the availability of large areas of fairly good agricultural land, the Far East is not self-sufficient in basic foodstuffs.

The low stage of development in this region is due in part to the small size of its population but also to a certain lack of attention to its problems by Soviet planners. The wave of intensive industrial development which has swept across the Urals into Siberia has barely touched the Far East. In view of past tensions between the Soviet Union and China, the underdeveloped nature of this region must be viewed with disquiet from Moscow and we may expect to see a more determined effort directed towards the development of the Far East in the coming decade. There is some evidence of this in the recent construction of the Baykal–Amur Mainline (BAM) railway. Running from Bratsk on the Angara to Komsomolsk on the Amur, some 300 km north of the Trans-Siberian, the BAM project was first mooted in the 1930s but construction did not begin until the mid-1970s and the line will not be fully operational before 1990. This 3000 km long railway is intended to lessen dependence on the Trans-Siberian, which runs uncomfortably close to the Chinese border, to strengthen the links between the Pacific coastlands and the rest of the country and to assist in the development of the industrial resources which lie along its route. Once this project is completed, we may expect the Far East to play a more important role in the Soviet economy than hitherto.

(12) The Donets–Dnepr Region

This is the most densely populated Major Economic Region and has one of the highest levels of urbanisation, yet at the same time is second only to the Black Earth Centre in the proportion of land under crops. The Donets-Dnepr region plays a leading role in the industrial economy as the Soviet Union's 'first metallurgical base', a function which it has now performed for more than 100 years, and produces more than a third of the country's coal, iron ore, pig iron and steel. The iron and steel industry became important in the second half of the nineteenth century on the basis of Donbas coal and carboniferous iron ores, but the Donbas was soon linked by rail to the Krivoy Rog iron ore field, thus permitting the establishment of a second iron and steel district in the Dnepr bend. A similar though smaller scale interchange, between Donbas coal and iron ore from Kerch in the Crimea, resulted in the building of steel plants at the coastal site of Zhdanov on the Sea of Azov. The Donbas and Dnepr bend areas still form two geographically separate industrial complexes, with a zone of mainly rural territory between them, but are completely interdependent economically, linked as they are by the vital interchange of fuel and ore. Each of the districts has additional advantages. The Dnepr bend, now a major source of hydro-electric power, with some oil nearby, has attracted a wide range of chemical and engineering activities. Chemicals are also important in the Donbas on the basis of local salts, coke-oven by-products and oil and gas piped in from the Caucasus and the Volga region.

The Donets–Dnepr region not only supplies steel, heavy engineering products and chemicals to a wide area of the European USSR, but also has large surpluses of raw materials and semi-finished products, including coal, coke, iron ore and pig iron, which it sends to regions deficient in those commodities, notably the Centre. In return it receives light industrial and consumer goods.

While industrial activity is most intensive in the Donbas–Dnepr bend zone, the region's largest city is Kharkov (1.6 million) which lies outside this zone but stands at the junction of routes between the Donbas, Dnepr bend and Centre. The city is now the centre of a rapidly growing engineering and chemical complex based on Shebelinka natural gas and its contacts with the older industrial districts to the south.

Although the region is the scene of quite intensive and productive agriculture, this does not produce enough to support the large urban population and foodstuffs are brought in from the North Caucasus and western Ukraine.

(13) The South-west

In contrast to the Donets–Dnepr region, the South-west, which is considerably larger, is still largely rural, with only 54% of its population living in towns. Until the re-organisation of Major Economic Regions in the 1960s, the South-west along with Moldavia and the present Southern region, were part of a single region, balancing the predominantly industrial eastern part of the Ukraine. Despite its low level of urbanisation, the South-west is among the most densely populated parts of the European plain, a fact which reflects the particularly favourable conditions for agriculture provided by its rich chernozem soils and relatively mild, humid climate.

The South-west is not without its industrial resources. Oil, coal and natural gas are all produced in the Carpathian hill-foot zone, which is also a source of sulphur and potash, and there are major hydro-electric plants on the Dnestr and Dnepr rivers. The major cities have a variety of engineering, food-processing, chemical and other industries. All these are expanding and the large rural population provides an adequate labour supply.

Despite these developments, industry still takes second place to agriculture which supplies grain, sugar beet, potatoes and livestock products to the eastern Ukraine and the regions of the forest zone to the north. This is a function which the region has performed since the nineteenth century, aided by a relatively dense network of communications linking it to its neighbours and to the Black Sea ports.

(14) The South

This is the smallest of the Ukrainian regions and is a good deal less densely populated than the other two. It comprises the drier steppelands of the southern Ukraine and Crimea, where agriculture has been intensified in recent years with aid of irrigation, and the high-value crop zone of the Crimean coast. Its main products are thus agricultural, including grain, livestock products, vines, fruit, vegetables and tobacco. The main industrial resource is the Kerch iron ore, which contributes to the Donets–Dnepr steel complex.

A kolkhoz market in Dushanbe. Collective farm markets provide outlets for the farms and fresh produce for consumers throughout the Soviet Union.

The region gains added importance from its ports, particularly Odessa, which handles a large share of Soviet foreign trade, and from the health and holiday resorts of the Crimea.

(15) The Baltic Region

The three small Baltic republics of Estonia, Latvia and Lithuania, together with the Kaliningrad oblast of the RSFSR (formerly part of East Prussia) constitute the Baltic economic region. This is a region with a particularly poor resource base. Lying entirely within the forest zone and the podzol soil belt, with extensive tracts of poorly drained lowland, the Baltic region presents limited opportunities for agriculture. Although more than a quarter of the land is under crops, foodstuffs, particularly grain, have to be brought in from the Ukraine. In recent years, agriculture has become increasingly specialised on the livestock side, which would appear to be the most suitable agricultural activity for this region.

Industrial resources are extremely limited. Apart from timber and flax, which are the main items, the only materials available in quantity are peat and oil shale (the latter in Estonia), both used for the generation of electricity. The three republics were detached from the USSR between 1917 and 1940 and thus were not affected by Soviet programmes of economic development, so that in the early post-war years, the region lagged behind the rest of the country. Since the mid-1950s, the significance of the Baltic region to the Soviet economy has increased. Power supplies have been augmented by the extension of pipelines into the region and by the development of hydro-electricity on the Western Dvina, and the Baltic ports have benefited from the growth of foreign trade. The problem of poverty in industrial raw materials remains, but the larger towns have their ship-building, engineering, chemical, food-processing and timber industries. Despite these developments, the Baltic region seems destined to remain of rather limited importance to the Soviet economy, not only because of its poor resource base but also as a result of its peripheral location with respect to the more highly developed parts of the country.

(16) The Transcaucasian Region

As its name suggests, this region, which unites the three republics of Georgia, Armenia and Azerbaydzhan, lies to the south of the main Caucasian range. This location is largely responsible for the highly distinctive character of Transcaucasia, seen in its physical environment, history, culture and economy.

Although the cultivated area is small, the agriculture of the region is of special importance, since the local climates permit the growth of crops which can be produced in only a few parts of the Soviet Union. These include tea, citrus fruits, vines and tobacco, grown mainly in the humid lowlands of western Georgia, and cotton, grown under irrigation in the drier lowlands of Azerbaydzhan. These high-value crops, together with livestock products from the mountains, are sent to other parts of the USSR, but agricultural specialisation has led to a deficiency in basic foodstuffs, particularly grain, which has to be brought into the region.

Transcaucasia has a variety of mineral resources. The most important of these is oil, which, together with natural gas, occurs mainly in the Baku district of Azerbaydzhan. Baku oil was the region's main contribution to the national economy from the nineteenth century until the Second World War and, until the development of the Volga–Ural field, Baku and the North Caucasus together produced three-quarters of all Soviet oil. The decline in production which set in in the 1940s continues, despite the development of reserves beneath the Caspian, and Baku must now be considered only a minor producer. Much of the oil and most of the natural gas are now consumed within the region. Transcaucasia has a large hydro-electric potential. Until the 1950s this was developed on a limited scale, but over the past 30 years a number of important stations have been built, notably in Armenia. Coal and iron are both present, though in relatively small quantities, and there is an iron and steel plant at Rustavi. There are a variety of other metallic ores, including copper, lead, zinc, molybdenum and tungsten, but the only item of major importance is the Chiatura manganese deposit, one of the largest in the world.

The industrial resource base of Transcaucasia is varied but modest in size and the region is unlikely to become a major industrial zone. At present, its main contributions to the Soviet economy are in the form of foodstuffs and raw materials. Large-scale industry is confined to a few major cities, notably Baku, Tbilisi and Yerevan, which have engineering, chemical and food-processing plant. In this densely populated region, labour-intensive light industries have developed quite rapidly, but remoteness from major markets is a disadvantage.

This is a region in which the varied resource base and an isolated location would suggest all-round development aimed at regional self-sufficiency. This has not so far taken place and attention has been concentrated on the production of items for which the region has special advantages, namely oil and high-value crops, and Transcaucasia is heavily dependent on other regions for both foodstuffs and manufactures. The industrial economy is being diversified, but the importance of this region to the USSR is appreciably less great than in the inter-war period.

(17) The Central Asian Republics
The four Central Asian republics (Kirgiz, Tadzhik, Uzbek and Turkmen) show certain similar-

The Registan square in Samarkand, one of the oldest cities in Central Asia

ities to the Transcaucasian region, particularly as regards their cultural distinctiveness and the level and nature of their economic development. The physical environment is extremely diverse, the major contrasts being between the desert, the mountains and the intervening hill-foot zone. As a result, Central Asia contains some of the most densely settled as well as some of the most thinly inhabited parts of the Soviet Union; the average density figure of 24 per km^2 given in Table 12.2 is very misleading—most areas are either well above or well below average.

The bulk of the population is associated with the intensive irrigated agriculture of the hill-foot zone and the river valleys. Agriculture is the region's main economic activity and Central Asia's leading contribution to the USSR's economy is its cotton, which constitutes more than 80% of the Soviet crop. Fruit, vegetables and livestock products are also important. Concentration on these high-value crops has, as in Transcaucasia, led to a grain deficit, which is made up by 'imports' from Kazakhstan.

Central Asia has a broad industrial resource base, with coal, iron, oil, natural gas, non-ferrous metals and a large hydro-electric potential, but has experienced only a modest degree of industrialisation. Industrial development has been selective and mainly post-Second World War. A small steel plant was built in the 1930s, but pig-iron and steel are still brought in from other regions. The oil, most of which occurs in western Turkmenia, is mainly shipped out via the Caspian, and ores and non-ferrous metals are also 'exported'. Over the past two decades, there has been more rapid industrial expansion, supported by natural gas and hydro-electricity, and the engineering, chemical, textile and food-processing branches have all expanded. Despite these developments, Central Asia's main function continues to be that of a supplier of primary products to the more developed parts of the country. The continuation of this traditional role is exemplified by the development of the large natural gas deposits in the Uzbek and Turkmen republics which now produce about 20% of all Soviet gas. Although these have certainly been used to strengthen the energy base of the Central Asian region, a large part of the output is piped to the energy-deficient Urals and Centre. Also as in Transcaucasia, a rapidly growing population supports industrial expansion, indeed Central Asia has a labour surplus, but there is a continuing reliance on other regions for manufactured goods. This is another region in which the possibility of self-sufficiency appears to have been sacrificed to the requirements of the more developed regions.

(18) Kazakhstan

Second in size only to the RSFSR and much bigger than the whole of western Europe, the Kazakh republic is also one of the largest economic regions and contains several contrasting elements. Agricultural patterns reflect the varied physical conditions. In the north is a belt of steppe-land which runs across the republic from the southern Urals to the Altay. Along with the adjacent part of the West Siberian region, this zone was affected by the virgin lands scheme of the 1950s and now takes second place only to the Ukraine in the production of wheat which, together with livestock products, is 'exported' to the European regions, to Central Asia and to eastern Siberia. The steppe gives way southwards to the desert zone, where extensive stock-rearing is the dominant activity, supplying both meat and wool. Beyond the desert is a hill-foot zone, similar to that of the Central Asian region, where intensive irrigated farming is carried on. Thus Kazakhstan is a major source of both basic foodstuffs and high-value crops.

The republic has also been the scene of major industrial developments in the Soviet period. Among the earliest was the exploitation of the Karaganda coalfield as a source of coking coal for the Urals, which led to the establishment of an iron and steel industry at Karaganda itself. This no longer depends on the Urals for its iron ore, which now comes partly from the Atasu deposit, 150 km to the south-west. The significance of northern Kazakhstan to the Soviet iron and steel industry has been greatly enhanced by the discovery of large deposits of iron ore, mainly in the Kustanay oblast, which have been developed to supply the steelworks of the Urals and to support expansion at Karaganda. Another major development in this zone has been the opening up of the Ekibastuz coalfield, 250 km north-east of Karaganda, which now produces more coal than Karaganda itself, though of a lower quality. Ekibastuz coal is mined open-cast and is fed into large thermal generating stations nearby. These will eventually be linked to a high voltage grid connecting the Kuzbas, the Urals and the

European regions. Thus northern Kazakhstan is becoming ever more closely linked to the West Siberian and Ural regions.

Kazakhstan also plays a major role in the production of ferro-alloy and non-ferrous metals. Major items include the lead and zinc of the upper Irtysh valley, the copper of Dzhezkazgan and Balkhash and several sources of chrome, manganese and bauxite. The development of these resources has placed Kazakhstan on an equal footing with the Urals as a producer of non-ferrous metals. In addition, the western part of the republic lies in the oil-rich zone around the Caspian. The Guryev (Emba) field has been in operation since the nineteenth century and the 1960s saw the development of oil and gas further south, in the Mangyshlak peninsula. The latter has, however, proved less rich than anticipated, and Kazakhstan produces less than five per cent of Soviet oil and gas.

Thus Kazakhstan has a rich and varied resource base and is increasing in importance to the Soviet economy. As an economic region it clearly lacks unity and linkages between the various industrial zones are weak. A more rational approach would be to link northern Kazakhstan to the Urals and West Siberia and the south to the Central Asian region. The policy of maintaining intact the constituent republics of the Union, rather than any real economic unity, would seem to be the main reason for the continued existence of Kazakhstan as a Major Economic Region.

(19) Belorussia

Belorussia shows many similarities to the Baltic republics. Like the latter, it has a poor resource base and, because much of present-day Belorussia lay within Poland until 1939, experienced little development before the Second World War. Agriculture is hindered by poor soils and bad drainage and is concerned mainly with potatoes, flax and livestock. The energy base is particularly weak, having been confined to peat, which is used in the generation of electricity. Oil and natural gas were discovered in the 1960s, but reserves appear to be very small. There are few minerals. Over the past two decades there has been some industrial development in the main towns, particularly Minsk (1.5 million), the capital, which have expanded their light engineering and chemical industries on the basis of electric power, oil and gas piped into the region, and local potash salts. Belorussia has long suffered from a degree of over-population and attempts have been made to offset this by the expansion of labour-intensive manufacturing industries. However, the republic continues to lie outside the mainstream of the Soviet economy, though clearly benefiting to some extent from the development of COMECON links.

(20) Moldavia

Under the present system, this small republic is not attached to any major economic unit, though in the past it was included in the Ukrainian economic region. Moldavia is characterised by a very low level of urbanisation accompanied by very high rural population densities, a situation reflecting favourable conditions for agriculture. Thus the republic specialises in high-value crops, notably vines, fruit and tobacco. Its main industries are concerned with processing agricultural products. In recent years there has been a marked expansion of light engineering activities, but Moldavia remains predominantly agricultural.

Conclusion

In the preceding section, an attempt has been made to indicate the characteristic features of the 20 regions rather than to give a complete systematic description of each. The various regions can be grouped, according to the nature of their contribution to the Soviet economy, into a number of types. One such group includes regions whose main function is as heavy metallurgical bases, providing steel and heavy engineering products—the Donets-Dnepr and Ural regions are clearly in this category whilst West Siberia performed the same function on a smaller scale but is now most significant as an energy source. One region, the Centre, stands in a class of its own, making a major contribution to the Soviet industrial economy owing to its advantages of location, despite its poverty in resources. Another, the Volga region, is primarily a source of energy and the same can be said of West and East Siberia, though all three are important in other respects as well. A further set of regions are important mainly for their agricultural production: the Black Earth Centre, the South-west, Moldavia and the South fall into this category. In contrast, Belorussia and the Baltic republics have very poor

resource bases and are of limited importance in either agriculture or industry. Other regions are much less easy to allocate to a particular group. Kazakhstan, for example, is of major significance in both agriculture and industry. The North, East Siberia and the Far East contain zones of rapid development and large stretches of undeveloped territory, while Transcaucasia and Central Asia have a varied resource base and are now developing quite rapidly, but remain somewhat peripheral to the Soviet economy as a whole.

The matters of the location of each region, the degree of development each has achieved and its contribution to the economy of the USSR are neatly summarised by Hooson's identification (Hooson, 1966, p. 121) of three types of region: the 'established core', embracing the Centre, the Ukraine and certain adjacent districts, the 'recent expansion funnel' covering a zone from the Volga to Lake Baykal, and the 'relatively marginal' or 'peripheral' areas of the Far East, the Arctic and Sub-Arctic zones, the Caucasus, Central Asia, Belorussia and the Baltic republics.

The difficulty of allocating the existing Major Economic Regions to groups with common characteristics draws our attention to the questionable nature of some of the present boundaries. The illogicalities which result from the insistence that republican boundaries should remain inviolate have already been commented upon with respect to the splitting of the Donbas between the Donets–Dnepr and North Caucasus regions. In addition, there are several cases, of which Kazakhstan is the best example, but which also include the North Caucasus, Volga–Vyatka and others, of a combination of diverse and unconnected elements within a single region and the ignoring of clearly established inter-regional linkages. Many of these problems can still be attributed to the conflicting ideas of regional self-sufficiency and regional specialisation, neither of which can be wholly achieved. As the Soviet economy develops and new inter- and intra-regional linkages are established, the present set of economic regions, now in existence for more than twenty years, seems likely to become even less satisfactory as a basis on which to plan the economy. There is increasing need for a radical reorganisation of the economic regionalisation of the whole country.

BIBLIOGRAPHY

Alampiyev, P. M. (1960), 'Problems of general economic regionalization at the present stage,' *Soviet Geography*, **1**, pp. 3–15.

Alampiyev, P. M. (1963), 'Economic regionalization and its place in economic geography,' *Soviet Geography*, **4**, pp. 60–67.

Alampiyev, P. (1964), *Economic areas in the USSR*, Progress Publishers, Moscow.

Altman, L. P. (1965), 'Economic regionalization of the USSR and new methods in economic-geographic research,' *Soviet Geography*, **6**, pp. 48–55.

Bandera, V. N. and Melnyk, Z. L. (1973), *The Soviet Economy in Regional Perspective*, Praeger, New York.

Bone, R. M. (1967), 'Regional planning and economic regionalization in the Soviet Union,' *Land Economics*, pp. 347–354.

Chambre, H. (1959), *L'aménagement du territoire en URSS*, Mouton, Paris.

Dewdney, J. C. (1967), *Patterns and problems of regionalisation in the USSR*. Research Paper No. 8, Dept. of Geog., University of Durham.

Hooson, D. (1966), *The Soviet Union*, ULP (now Hodder and Stoughton), London.

Kalashnikova, T. M. (1969), *Ekonomicheskoye rayonirovaniye*, Moscow University.

Kazanskiy, N. N. and Khorev, B. S. (1976), 'Problems of economic regionalisation at the present stage,' *Soviet Geography*, **17**, pp. 637–645.

Kistanov, V. V. (1960), 'Aspects of the formation of economic regions in the eastern USSR,' *Soviet Geography*, **1**, pp. 52–59.

Kolosovskiy, N. N. (1961), 'The territorial-production combination (complex) in Soviet economic geography,' *Journal of Regional Science*, **3**, pp. 1–25.

Koropeckij, I. S. and Schroeder, G. E. (eds) (1981), *The Economics of Soviet Regions*, Praeger, New York.

Kurakin, A. F. (1962), 'Economic-administrative regions: their specialization and their integrated development,' *Soviet Geógraphy*, **3**, pp. 29–39.

Linge, G. J. R., Karaska, G. J. and Hamilton, F. E. I. (1978), 'An appraisal of the Soviet concept of the territorial production complex,' *Soviet Geography*, **19**, pp. 681–697.

Lydolph, P. E. (1977), *Geography of the USSR*, 3rd edn, Wiley, New York.

Melezin, A. (1968), 'Soviet regionalization: an attempt at the delineation of socio-economic integrated regions,' *Geographical Review*, **58**, pp. 593–621.

Mellor, R. E. H. (1959), 'Trouble with the regions: planning problems in Russia,' *Scottish Geographical Magazine*, **75**, pp. 44–48.

Mieczkowski, Z. (1965), 'The major economic regions of the USSR in the Khrushchev era,' *Canadian Geographer*, **9**.

Mieczkowski, Z. (1967), 'The economic-administrative regions in the USSR,' *Tijdschrift voor Econ. en Soc. Geografie*, **58**, pp. 209–19.

Moshkin, A. M. (1962), 'What is a territorial-production complex?,' *Soviet Geography*, **3**, pp. 49–55.

Moshkin, A. M. (1977), 'A typology of regional territorial-production complexes,' *Soviet Geography*, **18**, pp. 60–67.

NATO (1979), *Regional Development in the USSR*, Oriental Research Partners, Newtonville, Mass.

Nikolskiy, I. V. (1975), 'Economic regions, administrative regions and territorial production complexes,' *Soviet Geography*, **16**, pp. 374–381.

Pallot, J. and Shaw, D. B. (1981), *Planning in the Soviet Union*, Croom Helm, London.

Parkhomenko, I. I. (1966), 'Detailed (intra-oblast and lower level) economic regionalization in the USSR,' *Soviet Geography*, **7**, pp. 33–47.

Pokshishevskiy, V. V. (ed) (1964), *Geograficheskiye problemy krupnykh rayonov SSSR*, Mysl, Moscow.

Pokshishevskiy, V. V. (1966), 'Economic regionalization of the USSR: a review of research during 1962–64,' *Soviet Geography*, **7**, pp. 4–32.

Pokshishevskiy, V. V. (1975), 'On the Soviet concept of economic regionalisation,' *Progress in Geography*, **7**, pp. 1–52.

Privalovskaya, G. A. (1979) "General tendencies of development of territorial production complexes", *Soviet Geography*, **20**, pp. 82–96.

Probst, A. Ye. (1966), 'Territorial production complexes in the USSR,' *Soviet Geography*, **7**, pp. 47–55.

Probst, A. Ye. (1977), 'Territorial production complexes,' *Soviet Geography*, **18**, p. 195.

Rodoman, B. B. (1972), 'Principal types of geographical regions,' *Soviet Geography*, **13**, pp. 448–454.

Rodoman, B. B. (1973), 'Territorial systems,' *Soviet Geography*, **14**, pp. 100–105.

Saushkin, Yu. G. (1976), 'Economic regionalisation,' *Soviet Geographical Studies*, Academy of Sciences of the USSR, Moscow, pp. 57–73.

Saushkin, Yu. G. and Kalashnikova, T. M. (1960), 'Current problems in the economic regionalisation of the USSR,' *Soviet Geography*, **1**, 6, pp. 50–60.

Sdasiuk, G. (1962), 'The history of regionalisation of the USSR,' *Nat. Geog. Journ. India* (Varanasi), **8**, pp. 145–156.

Shabad, T. (1946), 'Political administrative divisions of the USSR in 1945,' *Geographical Review*, **36**, pp. 303–312.

Shabad, T. (1953), 'The Soviet concept of economic regionalization,' *Geographical Review*, **43**, pp. 214–222.

Shimkin, D. (1952), Economic regionalization in the USSR, *Geographical Review*, **42**.

Soviet Geography (1966), 'Bibliography on economic regionalization,' *Soviet Geography*, **7**, pp. 65–96.

Taaffe, R. N. L. (1980) "Soviet regional development" *in* S. F. Cohen *et al* (eds) *The Soviet Union since Stalin*, Macmillan, London.

Zawadskiy, S. (1973), *Osnovy regionalnogo planirovaniya*, Progress Publishers, Moscow

Index